HISTORY OF
SCIENCE AND
THE EMOTIONS

EDITED BY

Otniel E. Dror, Bettina Hitzer,
Anja Laukötter, and Pilar León-Sanz

O S I R I S | 31

A Research Journal Devoted to the
History of Science and Its Cultural Influences

Osiris

Series editor, 2013–2018

ANDREA RUSNOCK, *University of Rhode Island*

Volumes 28 to 32 in this series are designed to connect the history of science to broader cultural developments, and to place scientific ideas, institutions, practices, and practitioners within international and global contexts. Some volumes address new themes in the history of science and explore new categories of analysis, while others assess the "state of the field" in various established and emerging areas of the history of science.

Series editor, 2002–2012

KATHRYN OLESKO, *Georgetown University*

Cover Illustration:

Spike Walker, *Dried tears from watery eyes*, Light Microscopy, Wellcome Images.

Acknowledgments

This volume emerged from two workshops. We would like to thank the Emotional Culture and Identity Project of the Institute for Culture and Society (University of Navarra) and the Center for the History of Emotions at the Max Planck Institute for Human Development (Berlin) for their generous support of the two workshops.

An Introduction to *History of Science and the Emotions*

by Otniel E. Dror,* Bettina Hitzer,**
Anja Laukötter‡, and Pilar León-Sanz§

ABSTRACT

This essay introduces our call for an intertwined history-of-emotions/history-of-science perspective. We argue that the history of science can greatly extend the history of emotions by proffering science qua science as a new resource for the study of emotions. We present and read science, in its multiple diversities and locations, and in its variegated activities, products, theories, and emotions, as constitutive of the norms, experiences, expressions, and regimes of emotions. Reciprocally, we call for a new reading of science in terms of emotions as an analytical category. Assuming emotions are intelligible and culturally learned, we extend the notion of emotion to include a nonintentional and noncausal "emotional style," which is inscribed into (and can reciprocally be generated by) technologies, disease entities, laboratory models, and scientific texts. Ultimately, we argue that emotional styles interrelate with broader emotional cultures and thus can contribute to and/or challenge grand historical narratives.

> Descartes would have been more nearly right in saying, "I feel, therefore I am." —Ralph W. Gerard, 1941[1]

Over the past two or three decades, the study of emotions has revolutionized our conceptions of human nature. What we now call the "Emotional Turn" challenged earlier scientific understandings of humans—our brains, our bodies, and the laws that govern their functions within and between individuals—and of society as a whole. Research on emotions has contributed to the invention of new scientific techniques and instruments, to the development of new ways of seeing and making visible, to the reorganization of the hierarchy of disciplines in science, as well as to the public visibility of and political interest in science.

This turn to emotions is now visible in the neurosciences, economics, sociology, anthropology, criminology, philosophy, and literary and media studies.[2] In history, the

---

* History of Medicine, Hebrew University Medical Faculty, P.O. Box 12272, Jerusalem 91120, Israel; otnield@ekmd.huji.ac.il.
** Max Planck Institute for Human Development, Berlin, Germany; hitzer@mpib-berlin.mpg.de.
‡ Max Planck Institute for Human Development, Berlin, Germany; laukoetter@mpib-berlin.mpg.de.
§ School of Medicine, University of Navarra, 31008 Pamplona, Spain; mpleon@unav.es.
[1] Ralph W. Gerard, *The Body Functions: Physiology* (New York, 1941), 256.
[2] "In the past thirty years, the dramatic explosion of interest in emotions has become evident throughout a wide variety of disciplines as well as around the world of research and scholarship." In-

study of emotions is no longer the lonely enterprise of a marginal avant-garde but a burgeoning, increasingly respected, and established field within historiography. Numerous conferences, dedicated monograph series, and new journals and centers for the study of the history of emotions testify to the success of this new field.[3]

But should one follow this new historiographical trend and add a history-of-emotions perspective to the history of science and a history-of-science perspective to the history of emotions? This question is as legitimate as it is rhetorical since this volume testifies to our conviction that the history of emotions can indeed contribute a valuable and challenging perspective to the history of science and that the history of science can significantly enrich the history, sociology, and anthropology of emotions, as well as the study of emotions in the neurosciences.

One compelling argument in favor of such a perspective lies in the necessary intellectual critique of the present-day viewpoint of science, the need to historicize emotion concepts, to detect alternative but oftentimes marginalized scientific routes into emotion research, and, ultimately, to articulate the complex genealogy of the emotional turn from a history-of-science perspective.[4] This latter history of the science of the present is no less significant for the humanities than it is for the neurosciences because numerous affective theorists and emotion researchers in history, sociology, anthropol-

---

ternational Society for Research on Emotion, "ISRE: Past and Present," http://isre.org/isre.php (accessed 4 March 2016).

[3] For an overview of the history of emotions, see Jan Plamper, *The History of Emotions: An Introduction* (Oxford, 2015); Bettina Hitzer, "Emotionsgeschichte—ein Anfang mit Folgen," *H/Soz/Kult*, 23 November 2011, http://hsozkult.geschichte.hu-berlin.de/forum/2011-11-001.pdf (accessed 4 March 2016); Susan J. Matt and Peter N. Stearns, eds., *Doing Emotions History* (Urbana, Ill., 2014); Matt, "Current Emotion Research in History: Or, Doing History from the Inside Out," *Emotion Rev.* 3 (2011): 117–24; Frank Biess, Alon Confino, Ute Frevert, Uffa Jensen, Lyndal Roper, and Daniela Saxer, "History of Emotions: Forum," *Germ. Hist.* 28 (2010): 67–80; Plamper, "The History of Emotions: An Interview with William Reddy, Barbara Rosenwein, and Peter Stearns," *Hist. & Theory* 49 (2010): 237–65; Anna Wierzbicka, "The 'History of Emotions' and the Future of Emotion Research," *Emotion Rev.* 2 (2010): 269–73; Barbara Rosenwein, "Problems and Methods in the History of Emotions," *Passions in Context: J. Hist. Phil. Emotions* 1 (2010), http://www.passionsincontext.de/uploads/media/01_Rosenwein.pdf (accessed 4 March 2016); William M. Reddy, "Historical Research on the Self and Emotions," *Emotion Rev.* 1 (2009): 302–15; Rosenwein, "Worrying about Emotions in History," *Amer. Hist. Rev.* 107 (2002): 821–45. Moreover, research centers have been established around the world: Centre for the History of Emotions, Queen Mary University of London; Center for the History of Emotions at the Max-Planck-Institute for Human Development, Berlin; and the Australian Research Council, Centre of Excellence for the History of Emotions, which is composed of multiple units in numerous Australian universities. Its main center is in Perth. Another example is EMMA: Emotions au Moyen Age, a research program on emotions in the Middle Ages in France and Québec, emma.hypotheses.org (accessed 4 March 2016). Also, specific publications have been established: e.g., *Emotion Review* and the website *History of Emotions—Insights into Research*, www.history-of-emotions.mpg.de/en (accessed 4 March 2016). Moreover, several new series focus on the history of emotions: Thomas Dixon and Ute Frevert, eds., Emotions in History, 1500–2000, Oxford University Press; Susan J. Matt and Peter Stearns, eds., Studies in the History of the Emotions, University of Illinois Press; William M. Reddy and David Lemmings, Palgrave Studies in the History of Emotions, https://www.palgrave.com/us/series/14584 (accessed 4 March 2016).

[4] See John Deigh, "William James and the Rise of the Scientific Study of Emotion," *Emotion Rev.* 6 (2014): 4–12; Otniel E. Dror, "The Cannon-Bard Thalamic Theory of Emotions: A Brief Genealogy and Reappraisal," *Emotion Rev.* 6 (2014): 13–20; Frank Biess and Daniel Gross, eds., *Science and Emotions after 1945: A Transatlantic Perspective* (Chicago, 2014); Rhodri Hayward, *The Transformation of the Psyche in British Primary Care, 1880–1970* (London, 2014); Plamper, *History of Emotions* (cit. n. 3); Daniel M. Gross, *The Secret History of Emotion: From Aristotle's "Rhetoric" to Modern Brain Science* (Chicago, 2006); Claudia Wassmann, "The Science of Emotion: Studying Emotions in Germany, France, and the United States, 1860–1920" (PhD diss., Univ. of Chicago, 2005); Thomas Dixon, *From Passions to Emotions: The Creation of a Secular Psychological Category* (Cambridge, 2003).

ogy, and philosophy draw deliberately or unselfconsciously on the recent neurosciences of emotions in formulating their conceptions and assumptions regarding the
emotions.[5] While some criticize the explicit adoption of scientific models of emotions
in the humanities and social sciences,[6] others present it as a positive and progressive
development in overcoming the abstractions and reductionisms of discourse, incorporating the body into history, and bypassing the Cartesian divide between reason
and emotion (mind and body).[7] A critical history of the modern sciences of emotions
is thus essential for reflecting on and contextualizing our own assumptions and scholarship, as several contributions to this volume exemplify: be it the recent history of
motherly love going schizogenic, as Anne Harrington shows, or the neurophysiological discovery of a "new" post-1945 "supramaximal" "super-pleasure," which Otniel
E. Dror examines.[8]

But the history of science and the history of emotions can offer each other much
more than this important historical inquiry into their respective presents. The history
of science can extend the history of emotions by proffering science qua science as a
new, even if at first blush unlikely, resource for the study of emotions. It can also expand the very notion of what an emotion is (and what the study of the history of emotions is about) and it can enrich and challenge narratives of emotions, as Piroska Nagy
and Damien Boquet demonstrate in this volume by calling into question the grand narrative of scientific emotions as a specifically modern phenomenon. In a reciprocal
manner, the history of emotions can provide historians of science with what has been
repeatedly demanded during the last decade: a broader perspective on history beyond
the scope of micro- or case studies, as well as a more tightly knit integration of the
history of science into general history.[9] As Robert E. Kohler and Kathryn M. Olesko
suggested in a recent *Osiris* volume, instead of calling for a big-picture history, historians of science should move toward big thinking. Big thinking is a call to "develop the
generalities that grow out of and remain rooted in specialized research."[10] Kohler and

[5] Otniel E. Dror, "What Is an Excitement?" in Biess and Gross, *Science* (cit. n. 4), 121–38.

[6] Constantina Papoulias and Felicity Callard, "Biology's Gift: Interrogating the Turn to Affect," *Body & Soc.* 16 (2010): 29–56; Ruth Leys, "The Turn to Affect: A Critique," *Crit. Inq.* 37 (2011): 434–72; Rafael Mandressi, "Le temps profond et le temps perdu: Usages des neurosciences et des sciences cognitives en histoire," *Rev. Hist. Sci. Hum.* 25 (2011): 165–202.

[7] Above all William M. Reddy, "Against Constructionism: The Historical Ethnography of Emo-tions," *Curr. Anthropol.* 38 (1997): 327–51; Reddy, *The Navigation of Feeling: A Framework for the History of Emotions* (New York, 2001); Reddy, "Historical Research on the Self and Emotions," *Emotion Rev.* 1 (2009): 302–15; but see also Reddy, "Neuroscience and the Fallacies of Functional-ism," *Hist. & Theory* 49 (2010): 412–25. See also Sara Ahmed, *Cultural Politics of Emotion* (New York, 2004); Luc Ciompi, *Die emotionalen Grundlagen des Denkens: Entwurf einer fraktalen Affekt-logik* (Göttingen, 1997). An example of this type of argument is provided by Carolyn Pedwell. She has noted that against the backdrop of the "'science of empathy' in the wake of the discovery of mirror neurons," one can identify a recourse to biology, neuroscience, and genetics as well as to different evo-lutionary scenarios to explicate "empathic circuits of feeling within the human body" and current "emo-tional politics of contemporary societies internationally." See Pedwell, *Affective Relations: The Trans-national Politics of Empathy* (Basingstoke, 2014).

[8] On post-1945 science and emotions, see also Biess and Gross, *Science* (cit. n. 4).

[9] Casper Hakfoort, "The Missing Syntheses in the Historiography of Science," *Hist. Sci.* 29 (1991): 207–16; James A. Secord, ed., "The Big Picture," *Brit. J. Hist. Sci.* 26 (1993): 387–483; Robert E. Kohler, ed., "Focus: The Generalist Vision in the History of Science," *Isis* 96 (2005): 224–51; Josep Simon and Nestor Herran, "Introduction," in *Beyond Borders: Fresh Perspectives in History of Sci-ence*, ed. Josep Simon and Nestor Herran (Newcastle, 2008), 1–26.

[10] Robert E. Kohler and Kathryn M. Olesko, "Introduction: Clio Meets Science: The Challenges of History," *Osiris* 27 (2012): 1–16, on 4.

Olesko rightfully argued that this kind of midscale history entailed a critical reexamination of three historiographical fundamentals: namely, the categories of analysis, the subjects, and, eventually, the periodizations.[11] The history of emotions invites historians to rethink and reframe these three historiographical fundamentals.

* * *

What is the history of emotions about? Are emotions a new analytical category like gender or class (even if very different in kind), or are they a subject of historical analysis in and of themselves, like the senses or pain?[12] In the title of his famous 1884 *Mind* article, William James asked, "What Is an Emotion?" This question is open to historical investigation.[13] Historians of emotions reject essentialist definitions of emotion(s). "Emotion" basically indicates that humans at different times and from different regions, classes, and genders used concepts with many different names to describe a human (sometimes also an animal) faculty that was in some way or another related to the bodily sensations of feeling, touching, and moving, that was more or other than just thoughts, and that was part of the human experience as shared experience and communication. Emotion, in this basic sense, was for most of history's course regarded as fundamental to human existence and to human action.[14] Recently, scholars have begun to discuss the possible differences between and assumed political significance of "emotions" and "affects." "Affects"—as theorists like William E. Conolly and Nigel Thrift assume—are bound more closely to the body than emotions, cannot be reduced to discourses, are preconscious and nonintentional, and are thus incompletely understood in rational terms.[15]

But in using emotion(s) as an analytical category, historians have to go beyond this basic definition without losing their historical sensitivity to the equivocality of and the conceptual shifts in emotion concepts. Thus, even within a single scientific discipline, various conflicting concepts of "emotions" might have coexisted, as Bettina Hitzer and Pilar Léon-Sanz demonstrate in this volume in their study of psychosomatic cancer medicine in the twentieth century.[16] If emotion is to be as useful a category of historical analysis as gender currently is, it has to be regarded as intelligible; if it is to be historicized, it has to be, at least in part, learned and culturally specific. If emotions

[11] Ibid., 10–1.

[12] Quentin Deluermoz, Emmanuel Fureix, Hervé Mazurel, and M'hamed Oualdi, "Écrire l'histoire des émotions: de l'objet à la catégorie d'analyse," *Rev. Hist. XIXe Siècle* 47 (2013): 155–89.

[13] William James, "What Is an Emotion?" *Mind* 9 (1884): 188–205.

[14] We note that any provisional definition must be historicized and can be challenged.

[15] See, e.g., Nigel Thrift, "Intensities of Feeling: Towards a Spatial Politics of Affect," *Geogr. Ann. Ser. B Human Geogr.* 86 (2004): 57–78. For a lively discussion on affect and its intention between scholars Ruth Leys and William E. Connolly, see Leys, "The Turn to Affect: A Critique," *Crit. Inq.* 37 (2011): 434–72; Connolly, "The Complexity of Intention," *Crit. Inq.* 37 (2011): 791–8; and Leys, "Affect and Intention: A Reply to William E. Connolly," *Crit. Inq.* 37 (2011): 799–805. Part of this discussion harks back to the earlier work by Brian Massumi, "The Autonomy of Affect," *Cult. Critique* 31 (1995): 83–109.

[16] For the historical change of emotion concepts, see Dixon, *Passions* (cit. n. 4); Philip Fisher, *The Vehement Passions* (Princeton, N.J., 2002); David Thorley, "Towards a History of Emotion, 1562–1660," *Seventeenth Cent.* 28 (2013): 3–19; Ute Frevert, Monique Scheer, Anne Schmidt, Pascal Eitler, Bettina Hitzer, Nina Verheyen, Benno Gammerl, Christian Bailey, and Margrit Pernau, *Emotional Lexicons: Continuity and Change in the Vocabulary of Feeling 1700–2000* (Oxford, 2014); Jutta Stalfort, *Die Erfindung der Gefühle: Eine Studie über den historischen Wandel menschlicher Emotionalität (1750–1850)* (Bielefeld, 2013); and Joanna Bourke, *The Story of Pain: From Prayer to Painkillers* (Oxford, 2014).

remain irrational and arbitrary, they are not fully accessible for historical analysis.[17] In this case, historians can only describe in which way they triggered historical events or decisions. The emotion itself, its internal dynamic, its implications within processes of subjectification, as well as its role in shaping practices and material cultures, would remain opaque. If we regard emotions not as universals, then we might ask how emotions evolve, how they are learned, why they change, and what effects these changes have, a challenge that Felicity Callard takes up in investigating the history of panic disorder in this volume. Only this understanding of emotions allows us to include questions pertaining to historical developments and changes on the individual, social group, and societal levels in the analysis.[18]

As a category of analysis sui generis, emotion is not reducible to other categories or forms of historical explanations, and it offers a distinctive prism through which to (re)interpret historical events. It partly constitutes historical developments rather than being (only) derivative. It is, in other words, "a force giving shape to politics, society and culture, to beliefs and values, and to everyday life, institutional settings, and the processes of decision making."[19]

Emotion is not necessarily intentional, nor is it limited to explaining human actions or motivations.[20] But this is, of course, one possibility in using the category "emotion." It was in this first sense that the history of emotions shifted the historiographical ground by emphasizing agency over causation in historical analysis. What scared people, "what made them joyful or proud or disgusted or compassionate [became] . . . fundamental to understanding what those same people decided and did."[21] Taking this route into the history of emotions does not dismiss the possible significance of other elements that determine human agency, since emotions are often part of a complex interplay with political, economic, scientific, or other arguments, as well as pragmatic considerations. The analysis of this complex interweaving is central to many contributions to this volume. Anja Laukötter, for example, highlights how the very constitution of the modern human subject and his or her emotions was negotiated when discussing the psychological underpinnings of health education films.

In contrast to psychohistorical narratives, which also regarded emotions as decisive, the new history of emotions presents individual and collective emotions as embedded in social structures and relations and as part of some larger emotional world. Different terms have been coined to exemplify this embeddedness: "structures of feelings" (Raymond Williams), "emotionology" (Peter N. Stearns and Carol Z. Stearns), "emotional

---

[17] See Rosenwein, "Worrying" (cit. n. 3), 821–45.

[18] See Ute Frevert, Pascal Eitler, Stephanie Olsen, Uffa Jensen, Margrit Pernau, Daniel Brückenhaus, Magdalena Beljan, et al., *Learning How to Feel: Children's Literature and the History of Emotional Socialization, 1870–1970* (Oxford, 2014).

[19] Biess et al., "History of Emotions" (cit. n. 3), 71.

[20] The role of the subject and his or her emotional agency has been widely debated in recent years between affect theorists such as William E. Connolly and Couze Venn on the one hand and neo-universalists such as Michael Hochgeschwender and William M. Reddy on the other hand—although both share in the attempt to transgress the poststructuralist focus on language. While neo-universalists hold on to the intentional subject, affect theorists want to dissolve the intentional subject in favor of trans-individual and -subjective phenomena. This has been criticized by Ruth Leys, among others, as a form of affect determinism; see also Benno Gammerl and Bettina Hitzer, "Wohin mit den Gefühlen? Vergangenheit und Zukunft des Emotional Turn in den Geschichtswissenschaften," *Berliner Debatte Initial* 24 (2013): 31–40, on 35–6.

[21] Sophia Rosenfeld, "Thinking about Feeling, 1789–1799," *French Hist. Stud.* 32 (2009): 697–706, on 698. See also the argument by Sophie Wahnich, *La longue patience du peuple: 1792, naissance de la République* (Paris, 2008).

regimes" (William M. Reddy), and "emotional communities" (Barbara H. Rosenwein) to name the most important at this point.[22]

Nevertheless, emotions are not determined by these larger frameworks. They leave room for maneuvering, as William Reddy noted in proposing the term "emotive" to explain that for one person emotion statements might serve as a choice between different possible emotional reactions, and for another they are transformative insofar as they change their objects and reflect back on themselves. This performative dimension of emotions also allows us to transcend old dichotomies like "body and language," privileging dimensions of bodily sensations and senses while suggesting new ways of thinking about the relationship between the individual and the "larger world."[23]

* * *

Against this preliminary background, we pose the question: What were the historical interrelationships between science and its cultures on the one hand and cultures of emotions on the other? This key historical question harks back to a grand Western narrative that still frames contemporary historiography of science and the emotions. From at least the eighteenth century, numerous commentators observed that science transformed Western emotions. Framing their argument in terms of a progressive narrative of diminishing fear, pain, and wonder in Western societies, they contended that science dispelled "supernatural fears . . . by the discoveries that reveal the laws of the world," and brought "man more or less to the position of 'nil admirari,'" as Alexander Bain argued in *The Emotions and the Will* (1859).[24] This was the meaning of the Enlightenment, according to Max Horkheimer and Theodor W. Adorno. In *Dialektik der Aufklärung*, as Michael Hagner has argued, Adorno and Horkheimer "defined the central aim of the enlightenment as 'delivering humans from fear and installing them as masters.' . . . The goal was reached 'if there is nothing unknown left.'" The Enlightenment established a direct link between the accumulation of knowledge and "overcoming fear."[25]

---

[22] The term "structure of feelings" describes feelings as a form of reorganization of social experiences, a set of perceptions and values that is common to a generation or social group at a specific moment in history. One can discern these structures of feelings above all in artistic productions; see Raymond C. Williams, *Marxism and Literature* (Oxford, 1977). The concept of "emotionology" conveys the idea that expressions of emotion, as well as both their intensity and their duration in societies, are guided by certain norms. Conflict might arise when the emotions experienced by individuals or groups differ from these collective emotional standards. See Peter N. Stearns and Carol Z. Stearns, "Emotionology: Clarifying the History of Emotions and Emotional Standards," *Amer. Hist. Rev.* 90 (1985): 813–30. Using anthropological and cognitive psychological research as his basis, William M. Reddy underlined the performative aspect of emotions in highlighting the reciprocity of the inner emotional state and its verbal expressions, thus his concept of "emotive." Emotional norms or "emotional regimes," as he calls them, are created by dominant groups or societies and might give rise to emotional suffering of other groups that will result in conflicts and even politico-emotional change if there is no space for taking emotional refuge. See Reddy, "Against Constructionism" (cit. n. 7), 327–51. Barbara Rosenwein has shown that "social groups" such as families, parties, and guilds create "systems of feeling." According to her, several "emotional communities," which are created along the lines of social and political groups and classes, can coexist. See Rosenwein, "Worrying" (cit. n. 3), 821–45.

[23] Rosenfeld, "Thinking" (cit. n. 21), 703.

[24] Alexander Bain, *The Emotions and the Will*, ed. Daniel N. Robinson (1859; repr., Washington, D.C., 1977), 72, 90.

[25] Hagner emphasizes the spatial metaphor that is at the heart of Adorno's and Horkheimer's argument. See Hagner, "Enlightened Monsters," in *The Sciences in Enlightened Europe*, ed. William Clark, Jan Golinski, and Simon Schaffer (Chicago, 1999), 175–217, on 178.

This grand historical narrative persisted throughout the nineteenth century and appeared in the writings of John Stuart Mill and William James, among others, and in the diatribes of the critics of science.[26] These critics endorsed the grand narrative of the inverse relationships between science and emotions, but instead of a progressive narrative of diminishing fear and pain, they disparaged science and medicine for diminishing feelings and for the triumph of the head over the heart.[27]

For historians of science and of emotions, this grand Western narrative raises a myriad of historical and methodological questions: To what extent did science, in its multiple diversities and locations, and in its variegated activities, products, theories, and emotions, constitute the norms, experiences, expressions, and regimes of emotions?[28] Is there a grand narrative of the interrelationships between the emotions and science, or between particular emotions and science?[29] Did discrete scientific developments structure the expression, experience, visibility, or nature of emotions? In this vein, Naama Cohen-Hanegbi's contribution to this volume examines how medical debates about emotions and their treatment by medical means shaped not only the science of medicine itself but also the status of the medical profession and its position in late medieval society. Rafael Mandressi's contribution addresses these issues in studying how affective cultures were linked to and negotiated by anatomical dissection in Early Modern Europe. Are there basic emotions (or sentiments) that undergird science?[30] What can the study of science, and its emotions, contribute to the history of emotions, beyond the study of emotions in literature, art, economics, and politics?

The first methodological challenge for the historian pursuing these questions is how to read emotions in, from, and through science. How does the historian identify emotions in a culture that deliberately and methodologically excluded emotions from its texts, narratives, practices, and practitioners, separated the factual-scientific from the personal-experiential, and admonished its members to suppress their blushes, control their disgust reactions, and abolish their tears, at least since the eighteenth century?[31] This latter difficulty partly explains the relative absence of histories of emotions and

---

[26] John Stuart Mill, "Civilization—Signs of the Times," *London Westminster Rev.*, April 1836, 1–28; and William James, *The Principles of Psychology* (London, 1891).

[27] John Gordon, "T. S. Eliot's Head and Heart," *ELH* 62 (1995): 979–1000; and Martin S. Pernick, "The Calculus of Suffering in 19th-Century Surgery," in *Sickness and Health in America: Readings in the History of Medicine and Public Health*, ed. Judith Walzer Leavitt and Ronald L. Numbers (Madison, Wis., 1985), 98–112.

[28] We emphasize at the outset for the nonhistorian of science the great diversity of the sites, meanings, and identities of science, each of which could present a unique subculture of emotions despite the progressive and general adoption of the modern laboratory as the ideal of modern science. On the laboratory as an ideal, see Robert E. Kohler, *Landscapes and Labscapes: Exploring the Lab-Field Border in Biology* (Chicago, 2002).

[29] Katharine Park and Lorraine Daston's study of the cognitive passion of "wonder" provides a grand narrative of the rise and decline of "wonders as objects of natural inquiry" and of "wonder as a passion of natural inquiry." Park and Daston study the coeval transformation in the nature of knowledge and in the passions that typified the knower over a period of several centuries. By the eighteenth century, the wonders of nature and wonder as a passion of scientific inquiry, both of which had typified natural enquiry, were excluded from nature and from natural philosophy. See Daston and Park, *Wonders and the Order of Nature, 1150–1750* (New York, 1998), 14.

[30] Lorraine Daston and Peter Galison, "Objectivity and Its Critics," *Victorian Stud.* 50 (2008): 666–77.

[31] This disappearance of overt feelings and experiences constituted one element in the long history of Western scientific objectivity. See Paul White, "Introduction," *Isis* 100 (2009): 792–7. Particularly from the eighteenth century, the experiences and personalities of scientists were progressively excluded from scientific texts. See Marie-Noelle Bourguet, "The Explorer," in *Enlightenment Portraits*, ed. Michel

histories of science that derive their history of emotions from science qua science—from the very fabric of science—and the penchant to seek the disappeared emotions of science in the archives: in the private correspondence, autobiographies, and journal entries of scientists.[32] Literature and art, on the other hand, have very often been presented as privileged sites for the study of cultures and histories of emotions, since they are, as the cultural materialist Raymond Williams put it, the social loci that best manifest the "actually lived through" and are closest to "original experience."[33]

Yet emotions suffused science and science suffused emotions. Fundamental aspects of science were undergirded by and can be explained only in terms of the emotions, and fundamental aspects of the emotions were undergirded by and can be explained only in terms of the sciences: the "fear" of the "epistemic fear" that motivated science writ large, the "sentimentalism" of the "sentimental empiricism" of eighteenth-century knowledge, the "sentiment" of the "sentiment of objectivity," the "wonder" of scientific epistemology, and the "laughter" or "seriousness" of science—all have a history of emotions.[34] Conversely, the "fear" of early twentieth-century Russian soldiers, the "excitement" of the football game, the category of survivor's "shame," the very concept of "emotion" (rather than "passion"), and even our contemporary scholarship of the history (anthropology and sociology) of emotions—all have a history of science.[35]

Vovelle, trans. Lydia G. Cochrane (Chicago, 1997), 257–315; and Mary Terrall, "Heroic Narratives of Quest and Discovery," *Configurations* 6 (1998): 223–42. On suppressing the blush, see William Osler, "Aequanimitas," in *Aequanimitas, with Other Addresses to Medical Students, Nurses and Practitioners of Medicine*, ed. William Osler (Philadelphia, 1904), 3–11. This was the valedictory address given at the University of Pennsylvania in 1889. But see also Thomas Dixon on a weeping judge and Paul White on Darwin's tears. Dixon, "The Tears of Mr. Justice Willes," *J. Victorian Cult.* 17 (2012): 1–23; and Paul White, "Darwin Wept: Science and the Sentimental Subject," *J. Victorian Cult.* 16 (2011): 195–213.

[32] In the case of Ludwik Fleck, even his scientific work and texts, in which he draws on the question of emotions and knowledge production, might be interesting for Fleck's emotional self-construction but could also be useful for other questions on emotional practices and emotional habitus in science; see Fleck, *Entstehung und Entwicklung einer wissenschaftlichen Tatsache: Einführung in die Lehre vom Denkstil und Denkkollektiv* (Basel, 1935), translated by Fred Bradley and Thaddeus J. Trenn as *Genesis and Development of a Scientific Fact*, ed. Thaddeus J. Trenn and Robert K. Merton (Chicago, 1979). Until fairly recently, historians of science have ignored the idea that emotions are integral to the practices and content of science, often relegating emotions to biographies of scientists. For early exceptions, see, e.g., Evelyn Fox Keller, *A Feeling for the Organism: The Life and Work of Barbara McClintock* (San Francisco, 1983); and Donna J. Haraway, *Primate Visions: Gender, Race, and Nature in the World of Modern Science* (New York, 1989). See also Otniel E. Dror, "The Affect of Experiment: The Turn to Emotions in Anglo-American Physiology, 1900–1940," *Isis* 90 (1999): 205–37.

[33] David Simpson, "Raymond Williams: Feeling for Structures, Voicing 'History,'" *Soc. Text* 30 (1992): 9–26, on 16–8.

[34] On "epistemic fear," see Daston and Galison, "Objectivity" (cit. n. 30). See also Lorraine Daston, "Fear and Loathing of the Imagination in Science," *Daedalus* 127 (1998): 73–95. On "sentimental empiricism," see Jessica Riskin, *Science in the Age of Sensibility: The Sentimental Empiricists of the French Enlightenment* (Chicago, 2002). On "sentiment of objectivity," see Paul White, "Darwin's Emotions: The Scientific Self and the Sentiment of Objectivity," *Isis* 100 (2009): 811–26. On "wonder," see Daston and Park, *Wonders* (cit. n. 29). On the decline of laughter and the emergence of seriousness as a defining feature of science from the seventeenth century, see Paula Findlen, "Between Carnival and Lent: The Scientific Revolution at the Margins of Culture," *Configurations* 6 (1998): 243–67.

[35] Jan Plamper, e.g., emphasized the role of science over other cultural forces in studying the significant shift to fear during the late nineteenth and early twentieth centuries. While previous historians, like Robert Nye or Joanna Bourke, argued that the valuation of and shift to fear in French and British science, respectively, reflected broader cultural concerns with courage and modernity, Plamper suggested that it was early twentieth-century psychiatrists who transformed fear from an object that was unspeakable to one that was overtly analyzed and discussed in Russia. Thomas Dixon's study *From Passions to*

These different and variegated emotions are not always visible or apparent in terms of ostensible "emotions." Conversely, ostensibly expressed "emotions" are not always (or have not always been) interpreted in terms of emotions. These ignored and dismissed emotions were often expressed by scientists. William James had written of the "thrill" of newly discovered ideas, Alexander Bain analyzed the "excitement" of "chasing" after knowledge, Elie de Cyon described the "ecstasy" of science, and Walter B. Cannon compared his mountaineering experiences to his experiences inside the laboratory.[36] Historians, however, have often interpreted these expressed emotions of scientists and physicians in ways that dismissed them as emotions.[37] These emotions, as Hilary Rose observed with respect to aggression and anger, have become so "normalized within science that the dominant discourse no longer recognizes them as emotions."[38] Other historians have argued that the ostensibly expressed emotions of scientists either depict the scientist as courageous, honorable, heroic, or masculine or present the tribulations, pain, and suffering that the scientist must endure in pursuing science and producing knowledge.[39] Still others have intellectualized the scientist's emotions.[40]

Emotions suggested an alternative interpretation of a significant cultural-linguistic shift by emphasizing the role of science. Dixon's monograph studied the shift from a vocabulary of "passions" to a vocabulary of "feelings," "moods," and "emotions" during the late eighteenth and early nineteenth centuries. This significant shift captured a more general secularization of the affects in the British context. This major cultural-linguistic shift, according to Dixon, was developed primarily in the context of the learned-scientific study of the emotions in Great Britain (by Thomas Brown, Charles Bell, and, later, Herbert Spencer and Charles Darwin), rather than primarily by seventeenth- and/or eighteenth-century philosophers, as other authors emphasized. In a similar vein, Ruth Leys had argued that it was US-based psychoanalysts and psychiatrists of the 1960s who initiated emotional shifts in retheorizing the concept of survivor guilt into shame. For courage and the preoccupation with fear, see Nye, "Medicine and Science as Masculine 'Fields of Honor,'" *Osiris* 12 (1997): 60–79. For the argument that fear was regarded as a hallmark of modernity during the early twentieth century, see Bourke, "The Emotions in War: Fear and the British and American Military, 1914–45," *Hist. Res.* 74 (2001): 314–30. For psychiatrists in Russia, see Plamper, "Fear: Soldiers and Emotion in Early Twentieth-Century Russian Military Psychology," *Slav. Rev.* 68 (2009): 259–83. Dixon's argument can be found in *Passions* (cit. n. 4). On the seventeenth-century history of the fluctuating meanings of "emotion," see Thorley, "History" (cit. n. 16). Dixon's emphasis on nineteenth-century thinkers-scientists is not without its critics. See, e.g., Max Rosenkrantz, review of *From Passions to Emotions*, by Thomas Dixon, *J. Hist. Phil.* 43 (2005): 214–5. See also Leys, *From Guilt to Shame: Auschwitz and After* (Oxford, 2009).

[36] James, "What Is an Emotion?" (cit. n. 13); Bain, *Emotions* (cit. n. 24); E. Cyon, *Methodik der physiologischen Experimente und Vivisectionen* (Giessen, 1876); and Walter Bradford Cannon, *The Way of an Investigator: A Scientist's Experience in Medical Research* (New York, 1945).

[37] For exceptions, see, e.g., Dixon, "Tears"; White, "Darwin" (both cit. n. 31); and Elizabeth A. Wilson, "'Would I Had Him with Me Always': Affects of Longing in Early Artificial Intelligence," *Isis* 100 (2009): 839–47.

[38] Hilary Rose, "Gendered Reflexions on the Laboratory in Medicine," in *The Laboratory Revolution in Medicine*, ed. Andrew Cunningham and Perry Williams (1992; repr., Cambridge, 2002), 324–42, on 336.

[39] Nye, "Medicine" (cit. n. 35); and Naomi Oreskes, "Objectivity or Heroism? On the Invisibility of Women in Science," *Osiris* 11 (1996): 87–113. Rebecca Herzig studied how scientific authority was associated with the suffering scientist, and Cohen et al., in an edited volume on *Knowledge and Pain*, presented numerous case studies in which knowledge and pain were linked in a variety of contexts over a period of several centuries. See Rebecca M. Herzig, *Suffering for Science: Reason and Sacrifice in Modern America* (New Brunswick, N.J., 2005); and Esther Cohen, Leona Toker, Manuela Consonni, Otniel E. Dror, eds., *Knowledge and Pain* (New York, 2012). On the "gendering" of emotions, see Ute Frevert, *Emotions in History—Lost and Found* (New York, 2011); and Manuel Borutta and Nina Verheyen, eds., *Die Präsenz der Gefühle: Männlichkeit und Emotion in der Moderne* (Bielefeld, 2010).

[40] On the emotion of "intellectual passion" in Michael Polanyi, see Stuart Jacobs, "Polanyi's Presagement of the Incommensurability Concept," *Stud. Hist. Phil. Sci.* 33 (2002): 105–20.

But should one accept this emotion-denying perspective, or should the historian study the ways and means through which scientists expressed their emotions, perhaps in other terms and/or through alternative mediums and/or in terms of other kinds of emotions?[41] These other kinds and mediums of emotions may not be immediately recognizable because they did not appear in terms of commonsense notions of emotions or in terms of psychological, literary, or artistic emotions. They were not necessarily manifest in overt actions, in ostensible locutions, expressions, or behaviors of emotions, or in terms of conation and motivation.

These emotions, for example, could be manifest in terms of the exchange, mutual exhibition, and circulation of graphs and/or numerical tables of one's own embodiments among members of the scientific community (who recognized them as emotions), hand in hand with the exclusion of any direct reference to one's nameable emotions.[42] These were the emotions of communities of scientists during the late nineteenth century.[43] Are they "less than" or "not" an emotion vis-à-vis a nameable emotion—such as "fear" or "anxiety"? What are the meaningful distinctions between manifesting emotions in terms of these graphs and mediums versus the nameable emotion (or the overt action and expression)? The inclusion of these emotions extends the identities of emotions and the possibilities for histories, anthropologies, and sociologies of (these) emotions. It also presents alternative readings of the history of emotions and a different arrangement of the historical periodization of the emotions.[44] Conversely, the nameable emotion in science could indicate something that we do not conceive of or identify as an emotion. During the late nineteenth and early twentieth centuries, the named emotions of "jealousy," "anxiety," or "apprehension," for example, encompassed and gave meaning to a variety of disruptive laboratory and clinical events—all of which were subsumed under nameable emotions.[45] Physiologists and clinicians invoked "emotion" in observing numerous unexpected and, from their perspective, disruptive events. These events included anything from an unex-

---

[41] During the nineteenth century, Claude Bernard presented scientific discovery in terms of three stages, the first of which was "sentiment" (followed by rationality and laboratory experiments). See Bernard, *Introduction à l'étude de la médecine expérimentale* (Paris, 1865). On the "emotive components of experimentation" of Bernard, see Atia Sattar, "The Aesthetics of Laboratory Inscription: Claude Bernard's *Cahier Rouge*," *Isis* 104 (2013): 63–85, on 84.

[42] As historians of emotions have noted, in some (male) cultures, fear is often expressed in other terms, e.g., by cursing rather than by saying "I am afraid." See Peter N. Stearns, *American Cool: Constructing a Twentieth-Century Emotional Style* (New York, 1994). For the argument that new technologies reconfigured the exchange of emotions between humans and between humans and animals, see Otniel E. Dror, "The Scientific Image of Emotion: Experience and Technologies of Inscription," *Configurations* 7 (1999): 355–401; and Dror, "Counting the Affects: Discoursing in Numbers," *Soc. Res.* 68 (2001): 357–78. See also Claire Brock, "Risk, Responsibility and Surgery in the 1890s and Early 1900s," *Med. Hist.* 57 (2013): 317–37, for the argument that new surgical technologies at the end of the nineteenth century increased fear among patients.

[43] For an additional approach to the exchange of emotions among scientists during the twentieth century, see Wilson, "'Would I Had Him with Me Always'" (cit. n. 37). On "the subdued emotional register of exchange among scientists [which] was distinct from the more intense personal investments in exchange between scientist and Fore," see Warwick Anderson, "Objectivity and Its Discontents," *Soc. Stud. Sci.* 43 (2013): 557–76, on 566.

[44] Otniel E. Dror, *Blush, Flush, Adrenalin: Science, Modernity and Paradigms of Emotions, 1850–1930* (manuscript, Department of the History of Medicine, Hebrew University of Jerusalem).

[45] Examples include failure to replicate an experiment, a deviation of the subject/animal from its habitual reactions, or an inexplicable observation. See Daniel P. Todes, *Pavlov's Physiology Factory: Experiment, Interpretation, Laboratory Enterprise* (Baltimore, 2002); Todes, "Pavlov's Physiology Factory," *Isis* 88 (1997): 205–46; and Dror, "Affect" (cit. n. 32).

pected shift in blood pressure (during a blood pressure measurement), to a variety of pathological-like clinical findings (such as an elevated glucose level), to failures to replicate experiments. These inexplicable and disturbing moments inside the laboratory and clinic were articulated in terms of the language of emotions—by arguing that an emotion had infiltrated—was present inside—the laboratory or clinic. These invocations of emotions extended the possible meanings of nameable emotions, as well as assumptions about their corporeal expression, about the impact they had on the body, and about the way one could influence, control, or manage them. These invocations extend the reach of the history of emotions.[46]

Scientific technologies also extend the reach of the history of emotions because scientists designed into scientific technologies aspects of the affective logic of their societies. Late nineteenth-century scientific instruments, which represented a mechanical objectivity that erased feelings, also implicitly embodied a hypersensitivity and reactivity. On the one hand, these scientific technologies embodied the contemporary masculine emotional culture of self-control and restraint, and, on the other hand, they embodied a contemporary feminine emotional culture of sensitivity and feelings—of the female *Sensitive*. The attributes that explained (away) the female Sensitive's ability to "read" people were designed into the instruments. The instruments, like the Sensitive, worked by touching the body of the subject and detecting minute movements, which indicated shifts in emotions. The instruments embodied and negotiated these masculine and feminine facets of the culture of emotions in their very design and mechanisms.[47] Pervasive moods or structures of feelings were also embodied in particular disease entities and laboratory models, as were, we suggest, scientific texts.[48] In turn, scientific texts, models, and technologies generated emotions in different communities, like the antivivisectionists, but no less so in the investigators and, ultimately, in those who were measured, tested, and treated using these instruments and therapies.[49]

By reading these technologies, disease entities, laboratory models, and scientific texts for their emotions, historians of emotions and of science can study the broader affective logic that was designed into these entities and that partly structured and constituted them, and they can interpret aspects of these technologies, disease entities, laboratory models, and texts in terms of the culture of emotions that partly shaped them. They can also study the emotions that were generated by these technologies, disease entities, laboratory models, and texts and their significance for science qua science and for cultures of emotions. These different readings can reveal how the inadequacies and limitations of laboratory models and research practices can be crucial in shaping a spe-

---

[46] Dror, "Affect" (cit. n. 32).

[47] Otniel E. Dror, "Seeing the Blush: Feeling Emotions," in *Histories of Scientific Observation*, ed. Lorraine Daston and Elizabeth Lunbeck (Chicago, 2010), 326–48.

[48] In suggesting this possibility, we draw on Philip Fisher's analysis of the Gothic novel but apply it to scientific texts. See Fisher, *Vehement Passions* (cit. n. 16). For the argument that during the second half of the twentieth century, disease models of stress embodied two distinctive and oppositional affective moods of modern culture—"comforting" vs. "panic sentiment"—see Otniel E. Dror, "From Primitive Fear to Civilized Stress: Sudden Unexpected Death," in *Stress, Shock and Adaptation in the Twentieth Century*, ed. David Cantor and Edmund Ramsden (Rochester, N.Y., 2014), 96–120.

[49] See, e.g., Francis Gano Benedict, "The Excretion of Nitrogen during Nervous Excitement," *Amer. J. Physiol.* 6 (1902): 398–410. The situatedness of emotions within their cultural boundaries links these phenomena to the concept of "framing disease" as developed by Charles E. Rosenberg and Janet Golden. See Rosenberg and Golden, eds., *Framing Disease: Studies in Cultural History* (New Brunswick, N.J., 1991).

cific scientific understanding of emotions, as Eric Engstrom argues in his study of the late nineteenth-century psychiatrist Emil Kraepelin's research on affective disorders.

The generation of emotions in and by scientific texts, models, and technologies also suggests that we should extend the cerebral concept of "virtual witnessing" to include a new concept: "virtual feeling."[50] This term conveys the re-creation of the feelings/experiences of the experimenter inside the laboratory and/or of the adventurer explorer in the readers of scientific texts. The emotions of the experimenter inside the laboratory or of the field scientist were crucial to these endeavors and encounters. They raise the further question of whether historians, in following and modifying the model set by Otto Sibum in a different context and with other objectives in mind, should attempt to re-create the emotions of past experiments and experimenters.[51]

* * *

Envisaging the emotions of past experiments and experimenters also entails taking the emotions of a different but not less important community into consideration whose emotions have often been obfuscated and invisible in histories of science or emotions: the emotions of animals. Historians of emotions have thus far provided meager tools for incorporating animal emotions into history: as part of our human histories and as part of their histories. Many of the dominant concepts and methodologies in the history of emotions—including emotional communities, emotional regimes, emotional styles, emotives, and emotionology—are strictly anthropocentric.[52] They exclude the emotions of animals and the history of animal emotions from histories of emotions. In the history of emotions writ large, animal emotions are arguably the most significant absent other in what is largely and predominantly a history of human emotions.[53] This absence is particularly evident to historians of science, since historians of science encounter animals and their emotions in numerous contexts and scientific texts. Animals and their emotions, moreover, provided the early foundations for our contemporary science of the neurophysiology of (human) emo-

---

[50] In virtual witnessing, one provides a type of description of the experiment that creates in the readers of the description the feeling that they "witnessed" the enacted experiment. On virtual witnessing, see Steven Shapin and Simon Schaffer, *Leviathan and the Air-Pump: Hobbes, Boyle, and the Experimental Life* (Princeton, N.J., 1986).

[51] In his historical study, Sibum replicated a nineteenth-century experiment, in reconstituting the practices and gestures of the original experiment. See Heinz Otto Sibum, "Reworking the Mechanical Value of Heat: Instruments of Precision and Gestures of Accuracy in Early Victorian England," *Stud. Hist. Phil. Sci.* 26 (1995): 73–106.

[52] Despite the rapidly expanding literature in animal studies, the history of emotions has not incorporated the challenges that animals pose for its basic concepts and methodologies. For a first historical approach, see Pascal Eitler, "The 'Origin' of Emotions: Sensitive Humans, Sensitive Animals," in Frevert et al., *Emotional Lexicons* (cit. n. 16), 91–117; Eitler, "Übertragungsgefahr: Zur Emotionalisierung und Verwissenschaftlichung des Mensch-Tier-Verhältnisses im Deutschen Kaiserreich," in *Rationalisierung des Gefühls: Zum Verhältnis von Wissenschaft und Emotionen 1880–1930*, ed. U. Jensen and D. Morat (Munich, 2008), 171, 188; and Rob Boddice, ed., *Anthropocentrism: Humans, Animals, Environments* (Boston, 2011).

[53] The history of non-Western emotions is also largely undeveloped and underexplored. For exceptions to this neglect of non-Western emotions, see, e.g., Margrit Pernau, *Ashraf into Middle Classes: Muslims in Nineteenth-Century Delhi* (New Delhi, 2013); Pernau, *Transnationale Geschichte* (Göttingen, 2011); Angelika C. Messner, "Aspects of Emotion in Late Imperial China: Editor's Introduction to the Thematic Section," *Asia. Stud./Étud. Asia.* 66 (2013): 893–913.

tions, which also partly shapes our own scholarship in the history (and anthropology and sociology) of emotions.[54]

* * *

Exploring animal emotions in scientific settings also opens up a different conceptual perspective on emotion in general, for emotions are not necessarily the reason for (not) doing or deciding (not) to do something. They are often an integral part of the (non) actions themselves, accompanying and molding them in a noncausal or nonintentional way. They do not have to give an answer to the question why something is done but to the question how something is done and what the action signifies. Using the concept of an "emotional habitus," emotions are understood as embodied or even as an emotional disposition. This more Bourdieuan approach to emotion history emphasizes the role of the body in producing and sustaining emotions through practices—and in doing so it links the bodily dimensions with the socially and culturally constructed dimension of emotions.[55] This methodological assumption underpins Dolores Martín-Moruno's essay in this volume. Martín-Moruno analyzes the nineteenth-century physiologist Paolo Mantegazza, whose study of pain included laboratory research, as well as photographic studies.

Especially with regard to the history of science, one could ask whether emotional practices could be conceived of as something similar to what Ludwik Fleck had defined as "styles of thinking" (*Denkstile*) and "collectivities of thinking" (*Denkkollektive*) that characterize and convey the way a certain scientific group constructs, investigates, and verifies its research.[56] Fleck himself had conceded that emotions cannot be regarded as separate from these styles of thinking. According to Fleck there is no "emotionless state" as there is no "rationality as such." The entire process of scientific research (from observation to explanation) and the epistemic interest of the scientist were regarded by Fleck as driven by emotions.[57]

But postulating the existence of emotional styles in science (and elsewhere) means more than the occurrence of specific emotions at specific moments in the process of scientific research and theory construction.[58] Thus, it is not only about modifying a

---

[54] The emotions of animals were self-evident to major twentieth-century investigators of the emotions. As one of the leading interwar investigators of the emotions put it: "That various animals have subjective experiences is a supposition that is difficult to deny. . . . With those of us who perform experiments on animals as well as with the antivivisectionists this is an assumption that is acted upon with conviction." See Philip Bard, "On Emotional Expression after Decortication with Some Remarks on Certain Theoretical Views: Part I," *Psychol. Rev.* 41 (1934): 309–29, on 320.

[55] Deborah B. Gould, *Moving Politics: Emotion and ACT UP's Fight against AIDS* (Chicago, 2009); Eva Illouz, *Gefühle in Zeiten des Kapitalismus* (Frankfurt am Main, 2006); Pascal Eitler and Monique Scheer, "Emotionengeschichte als Körpergeschichte: Eine heuristische Perspektive auf religiöse Konversionen im 19. und 20 Jahrhundert," *Gesch. Gesell.* 35 (2009): 282–313; Scheer, "Are Emotions a Kind of Practice (and Is That What Makes Them Have a History)? A Bourdieuan Approach to Defining Emotion," *Hist. & Theory* 51 (2012): 193–220; Sabine Flach and Jan Söffner, *Emotionaler Habitus: Gefühle und Sinne zwischen Subjektivität und Umweltrelation* (Munich, forthcoming); Fay Bound Alberti, ed., *Medicine, Emotions and Disease, 1700–1950* (New York, 2006).

[56] Fleck, *Genesis* (cit. n. 32). See also Uffa Jensen and Daniel Morat, "Die Verwissenschaftlichung des Emotionalen in der langen Jahrhundertwende (1880–1930)," in *Rationalisierungen des Gefühls: Zum Verhältnis von Wissenschaft und Emotionen 1880–1930*, ed. Uffa Jensen and Daniel Morat (Munich, 2008), 11–34, on 12–3.

[57] Fleck, *Genesis* (cit. n. 32), 48.

[58] Dwight R. Middleton, "Emotional Style: The Cultural Ordering of Emotions," *J. Soc. Psychol. Anthropol.* 17 (1989): 187–201; William M. Reddy, "Emotional Styles and Modern Forms of Life," in

perspective or highlighting a previously underestimated step within the *Denkstil* concept. Nor is it solely about including emotions as one more or less decisive element within researching and explaining the production of knowledge. Certainly, the role of emotions within this process can and should be approached by investigating the emotional style that is dominant within a certain scientific field or setting and asking whether and how this style constituted knowledge production.

The concept of emotional style that is propagated here goes beyond this rather limited accessory role of emotions within the production of knowledge. It aims at a specific emotional *haltung*—an ethos, mindset, and attitude—that the researcher of a certain group, discipline, or science in general had internalized and is reenacting every time he or she is doing research. This emotional style can be multifaceted and can determine how the researcher deals with emotions in general or with specific ones, how he or she acts and interacts, and what techniques or instruments he or she uses. It is tied to and trained through practices and techniques that are part of his or her field. It is usually not limited to the narrow confines of a specific scientific setting. The emotional style, moreover, is not emotional in the conventional meaning of the word; that is, it is not necessarily emphatically emotional nor does it necessarily emphasize certain expressive emotions or emotionality as such. Emotional styles are thus not only intersubjective phenomena but—one could argue—they are part of an interobjectivity that is related to the interplay of humans, artifacts, and spaces.[59] Thus, Jessica Riskin's "sentimental empiricism," Lorraine Daston and Peter Galison's "objectivity," and Amir Alexander's "tragic" mathematics can all be interpreted as emotional styles.[60]

These emotional styles interrelate with broader emotional cultures in multitudinous ways. They might conflict or coexist with, be influenced by, or impact other coeval emotional styles—both scientific and extrascientific. They might even shape or con-

---

Sexualized Brains: Scientific Modeling of Emotional Intelligence from a Cultural Perspective, ed. Nicole C. Karafyllis and Gotlind Ulshöfer (Cambridge, Mass., 2008), 81–100; Eva Illouz, *Cold Intimacies: The Making of Emotional Capitalism* (Malden, Mass., 2007); Benno Gammerl, "Emotional Styles: Concepts and Challenges," *Rethink. Hist.* 16 (2012): 161–75.

[59] Andreas Reckwitz, "Affective Spaces," *Rethink. Hist.* 16 (2012): 241–58.

[60] Lorraine Daston and Peter Galison have explicitly wedded the emotion of "fear" to epistemology in arguing for an underlying "epistemic fear" as a basic and pervasive dimension of science in general— "all epistemology is driven by fear." Epistemic fear, they argue, is neither psychological nor sociological but straddles the boundaries between the sociological, the psychological, and the epistemic. It presents the authors' broader claim that all "epistemology is rooted in an ethos, which is at once normative and affective—or affective *because* normative." See Daston and Galison, "Objectivity" (cit. n. 30), 669, 671, 676; emphasis in the original. Amir Alexander's study of "tragic" mathematics and mathematicians charted the coeval shifts in the nature of mathematics and in the emotional nature of mathematicians during the late eighteenth and early nineteenth centuries, demonstrating changes in emotional styles. With the shift from Newton, Euler, d'Alembert, and Lagrange to Galois and Abel, the mathematician was transformed from a nonemotional to an extremely emotional being. This transformation in the nature of the mathematician went hand in hand with a major shift in the way that mathematics was conceived and practiced. During the eighteenth century (and before), the subject matter of mathematics was the physical world and its hidden harmonies. Mathematicians were those closest to the physical world and with the deepest understanding of it. They were literally "worldly men," and their successful public image reflected that. But in the nineteenth century mathematics broke with the physical world and became a separate universe of pure reason governed by mathematical laws alone. Mathematicians became like prophets, those gifted with a special vision that enables them to peer into this alternate reality. Extremism and emotionality were henceforth taken as paradigmatic of the true mathematician. See Alexander, *Duel at Dawn: Heroes, Martyrs, and the Rise of Modern Mathematics* (Cambridge, Mass., 2010).

tribute to shaping a new emotional style of Science in general, and they might explain shifts and changes in emotional styles in other arenas of society.

One brief example illustrates this latter possibility and the potential for rewriting aspects of the history of science and of the history of emotions by examining the interrelations between emotional styles in science and broader cultures of emotions, in proposing an integrated history of a science-emotions perspective. Over a decade ago, historian of emotions Peter Stearns identified a broad shift in the emotionology— "attitudes or standards that a society, or a definable group within a society, maintains toward basic emotions and their appropriate expression"—of late Victorian (mostly US) society.[61] One major feature of this emergent emotionology was a significant shift to an emphasis on emotional control and discipline. Stearns demonstrated that this "cool"-restraint emotionology dominated a wide range of social loci, one of which was science. The control and discipline of emotions in late nineteenth-century science was thus a product of the internalization of this broader cool emotionology of Western society. Historians of science, on the other hand, have demonstrated that the control over emotions in science was one element in a longer history of scientific objectivity. This history of scientific objectivity harked back to the seventeenth century. It preceded, rather than followed, the late Victorian shift to a cool emotionology.

These two divergent narratives, which draw on the history of emotions and on the history of science, suggest several new hypotheses that revise both narratives. One hypothesis privileges the history-of-science perspective but integrates it into recent models of historical change that draw on the history-of-emotions literature. According to this first interpretation, the shift to a cool emotionology during the late nineteenth century reflected the progressive dominance of the "emotional community" of scientists in late Victorian society.[62] It exemplifies one mechanism of change, in which one emotional community—science in this case—comes to dominate in a particular society at a particular moment in time. This first alternate reading of late nineteenth-century emotions can enrich and/or offer a possible alternative to Stearns's history-of-emotions narrative of late Victorian society.

Stearns's identification of a cool emotionology as a general feature of late Victorian society in turn complicates the history-of-science narrative. It challenges historians of science to heed the synchronic emotional contexts of the relationships between emotions and science, rather than to heed primarily the diachronic history-of-scientific-objectivity narrative. It suggests that historians of science have omitted a major factor in the history of scientific objectivity and that the history of objectivity should be rewritten to include the perspective of the history-of-emotions narrative and the emotional contexts of science.

---

[61] For the definition of "emotionology," see Stearns and Stearns, "Emotionology" (cit. n. 22), 813. On the shift to a cool emotionology, see Stearns, *American Cool* (cit. n. 42). According to Stearns, the changes in emotions went hand in hand with major social and economic changes. Other historians of emotions, like Susan Karant-Nunn, have maintained a stance that privileges emotions as primary movers, over economic and social changes. See Karant-Nunn, *The Reformation of Feeling: Shaping the Religious Emotions in Early Modern Germany* (New York, 2010).

[62] Barbara Rosenwein introduced the concept of "emotional community" in defining "social groups whose members adhere to the same valuations of emotions and their expression . . . the emotions that they value, devalue, or ignore; the nature of the affective bonds between people that they recognize; and the modes of emotional expression that they expect, encourage, tolerate, and deplore." See Rosenwein, "Problems" (cit. n. 3), 1–4.

For both historians of emotions and historians of science, this history indicates that late nineteenth-century science is the ideal locus for studying and characterizing the dominant culture of emotions during this particular period. Science, maybe even more than art or literature, best manifested (and was perhaps the harbinger of) the cool emotionology. The study of the history of science is the study of the history of emotions, and vice versa.

* * *

Each contribution to this volume exemplifies the historiographical work of emotion—what can a history-of-emotions approach offer historians of science/medicine/technology? What new insights become available with the inclusion of "emotion" as an analytical category? What is a history of emotions in science?

The essays are arranged under three major headings: Situating Emotions, Emotions into Practice, and New Emotions–New Knowledge–New Subjectivities. Many of the essays exemplify multiple dimensions of emotions and could fit under more than one major heading. The essays study developments in the history of technology and the human sciences (medicine, physiology, and psychiatry) on two different continents (North America and Europe) between the twelfth and the twentieth centuries. They exemplify the work that emotions can do for historians studying different periods, geographical areas, and disciplines.

SITUATING EMOTIONS

The contributions by Piroska Nagy and Damien Boquet, Naama Cohen-Hanegbi, Bettina Hitzer and Pilar León-Sanz, and Anne Harrington explore the ways in which medical and scientific communities approached and explained the functions of emotions and the concomitant positioning of emotions in and/or between body-mind intersubjectivity. In Nagy and Boquet's contribution, emotions serve as a common platform for bridging the medieval history of Christianity and the modern history of science. By drawing on the "religious" history of "passions," Nagy and Boquet address and redress the "scientific" history of modern emotions. They examine how Western thinking about emotions from the beginning of the twelfth century engaged new questions and received a new form and how emotions became an integral part of "human nature." Cohen-Hanegbi argues that an attention to emotions as an analytical category promotes a more contextualized reading of science, particularly in those historiographical enclaves where contextualization is still very much lacking. She fills a gap in late medieval history by approaching emotions in terms of medicine. She charts the contributions of scientific and medical thought to the formation of the late medieval culture of emotions. In this context, she also examines shifts in medical authority and the tensions between medicine as art and medicine as science in relation to the new authority over emotions. The contribution by Hitzer and León-Sanz presents a cross-Atlantic study of psychosomatic medicine in Germany and the United States in terms of distinct shifts in emotions. They study the introduction of new ideas with respect to the emotional body and the emotional self in psychosomatic medicine, which emerged between 1920 and 1960, and through a case study of cancer. They trace a significant shift in cancer: from an early twentieth-century organic conception of malignant cells, which was impervious to the influence of emotions, to a psychosomatic conception, which took off from emotions, over the course of the twentieth century. Their essay illuminates how and why this change occurred rather late in Germany and how the shift in Germany was influenced by the earlier turn toward framing cancer psychosomatically in the United States. Harrington's contribution charts a grand historical narrative of motherly love and of schizophrenia during the twentieth century. Her study weaves the history of motherly love and schizophrenia into a genealogy of major shifts in the conceptualization of emotions. She demonstrates how, during the period after World War II, the meandering history of motherly love and that of schizophrenia converged, creating the emblematic figure of the "Schizophrenogenic Mother."

EMOTIONS INTO PRACTICE

The contributions by Rafael Mandressi, Dolores Martín-Moruno, Eric J. Engstrom, and Anja Laukötter illustrate ways in which emotions infused practices and practices generated emotions. Mandressi examines how an attention to emotions affords new interpretations of scientific developments that are beyond the study of "emotions." Mandressi draws on the concept of "emotional communities" in explicitly challenging the history of medical professionalization during the early modern period. He examines how the shift to the dead body, with the emergence of the practice of anatomical dissection during the thirteenth century, had significant affective implications. The attention to the administration of affects, in turn, was important for the professional identity of practitioners and for their social status. Anatomy in early modern Europe was a significant locus "where affective cultures were produced and negotiated among several professional and social groups." Martín-Moruno's contribution juxtaposes three interrelated developments that converged on pain. By positioning pain as her organizing principle, she links between shifts in cultures of pain, the development of a science of pain, and the changing persona of the scientist during the nineteenth century. Martín-Moruno's contribution suggests how an attention to pain closes the gap between macro transformations in cultures of pain and micro productions of knowledge inside the laboratory. Engstrom's article emphasizes how the intractability of emotion as an object of knowledge and praxis was instrumental for the emergence of new knowledge and new emotions. Engstrom studies Emil Kraepelin's efforts to account for the significance of emotions in psychiatric illnesses. The failure to scientize emotions drove the research on affective disorders and decisively shaped one of the most influential psychiatric nosologies of the twentieth century. Laukötter's article examines how films on health education evolved as a specific "emotional engineering technique" to shape not only adults' emotions and behaviors but also the emotional and intellectual development of pupils in the classroom. Focusing on Germany and the United States in the first decades of the twentieth century, the essay explores how this practice of influencing through the visible was generated through various scientific fields, such as medicine and hygiene as well as psychology and pedagogy.

NEW EMOTIONS–NEW KNOWLEDGE–NEW SUBJECTIVITIES

The contributions by Felicity Callard and Otniel E. Dror study how new and emerging identities of and criteria for emotions created new knowledge and new subjectivities, and vice versa. Callard's contribution examines the emergence of a new category of emotion within the study of emotions—the "panic disorder." Callard draws on this new emotion in order to extend the historiographical focus on organizational, institutional, societal, and professional changes in the history of psychiatry to a focus on how and why the very shape, texture, and experience of symptoms, affects, and behaviors variously positioned within the purview of psychiatry and its proximate domains changed. Dror's contribution studies the discovery of a "new" post–World War II "supramaximal" "super-pleasure." Dror studies the laboratory enactments that constituted this new pleasure as "supramaximal," instant, and insatiable and suggests several postwar contexts that situate the new pleasure. He also reflects on—in introducing—an approach that "sides with" emotion, presents the notion of a "missed" emotion, and considers the meanings of "repetitions"—for science and for pleasure.

These diverse case studies argue for and provide exemplifications of the benefits and insights that become available in the dialogue between the history of emotions and the history of science. The reciprocal interchange between these histories can lead us to rethink our categories of analysis, our subjects, and our periodizations. The ten studies assembled here, which span several centuries, continents, and scientific disciplines, provide a preliminary map and guide for future endeavors. We hope that this volume will stimulate further research that combines the history of science and the history of emotions.

SITUATING EMOTIONS

Medieval Sciences of Emotions during the Eleventh to Thirteenth Centuries:

An Intellectual History

by Damien Boquet* and Piroska Nagy‡

ABSTRACT

The standard narrative of the development of Western thinking about emotions is that the concept of emotions emerged alongside the secularization of European society and thought and was linked to the emergence of psychology as a discipline. This essay argues that a systematic psychology of affectivity emerged far earlier and can be found in Western Christian thought. In the context of the cultural renewal of the eleventh to thirteenth centuries, Christian anthropology—the conception of the human being—was totally reshaped. We would like to trace the steps of this development. The shift took place gradually from the end of the eleventh century onward, both in the monastic context and in the new scholastic milieu. In earlier medieval times, emotions were understood in terms of the dual moral perspective of vices and virtues, as defined by the parameters of the Fall and Salvation: affective life was largely reduced either to negative disturbances of the soul that a Christian should resist or to a positive love of God that one should cultivate. From the beginning of the twelfth century onward, Western thinking about affective life began to engage new questions. Emotions both positive and negative came to be regarded as important aspects of a more complex picture of human nature and attracted growing attention as such. Without departing from the Christian framework, which remained the basis of the growing, psychologically oriented literature, emotions came to be described in relation to the powers of the soul, and their sensory and bodily dimensions, as well as their cognitive, rational, and volitional functions, were increasingly considered in an integrated way. Our essay analyzes the medieval psychology of emotions from a fresh perspective.

INTRODUCTION

The prevailing view of the development of Western thought on emotions is that it was only as late as the end of the nineteenth century, with the secularization of European

* Université Aix-Marseille, Département d'Histoire, 29, Avenue Robert Schuman, 13621 Aix-en-Provence Cedex 1, France; damien.boquet@gmail.com.
‡ Université du Québec à Montréal, Département d'Histoire, Case Postale 8888, Succursale Centre-Ville, Montréal, Québec, H3C 3P8, Canada; npmediumaevum@gmail.com.
A somewhat different treatment of the topic can be found in chapter 6 of our book, *Sensible Moyen Âge: Une histoire des émotions dans l'Occident médiéval* (Paris, 2015), entitled "La nature émotive de l'homme (XIe–XIIIe siècles)." This essay was enriched by the remarks of Laurence Moulinier, Monique Paulmier-Foucart, and Sylvain Piron, whom we thank warmly. Many thanks to Lauren Mancia and Curie Virag for their help in revising two successive versions of the English text and to Xavier Biron-Ouellet, Naama Cohen-Hanegbi, and Nicole Hochner, as well as to the anonymous readers, for their erudite comments.

society and thought, that a scientific psychology[1]—and with it, a scientific consideration of emotions—emerged. In his already classic book from 2003, *From Passions to Emotions: The Creation of a Secular Category*, Thomas Dixon proposed historicizing the category of emotion in psychology, which he assumes "did not exist until just under two hundred years ago."[2] He emphasizes that the rise of the term "emotion" corresponded to a new opposition to reason, which, in turn, resulted from the secularization of Western thinking. Dixon notes that, even before this turn, there was a lively scientific investigation of the moral and affective dimensions of human life, but earlier Western thinkers mostly considered the "passions" and "affects" of the soul as notions embedded in a network of Christian categories.[3]

Along the same lines, Fernando Vidal and Gary Hatfield have explored the emergence of the notion of psychology in the sixteenth century and its development as an independent discipline in the eighteenth.[4] Focusing on a later shift, Hatfield and Vidal do not deal with the Middle Ages but begin their narrative in the sixteenth century. Kurt Danziger, also concerned with the recent history of psychology as a discipline, devotes but one line to medieval developments, which are presumably a continuation of ancient thought.[5] Dixon's account, which deals specifically with emotions, reduces the medieval millennium to a passage from Augustine in the fifth century (who regarded the affects as part of the will) to Thomas Aquinas in the thirteenth (who would restore a theory of passions). Although their cursory treatment of medieval developments is understandable given their respective topics, all these scholars have surrendered to the *mala fama* of the Middle Ages while ignoring the interest and complexity of medieval thinking about affective life, with its nuanced meanings, manifold functions, and diverse social uses.[6] All of these dimensions were of crucial importance in many realms of medieval thought and culture, from spiritual speculations to hagiography and mystic discourse, from anthropology to theology, from medicine to natural philosophy.

In the limited space of this essay, we cannot embrace all of these fields. Concentrating only on theoretical developments, we aim to show that Western medieval intellectuals from the end of the eleventh century onward, without letting go of the Christian moral tradition that still constituted the anthropological framework of their thought,

---

[1] This idea, expressed in a popular form by Hermann Ebbinghaus, *Abriss der Psychologie* (Lepizig, 1908), at the beginning of the twentieth century, was challenged by Fernando Vidal in *The Sciences of the Soul: The Early Modern Origins of Psychology* (Chicago, 2011), 8–10. It is discussed in detail by Kurt Danziger, "Long Past, Short History: The Case of Memory," in *Problematic Encounter: Talks on Psychology and History*, http://www.kurtdanziger.com/Paper%2010.htm (accessed 25 August 2015); Danziger, "Psychology and Its History," *Theory Psychol.* 23 (2013): 829–39.

[2] Thomas Dixon, *From Passions to Emotions: The Creation of a Secular Category* (Cambridge, 2003), 1–11, quotation on 1.

[3] Ibid., 4.

[4] Vidal, *Sciences of the Soul* (cit. n. 1); Gary Hatfield, "Remaking the Science of Mind: Psychology as a Natural Science," in *Inventing Human Science: Eighteenth Century Domains*, ed. Christopher Fox, Roy Porter, and Robert Wokler (Berkeley and Los Angeles, 1994), 184–231.

[5] Danziger, "Long Past"; Danziger, "Psychology" (both cit. n. 1).

[6] For a glimpse into the vast scholarship on this subject, see Richard Sorabji, *Emotion and Peace of Mind: From Stoic Agitation to Christian Temptation* (Oxford, 2000); Bernard Besnier, Pierre-François Moreau, and Laurence Renault, eds., *Les Passions antiques et médiévales* (Paris, 2003); Simo Knuuttila, *Emotions in Ancient and Medieval Philosophy* (Oxford, 2004); Alain Boureau, *De vagues individus: La condition humaine dans la pensée scolastique* (Paris, 2008); Carla Casagrande and Silvana Vecchio, *Passioni dell'anima: Teorie e usi degli affetti nella cultura medievale* (Florence, 2015); Barbara H. Rosenwein, *Generations of Feeling: A History of Emotions (600–1700)* (Cambridge, 2015).

produced a psychology of emotions marked by a naturalistic approach and a rational, systematic method, which can be characterized as scientific in a broad sense. Taking into consideration the period of intense intellectual effervescence between the eleventh and thirteenth centuries may help us to understand the shift from Augustine's conception of passion to that of Thomas and in this way offer a new narrative of the emergence of a "scientific" discourse of emotions in the Christian West.

Before going further, we should provide a few clarifications on terminology. While we shall occasionally invoke the terms "science" or "scientific," we do not intend to discuss here at length the legitimacy of using them in the context of such a remote period, as the possible understandings of these words are diverse.[7] Historians of both ancient and medieval culture have always employed the term "science" to refer to what people of the period they studied understood by *scientia* in Latin (translating from *épistémé* in Greek and referring to "knowledge," as opposed to *doxa*, belief or opinion). Certainly, for medieval intellectuals—and for historians of medieval science—it does not possess the same meaning as it does for modern scientists—empirical knowledge produced by observation. Saint Augustine in the fifth century distinguished *scientia*, which covers the rational knowledge of temporal things, from *sapientia*, or the intellectual knowledge of eternal things.[8] In this sense, we use the term to designate the systematic, rationalized knowledge produced by medieval intellectuals, which aimed at discussing in a conscious and critical way the ancient and medieval written scientific tradition, especially in the context of urban schools that flourished from the eleventh century onward in the West. In the later scholastic context of the thirteenth century, the Augustinian distinction lost its validity when theology—the science of, or discourse on, God—came to be considered the "queen of sciences."

Similarly, we shall use the term "psychology"—a term coined far later,[9] and thus not used by medieval intellectuals, but used by medievalists—for speaking about medieval discourse on the human soul. We maintain that the period of the eleventh to the thirteenth centuries, due partly to the assimilation of ancient, Aristotelian knowledge, fostered the emergence of a scientific psychology of emotions—that is, methodical study of the ways by which affective movements and states operate in human life. Being part of a scientific discourse in the sense that medieval thinkers understood it, this lore on emotions remained mainly theoretical and was only partly based on observation. Still, its discussions show the practical problems that it was meant to resolve, and also the way it was integrated into the framework of a Christian anthropology or medieval "human science" program.[10]

Finally, our use of the term "emotion" for a period that did not know the word conforms to the present-day academic consensus.[11] The medieval nomenclature of emo-

---

[7] See David C. Lindberg, *The Beginnings of Western Science: The European Scientific Tradition in Philosophical, Religious, and Institutional Context, 600 B.C. to A.D. 1450* (Chicago, 1992), 1–4.

[8] Augustine, *De Trinitate libri quindecim*, 12.15.25: "Si ergo haec est sapientiae et scientiae recta distinctio, ut ad sapientiam pertineat aeternarum rerum cognitio intellectualis; ad scientiam vero, temporalium rerum cognitio rationalis, quid cui praeponendum, sive postponendum sit, non est difficile iudicare." See http://www.augustinus.it/latino/trinita/index2.htm (accessed 26 August 2015).

[9] Vidal, *Sciences of the Soul* (cit. n. 1), 21–57.

[10] See ibid., xi, 2, which, although focused on a much later period, has a similar aim.

[11] Reducing this note to medieval studies, in addition to the titles quoted in n. 6, see, e.g., Barbara H. Rosenwein, *Emotional Communities in the Early Middle Ages* (Ithaca, N.Y., 2006); Peter King,

tions in Latin was multiple and variegated, including terms from *motus animi/animae* to *perturbatio*, from *affectus* and *affectio* to *passio*, from *inclinatio* to *primus motus*. While every one of these notions has its own genealogy and cultural background that would necessitate a detailed discussion *in se*, it would be arbitrary to choose one while speaking in a synthetic way about a *longue durée* process, during which the phenomenon of affective movement (or habit) received various names. In addition, our word "emotion" refers to a phenomenon that is both cognitive and affective—an idea that has always been present in medieval reflections about affectivity. Finally, a third important reason for adopting the "E-word": our use of this umbrella term often designating any feature that belongs to the field of affectivity makes it possible for specialists of early and non-Western civilization and historians of contemporary society and science to understand one another. In this effort, some historians suggest the use of the terms "affect" and "affective," as they would help to acknowledge both the differences between affective movements, states, and long-lasting dispositions, on one hand, and the rise of a nuanced thinking about them, on the other. However, the recent success of the term in contemporary social sciences, through such expressions as "the affective turn" or "the turn to affect" and the correlated "affect theory," complicate the landscape, as they designate a turn to the neurosciences of emotions—an essentialist and biologically wired conception of emotion.[12] As far as there is consistency in the field,[13] "affect theorists," situated between the social and natural sciences, affirm the autonomy of affect from the realm of discourse, in a somewhat postmodern, if not posthuman way. Happily, for medievalists the problem is irrelevant: medieval texts impose upon us an integrated understanding of affective, sensitive, and rational discursive faculties and experiences.

According to Carla Casagrande and Silvana Vecchio, we can identify two types of discourse on affective life in medieval Western thought:[14] a theoretical discourse, which defines the place of emotions in human nature and in the ethical life; and a pedagogical discourse, in which emotions, understood as movements of the will, are analyzed in the framework of their management, use, or politics, in an endeavor to realize salvation. In this sense, while one can easily defend the idea that a spiritually oriented approach prevailed during the (earlier) medieval centuries in the West, one cannot argue that, at a given moment in the intellectual history of emotions in Chris-

---

"Emotions in Medieval Thought," in *The Oxford Handbook of the Emotions*, ed. Peter Goldie (Oxford, 2010), 167–88; Carla Casagrande, "Le emozioni e il sacramento della penitenza," in *La Penitenza tra I e II millennio. Per una comprensione delle origini della Penitenzieria Apostolica*, ed. Manlio Sodi and Renata Salvarani (Vatican City, 2012), 213–31; Martin Pickavé and Lisa Shapiro, eds., *Emotion and Cognitive Life in Medieval and Early Modern Philosophy* (Oxford, 2012).

[12] For such an approach, see Brian Massumi, "The Autonomy of Affect," *Cult. Critique* 31 (1995): 83–109; his article was perceptively criticized by Ruth Leys, "The Turn to Affect: A Critique," *Crit. Inq.* 37 (2011): 434–72; Rafael Mandressi, "Le temps profond et le temps perdu: Usages des neurosciences et des sciences cognitives en histoire," in "Les sciences de l'homme à l'âge du neurone," special issue, *Rev. Hist. Sci. Hum.* 25 (2011): 165–202; Brigitte Chamak and Baptiste Moutaud, eds., *Neurosciences et société: Enjeux des savoirs et pratiques sur le cerveau* (Paris, 2014), particularly their "Introduction: La vie sociale des neurosciences," 10–14.

[13] In their introduction to *The Affect Theory Reader*, Melissa Gregg and Gregory Seigworth enumerate eight very different trends in the field. See Gregg and Seigworth, eds., *The Affect Theory Reader* (Durham, N.C., 2010), 6–8.

[14] Carla Casagrande and Silvana Vecchio, "Les théories des passions dans la culture médiévale," in *Le Sujet des émotions au Moyen Age*, ed. Damien Boquet and Piroska Nagy (Paris, 2009), 107–22, esp. 112.

tian thought, a clear shift took place from a moral discourse to a scientific one. Already the Church Fathers, who included the passions in their system of vices and virtues, linked their descriptions to a vision of humankind (or anthropology) that was influenced by medical knowledge and natural philosophy. In the same way, instead of leading to the neglect of the moral aspects of emotions, the rise of an interest in nature in Western thought after the turn of the first millennium integrated moral questions with natural philosophy. The two perspectives were thus always entangled with one another and constituted two poles of the same cultural setting.

Acknowledging that Aquinas formulated the first synthetic "treatise of passions,"[15] our concern here is to trace the cultural and intellectual conditions in which the production of such an oeuvre became possible. We shall argue that, in the context of the cultural renewal of the West during the eleventh to thirteenth centuries,[16] the conception of human affectivity was significantly reshaped. In the earlier Middle Ages, affective movements in human life were understood in terms of the dual moral perspective of vices and virtues, as defined by the Fall and Salvation: affective life had been largely reduced either to negative disturbances of the soul that a Christian should fight or to a positive love of God that a Christian should cultivate. From the end of the eleventh century onward, in the context of a cultural dynamism to which we shall return later in this essay, Western Christian anthropology—that is, the Western Christian vision or conception of the human being—both produced and received new ideas, in part from medicine and the newly rediscovered natural philosophy.

It is at this moment that a medieval "scientific" psychology—understood here following Naama Cohen-Hanegbi[17] as the systematic, rational study of the soul and its faculties, in the framework of an anthropology tied to theology[18]—springs up. This development took place mainly in the cultural and intellectual context of the new urban schools, which would later become universities. In making this claim, we are proposing that the impetus for change was not the influx of "arabized" Aristotelian psychology, that is, the arrival in the West of the texts of Aristotle through translations from the Arabic and accompanied by the interpretations of Arabic commentators. Instead, it was an internal evolution of Western thought and *Weltanschauung*, which departed from early medieval Augustinian dualism. This evolution could already be discerned in the twelfth century, in the emergence of different approaches to anthropology, which stressed that affectivity was not only a spiritual and moral phenomenon, but also a bodily one. As a result, the affective life began to be recognized as an important and complex part of human nature and received growing attention as such. Through complex scholarly debates that frequently arose at least partly in the monastic context, this psychology of emotions took shape between the end of the eleventh and thirteenth centuries as part of an integrated—or interdisciplinary, as we would say today—field of Christian scholastic "human sciences" or anthropology.

---

[15] See Silvana Vecchio's introduction to Thomas Aquinas, *Le passioni dell'anima*, trans. Vecchio (Florence, 2002), 5–20.

[16] This context is dealt with in greater detail below in "The First Movements of Feeling and Sin: A Scholastic Debate."

[17] See Naama Cohen-Hanegbi, "A Moving Soul: Emotions in Late Medieval Medicine," in this volume.

[18] The study of man (or anthropology) was part of medieval theology. A good place to see this in detail is the *Summa theologiae* of Thomas Aquinas, where man is dealt with in the first part discussing God and his Creation, and more specifically in the other parts of the *summa*; see also below, "Twelfth-Century Monastic Anthropology."

This field included as many different disciplines and approaches to the human as those provided by medicine, theology, and natural philosophy. Without departing from the Christian perspective that remained a common framework for this new psychological literature, emotion, described in relation to the powers of the soul, itself became a power of the soul, so that the questions of senses, of bodily manifestations, and of links to cognition and rationality were all considered in an increasingly integrated way.

This essay intends to make four points about the advent of a medieval affective psychology—a *scientia de anima*, a scientific discourse. First, we shall examine the debate on the first movements of feeling, as discussed in early scholastic anthropology, which enables us to discern the anthropological and theological implications of diverse conceptions of emotions. Second, we shall consider the input of medical knowledge in the formulation of a vision of emotions informed by both physiology and psychology, that is, by knowledge of the body and soul. Third, we shall argue that the monastic anthropology of the twelfth century combines the new knowledge of the body with a vision of the human that remains spiritual in its orientation. Finally we shall analyze the steps of the development of what is called the "psychology of faculties," leading to the integration of emotions within a systematic science of the human, or human science, in the context of the thirteenth-century scholastic synthesis.

THE FIRST MOVEMENTS OF FEELING AND SIN: A SCHOLASTIC DEBATE

A complex and strange debate concerning the nature of the "first movements of feeling," one of the medieval terms for what we call emotions, is a good starting point for identifying the issues at stake and the orientations of one of the most important scientific debates in the twelfth and thirteenth centuries around the role of emotions in human life. What it reveals is the deeply ambivalent nature of emotions for medieval thinkers—before and after, and inside and outside the scholastic turn described above.

In the anthropology of the Fathers of the Church, the movements of the soul were directly dependent on the will. According to Gregory the Great (d. 604), who, along with Augustine (d. 430), was the most widely read of the Western Church Fathers, while the body possessed natural appetites like hunger or thirst, all feeling involving the soul entered the realm of the will. Thus, the pleasure felt while tasting a drink, for instance, potentially contained a sinful dimension, even though the fact of drinking was a natural, pressing need.[19] For a long time, these patristic ideas were well received; it was only at the beginning of the twelfth century that these questions started to be discussed again, in the framework of a renewed reflection on the role of the various faculties of the soul in the context of understanding sin.[20]

At the end of the 1130s, Abelard (d. 1142) argued in his *Ethics*, or *Scito te ipsum*, that there was no sin without the consent of consciousness: "the sin is not the desire

[19] Gregory the Great, *Moralia in Job*, ed. Marcus Adriaen (Turnhout, 1979–85), 30.18.61, quoted by Silvana Vecchio, "Il piacere di Abelardo a Tommaso," in *Piacere e dolore: Materiali per una storia delle passioni nel Medioevo*, ed. Carla Casagrande and Silvana Vecchio (Florence, 2009), 68n6.

[20] Odon Lottin, "Les mouvements premiers de l'appétit sensitif de Pierre Lombard à saint Thomas d'Aquin," in *Psychologie et morale aux XIIᵉ et XIIIᵉ siècles*, vol. 2, *Problèmes de morale* (Gembloux, 1948), pt. 1, 493–589; Knuuttila, *Emotions* (cit. n. 6), 185–95; Vecchio, "Il piacere" (cit. n. 19), 67–86.

for the wife of another . . . but rather to consent to this desire."[21] This argument, since referred to as the "morality of intention," led Abelard to reject the idea that *delectatio*, the suggestion of desire or spontaneous pleasure, could contain sin. A spontaneous feeling, without or before consent, was not sinful—even when it oriented the soul toward sin. In order for there to be sin, one needed to consent—to engage one's will, which was guided by reason.

At first sight, Abelard's position is not revolutionary. Augustine had already held that a movement engaging the will was a necessary condition for sin. Yet for Augustine, even the most spontaneous pleasure of temptation already contained a bit of consent.[22] This was a legacy of the idea of original sin. It was from this malediction, by which the will failed to exercise power over itself, that Abelard liberated the human subject. For Abelard, only a fully conscious will could be considered. At the same time, Abelard also liberated emotions from the weight of original sin by distinguishing vices of nature from vices of consent: the first was inclined toward evil but was not a sin. So, while a spiteful act perpetrated by anger was indeed a sin, being subject to anger and having a wrathful temperament was not, even if it was a natural defect of the soul:

> Mental vice of this kind is not, however, the same as sin; nor is sin the same as a bad action. For example, to be irascible—that is, prone or ready for the emotion of anger—is a vice and inclines the mind impetuously or unreasonably to do something, which is not at all suitable. However, this vice is in the soul, so that in fact it is ready to be angry. Even when it is not moved to anger, just as the limpness for which a man is said to be lame is in him even when he is not walking limply, because the vice is present even though the action is not. So, too, nature itself, or the constitution of the body, makes many prone to luxury just as it does to anger, yet they do not sin in this because that is how they are, but through this they have the material for a struggle so that triumphing over themselves through the virtue of temperance they obtain a crown.[23]

This doctrine had important consequences for the understanding of emotions: while preserving their moral engagement, they were subject to a more subtle and complex evaluation than before. The move that Abelard was making seems at first sight modest but was, in fact, of great consequence. Without neglecting the ethical dimension of emotions, Abelard assumed that emotional character fell within nature, not sin. People remained fully responsible for the acts that their emotions pushed them to commit, but emotionality in itself, as a constituent of human nature, was free from the weight of original sin.

Such a view, which was revolutionary from both the anthropological and the theological points of view, did not prevail in all milieus in Abelard's time. There was an important divide on this question—as on many others—between the very radical followers of the turbulent urban master Abelard and some more traditional circles, although, as we shall see, this divide was not a monastic versus secular one. The Cistercian monk William of Saint-Thierry (d. 1148) understood well what was at stake when he warned his friend, Bernard of Clairvaux (d. 1153), that this conception placed

---

[21] Peter Abelard, *Ethics*, ed. and trans. David E. Luscombe (Oxford, 1971), 1.3.25.

[22] Augustine of Hippo, *De Trinitate libri XV*, ed. William John Mountain, Corpus Christianorum Series Latina 50 (Turnhout, 1968), 12.12.372.

[23] Abelard, *Ethics* (cit. n. 21), 1.2.4–5; we revised the translation slightly.

moral responsibility at the level of consent. Indeed, the abbot of Clairvaux, whose position followed that of Gregory the Great,[24] included this thesis among the nineteen denounced before the council of Sens in May 1141.[25] Still, one cannot find a diametric opposition concerning this question between the monastic world as such (which would stick to a rather dark vision of the concupiscent will) and the urban world of schools (which would promote a neutral, natural conception of emotion). Belonging to the same monastic order as Bernard and William of Saint-Thierry, Aelred of Rievaulx (d. 1167) stated, when debating the question, that being troubled by these first movements was neither good nor bad.[26] But even in the world of the urban schools, there emerged a split in the second half of the twelfth century between two evaluations of the first movements. A majority of authors, including influential masters from Anselm of Laon (d. 1117) to Peter Lombard (d. 1160), whose *Sententiae* became the classical textbook for thirteenth-century students in theology, upheld the classic position inherited from the Fathers. Still, certain masters influenced by the school of Abelard, like Simon of Tournai (d. 1201) or Alan of Lille (d. 1202), argued that the first goad of desire [*titillatio carnis*] could not be a sin, since it escaped all control of the will. While the conservative option prevailed in the long term through the joint authority of Lombard and Aquinas, the argument defended by Abelard remained sporadically present in the first half of the thirteenth century.

At first sight, we are tempted to conclude that the attempt of Abelard and his school to naturalize the process of emotion in Christian anthropology was a failure. In reality, the discussion had a lasting influence on the cartography of sensitive powers of the soul in scholastic anthropology. The Patristic tradition had divided the appetites between sensual perception on the one hand, which engaged the body and the animal soul [*anima bestialis*], and an affective sensitivity on the other, made up of desires and repulsions, which fell under the competence of the rational soul [*anima rationalis*]. This conception sustained an ambiguity concerning the first movements, or emotions, as they could be found at both levels of the soul. To resolve the ambiguity, twelfth-century masters, using the scholastic method of distinction, identified a double sensitivity [*duplex sensualitas*]—the animal and rational. In this way, they cut the first movements in two, distinguishing, with Gilbertus Porretanus (d. 1154), "primary first movements," such as the sensation of hunger or thirst, which stimulated only the animal sensitivity, and "secondary first movements," or emotions proper, which fell under the rational sensitivity. Thus emotion, born unintentionally, involved the movement of the rational soul and was potentially controlled by reason, and so, ultimately, by the will. Irrepressible emotions fell within the realm of moral responsibility, since the soul was to step in to repress them if inappropriate. In this way, the weakness of nature coming from the malediction of original sin was transformed into an individual failure of reason, which was to exercise responsibility over the emotions. Albeit displaced, the

---

[24] "We have to squash the head of the serpent, that is, stifle in us the primary movement." Bernard of Clairvaux, "Sententia 107: Series tertia," in *Sancti Bernardi Opera*, ed. Jean Leclercq, Charles H. Talbot, and Henri Rochais (Rome, 1972), vol. VI-2, 173.

[25] On this affair and its chronology, see Pietro Zerbi, "Les différends doctrinaux," in *Bernard de Clairvaux: Histoire, mentalités, spiritualité* (Paris, 1992), 429–58; Constant J. Mews, *Abelard and Heloise* (Oxford, 2005), 233–43.

[26] Aelred of Rievaulx, *De speculo caritatis*, ed. Charles H. Talbot (Turnhout, 1971), 3.38.123. See also his Sermon XXXII, "For the Purification of the Virgin," in Aelred of Rievaulx, *Sermones I–XLVI: Collectio claraevallensis prima et secunda*, ed. Gaetano Raciti (Turnhout, 1989), 17–8.263.

original ambiguity of emotions survived, torn between the animal, moral, and rational natures, and between spontaneity and imputability.

Frustrating in its complex and technical nature, the debate on the first movements contributes to a partial naturalization of emotion, without depriving human beings of responsibility for the acts it incites. In this sense, and in another framework, one might investigate the extent to which the conception and the practice of emotions of some of these authors—such as Abelard or the Cistercian monk Aelred of Rievaulx, or, in another style, Bernard, all of whom left writings informing us about their lives—were consistent with each other. In other words, we could discuss how closely the history of science and the history of emotions were linked together on a microscale—and to what extent these major figures, writing about emotions in the twelfth century, did live in congruent emotional communities.[27]

If the theoretical inflections we have examined seem minute, they had long-lasting consequences. This dispute, which had high theological stakes, demonstrates how scholastic thinking on emotions evolved, already in the twelfth century, toward the integration of natural philosophy, which provided a new framework for human beings to perfect their nature as emotional creatures endowed with reason. The development of a science of emotions in the medieval West was, to this extent, inseparable from the renewal of medicine from the eleventh century onward, with the influx of Greek and Arabic knowledge,[28] as the essay by Cohen-Hanegbi in this volume soundly demonstrates.

A MEDICAL SCIENCE OF EMOTIONS BETWEEN PSYCHOLOGY AND PHYSIOLOGY[29]

While Abelard was not at all well informed concerning medicine, the naturalization of emotions that he proposed cannot be understood without this background. After a relatively dormant period in the early Middle Ages, when medical knowledge survived mostly in the monastic world, from about the second half of the eleventh century onward, a revival began in southern Italy, where Latin, Greek, and Arabic cultures had coexisted for several generations. Constantine the African (d. end of eleventh century), who arrived in southern Italy in 1077, most probably from a Christian community of the Maghreb, entered the monastery of Monte Cassino.[30] In the last decades of his life, he translated medical texts from Arabic to Latin, including the *Isagoge* of Ḥunayn ibn Isḥāq (Johannitius for the Latins), a standard textbook for medical stu-

---

[27] See Rosenwein, *Emotional Communities* (cit. n. 11); Otniel E. Dror, Bettina Hitzer, Anja Laukötter, and Pilar León-Sanz, "An Introduction to *History of Science and the Emotions*," in this volume.

[28] The bibliography is huge. For a first approach to medical science in the Middle Ages, see Mirko D. Grmek and Bernardino Fantini, eds., *Histoire de la pensée médicale en Occident*, vol. 1, *Antiquité et Moyen Âge* (Paris, 1995); Danielle Jacquart, *La Science médicale entre deux renaissances, XIIᵉ–XVᵉ siècle* (Aldershot, 1997).

[29] For the questions presented and raised here, see also Cohen-Hanegbi, "A Moving Soul" (cit. n. 17); she presents the story with a later focus, and in a slightly different light.

[30] On Constantine the African and his work and translations, see Charles Burnett and Danielle Jacquart, eds., *Constantine the African and 'Alī ibn al-'Abbas al-Maǧūsī: The "Pantegni" and Related Texts* (Leiden, 1994). An impressive number of translations have been attributed to Constantine. Specialists today think that certain Arabic texts had circulated earlier in Sicily and southern Italy and that the translations were progressive during the eleventh century. See Charles Burnett, "Physics before the *Physics*: Early Translations from Arabic Texts Concerning Nature in MSS British Library, Additional 22719 and Cotton Galba E IV," in *Arabic into Latin in the Middle Ages: The Translators and Their Intellectual and Social Context* (Farnham, 2009), 78–9.

dents until the end of the Middle Ages, and the *Pantegni*, a compendium composed mostly of translations from an encyclopedia written by the Persian doctor ʿAlī ibn al-ʿAbbas al-Maǧūsī at the end of the tenth century. This medical knowledge, arriving in the Christian world via the Mediterranean, was a synthesis of Hellenic and Arabic medicine, and was open to natural philosophy. While Constantine translated an important part of the Galenic corpus at Monte Cassino, not far from there, in the coastal town of Salerno, the bishop Alfano (d. 1085), formerly a monk of Monte Cassino, translated the *De natura hominis* of Nemesius of Emesa (active at the end of the fourth century), a treatise on human nature that contained important reflections on the passions of the soul. John of Damascus (d. 749), who integrated the ideas of Nemesius in his *De fide orthodoxa* in the eighth century, was himself translated by Burgundio of Pisa in the middle of the twelfth century. This interweaving of texts of different kinds demonstrates clearly that medical questioning was one component of global research on the nature of man that did not separate physiology from the psychology of emotions.

The general and cultural framework of this irrepressible curiosity is well known.[31] The period from the eleventh to the thirteenth centuries was one of a global expansion of the Christian West, fueled by demographic and economic growth and by the multiplication of contacts in the Mediterranean. For the written culture we are concerned with, the forms, methods, and content of learning were all renewed. The places in which knowledge was produced and circulated became increasingly diverse and interconnected with one another, while ways of dealing with questions became more open to rational treatment and to the integration of non-Christian input, both ancient pagan and heathen. This slow revolution, which took the form of a real project to produce a synthetic knowledge of the Creation, was specifically concerned with matters involving nature—among which was the constitution of the human being, made of body and soul in Christian terms.[32] At the end of the eleventh century and during the first half of the twelfth century, urban schools like those of Laon, Chartres, or Paris, but also Salerno and Bologna, gained the upper hand over the cloisters as the preferred places of study among lettered elites. As for urban masters, their interest turned increasingly toward logic and the arts of the *quadrivium*, as well as the sciences of nature and of the universe.[33] This movement, usually referred to as the "Renaissance of twelfth century,"[34] was part of an even larger shift in the medieval West, which Jacques Le Goff has referred to as the "descent of values from Heaven to Earth."[35] This put an end to Augustinian dualism: the growing consideration of earthly realities

---

[31] For a first approach, see, among others, Jacques Verger, *La Renaissance du XIIe siècle* (Paris, 1999); Verger, *Les Universités au Moyen Âge* (1973; repr., Paris, 1999); Alain de Libera, *La philosophie médiévale* (1993; repr., Paris, 1995). See also Cohen-Hanegbi, "A Moving Soul" (cit. n. 17).

[32] The period referred to here has been generally described as the "Renaissance of the twelfth century." On this last notion, see, first of all, Charles H. Haskins, *The Renaissance of the Twelfth Century* (Cambridge, Mass., 1972). See also Robert Louis Benson and Giles Constable, eds., *Renaissance and Renewal in the Twelfth Century* (Cambridge, Mass., 1982); Verger, *La Renaissance* (cit. n. 31).

[33] For an introduction, in addition to the works cited in n. 32, see Lindberg, "The Revival of Learning in the West," in Lindberg, *Beginnings* (cit. n. 7), 182–213; Jacques Le Goff, *Les intellectuels au Moyen Âge* (1957; repr., Paris, 2006).

[34] See n. 32.

[35] Jacques Le Goff, "Du ciel à la terre: la mutation des valeurs du XIIe siècle au XIIIe siècle dans l'Occident chrétien," in *Héros du Moyen Âge: Le roi, le saint au Moyen Âge* (Paris, 2004), 1263–87.

in and for themselves gave an important push to the integration of the ancient scientific tradition in Western culture and thought. Such a development, in turn, was linked to the rise of new contexts of everyday life: that of urban and commercial development, which took place alongside a pious turn toward the worldly life of the suffering Christ as part of the same cultural movement. From the twelfth and thirteenth centuries onward, an ever-growing number of men and women sought to incorporate the divine project of Salvation in their earthly lives. It was this endeavor that the transformation of schools and learning came to conceptualize and formalize. The human being as a whole came under investigation, and over the course of a few decades, a transition occurred in the conception of emotions from the early medieval psychomachia,[36] the battle of vices and virtues, to a psychosomatology, an integrated consideration of body and soul.

In the *Pantegni*, what we call emotions were present as movements of the breath of life, born in the heart and released in the whole body. Their dynamic was similar to the breathing of natural heat, driven by the blood, which converged from the whole body to the heart—the organ of emotions *par excellence*—and was diffused from the heart to the members. Yet the origin of emotions was in the soul: they were actions of the soul that led to somatic modifications like variations in cardiac rhythm, paleness and blushing, and so on. Emotions were considered active, which was why Constantine and the doctors who conceived the health *regimines* called them "accidents of the soul"—meaning that an emotion was a movement born from an external appeal—rather than "passions of the soul," an expression more rarely used by medical doctors.

Ancient and medieval medicine was holistic, based on the conviction that one's behavior and way of life had an impact on health. Resulting in the balance of humors that guaranteed good complexion, health could be restored or conserved by modifying the six nonnatural things. The six "nonnaturals," defined since Antiquity (air, food and drink, movement and rest, sleep and wakefulness, repletion and starvation, and the emotions—*accidentes animae, affectiones, passiones animae*),[37] were in the Galenic system opposed to the three "natural things" (members, physical spirits, and blood). In the fifth book of its theoretical part, the *Pantegni* dealt with the emotions following the six nonnaturals and defined the six accidents of the soul—that is, the six emotions (anger, joy, sadness, anxiety, fear, and shame)—situating them in relation to the movements of blood, and explaining the somatic manifestations that resulted from these movements. Emotions acted according to a mechanism of spirits and fluids, which brought about bodily reactions. So joy, for instance, was described as a process of the slow movement of the natural heat of the body toward the members. If joy was moderate, wrote Constantine, movement was slow and only brought about benefits to both the body and soul. However, a joy that was too intense resulted in a quick expulsion of the life substance and was risky for the heart; it could even provoke death. In the same way, excessive fear could also be fatal.[38] Based on the the-

---

[36] The term comes from the Greek Ψυχομαχία/*Psukhomakhia*, meaning "battle of/for the soul." It is the title of the allegorical poem of the late Antique poet Prudentius, *Psychomachia* (early fifth century), describing the conflict of (pagan) vices and (Christian) virtues.

[37] This list can contain slight variations. See Laurence Moulinier, "Magie, médecine et maux de l'âme dans l'œuvre scientifique de Hildegarde," in *"Im Angesicht Gottes suche der Mensch sich selbst": Hildegard von Bingen (1098–1179)*, ed. Rainer Berndt (Berlin, 2001), 545–59, on 551.

[38] Constantinus Africanus, *Pantegni, Theorica*, in *Opera omnia Ysaac*, vol. 2 (Lyon, 1515), 6.109, fol. 25v, and 4.7, fol. 16, quoted by Knuuttila, *Emotions* (cit. n. 6), 214.

ory of humors and temperaments, this dynamic scheme, which remained the foundation of the medical conception of emotions until the end of the Middle Ages, was enriched in the twelfth century by the contribution of the *Liber ad Almansorem* of Abū Babr ar-Rāzī (Rhazes) and the *Canon of medicine* and *De anima* of Ibn Sînâ (Avicenna).

From the end of the twelfth century onward, as social possibilities of access to medicine increased in medieval society, and as studying medicine in the framework of a university became the path to practicing medicine as a doctor, medical theories of the passions started to circulate more widely in scholastic thought and also well beyond, especially in the literate circles of society. The contribution of the Persian medical doctor and philosopher Avicenna (d. 1037) is, in this context, of great significance, the translation of his writings giving access for the first time in the Latin West to a complete psychosomatic theory of emotions, which reverberated in both medicine and philosophy.[39] For Avicenna, emotions were above all mental states provoked by external objects that contained, for the person who perceived them, an "intention." Sensual perception submitted this objective intention to the judgment of the force of estimation [*vis aestimativa*], localized in the brain. Emotion proper was the combination of this perceptive judgment and the instinctive, somatic reaction that accompanied it.[40] This consideration of emotion combined elements that human beings shared with animals—sensual perception, instinctive reaction, bodily movement—with elements that were specific to humans, since humans can make a rational decision to moderate, control, or even act against the inclinations of their emotions.

When the new medical science of emotions was aligned with the philosophical theories of the soul, linking psychology and physiology together in the framework of a global anthropology, there was an interesting practical consequence. Emotions could be taken charge of medically, either by curing emotional disorder with bodily remedies or by proposing a specific emotion as a form of therapy. From the eleventh century onward, philosophers and medical doctors did their best to find the physiological causes and remedies of psychological states. In the last centuries of the Middle Ages, as a consequence of this naturalization of affectivity, people began to believe that bad or excessive emotions could be treated by medicine, both by nutrition and by modifications in their way of life, and not only by correcting vices and amplifying virtues. Medical doctors, more and more present in the courts of princes and prelates and in the entourages of urban elites at the end of the thirteenth century,[41] started to write personalized recommendations [*consilia*] for their patients, prescribing general regimens of health based on the six nonnaturals.[42] In his *Regimen of Health*, written for the king of Aragon Jaume II El Just (James II; 1291–1327) and his *Mirror of Med-*

---

[39] See Avicenna, *Kitāb al-nafs*, ed. I. Madkour (Cairo, 1969), translated around 1160–70 by Dominicus Gundissalvi. This treatise contains a general classification of emotions: Avicenna Latinus, *Liber de anima seu Sextus de naturalibus*, pt. IV, chap. IV, ed. Simone Van Riet (Leuven, 1968), 54–62.

[40] Dag Nikolaus Hasse, *Avicenna's "De anima" in the Latin West: The Formation of a Peripatetic Philosophy of the Soul 1160–1300* (London, 2000), 139.

[41] Elisa Andretta and Maryline Nicoud, eds., *Être médecin à la cour, Italie, France, Espagne (XIIIᵉ–XVIIIᵉ siècle)* (Florence, 2013).

[42] For a first approach to regimens of health, see Pedro Gil-Sotres, "The Regimens of Health," in *Western Medical Thought from Antiquity to the Middle Ages*, ed. Bernardino Fantini and Mirko D. Grmek (Cambridge, Mass., 1998), 291–318, 395–8; Maryline Nicoud, *Les Régimes de santé au Moyen Âge: Naissance et diffusion d'une écriture médicale*, 2 vols. (Rome, 2007). On the *consilia*, see Jole Agrimi and Chiara Crisciani, eds., *Les "Consilia" médicaux* (Turnhout, 1994).

icine,[43] the well-known Catalan theologian and medical doctor Arnaldus of Villanova (d. 1311), for example, proposed a practical treatment of emotions based on Constantinus Africanus's model, which integrated Avicennian anthropology. Far from a speculation based on religious or moral considerations, Arnaldus wrote a sophisticated analysis of the causes of emotions. He explained that some were external, linked to a perceived object, and others were internal, simultaneously dependent on the estimated virtue and on the bodily predisposition someone might have to a specific emotion. He concluded by proposing to cure the harmful effects of certain emotions by a treatment that affected the complexion.[44] In other cases, the assessment of an emotion's value depended on complexion. Maino de Mainieri (d. 1368), for instance, claimed that anger did not always cause harm, recognizing that a fit of anger may be beneficial for persons of a cold and humid complexion as it diffuses sudden heat in the body.[45] Suggested remedies were varied, but this principle was widely adopted. John of Toledo (d. ca. 1275), for instance, recommended baths of salted water as a treatment for certain forms of sadness.[46] Modifications of nutrition were often recommended for people suffering from emotional troubles. For curing fear—which, according to medieval medicine, cooled the humors, causing paleness, trembling, and diarrhea—wine was considered a good remedy because it heated up the blood. Arnaldus of Villanova recommended a diet of white meat and fine dishes in order to boost moderate joy, a guarantee of good health.[47]

General recommendations concerning the way of life were also frequent. Doctors suggested avoiding strong emotions and devastating passions and instead recommended a tempered and sophisticated life, filled with beauty and good company, as in the regimen of health of Barnabas of Reggio (d. ca. 1365): "listening to the smooth songs and the delightful sounds of different instruments, and reading and listening [to] delightful books does not only calm down anger but helps also to fight pain and sadness."[48] According to ancient wisdom transferred to the Middle Ages by Boe-

---

[43] See *Arnaldi de Villanova Opera Medica Omnia*, X, 1. *Regimen sanitatis ad regem aragonum*, ed. Luis García-Ballester and Michael Rogers McVaugh (Barcelona, 1996); *Speculum medicine*, in *Arnaldi de Villanova. Hec sunt Opera* (Lyon, 1509), fols. 1–55. The monumental introduction of Pedro Gil-Sotres to the *Regimen*, 471–885, offers an excellent synthesis on the hygiene of emotions in the regimens of health on 803–27. Another redaction of this chapter is published by the author: "Modelo teórico y observación clínica: las pasiones del alma en la psicología medical medieval," in *Comprendre et maîtriser la nature au Moyen Âge: Mélanges d'histoire des sciences offerts à Guy Beaujouan* (Geneva, 1994), 181–204. See also Joseph Ziegler, *Medicine and Religion c. 1300: The Case of Arnau de Vilanova* (Oxford, 1998).

[44] Arnaldus de Villanova, *Speculum medicine*, chap. 80, fol. 27r–28v, commentary by Gil-Sotres; introduction to *Regimen*, 811–4 (both cit. n. 43).

[45] Maino de Mainieri, *Regimen sanitatis*, in *Praxis Medicinalis* (Lyon, 1586), II, chap. 9, 21; quoted by Gil-Sotres, introduction to *Regimen* (cit. n. 43), 827.

[46] John of Toledo, *Summa de conservanda sanitate*, Vienna, MS Mellcensis 728 (967), fol. 56v., quoted by Naama Cohen-Hanegbi, "Accidents of the Soul: Physicians and Confessors on the Conception and Treatment of Emotions in Italy and Spain, Late 12th–15th Centuries" (PhD diss., Hebrew Univ. of Jerusalem, 2011), 211.

[47] Gil-Sotres, introduction to *Regimen* (cit. n. 43), 819.

[48] "Audire cantos suaves et sonos delectabiles diversorum instrumentorum, non solum iram mitigat sed tristitiam et dolorem juvat et libros delectabiles legere et audire." Barnabas of Reggio, *Regimen sanitatis*, MS Paris, B.N., Lat., 16189 (XIV), fol. 10rb, quoted by Gil-Sotres, introduction to *Regimen* (cit. n. 43), 827. On Barnabas and his place in the diffusion of a practical medicine in lay society, see Maryline Nicoud, "L'adaptation du discours diététique aux pratiques alimentaires: l'exemple de Barnabas de Reggio," *Mélanges de l'École Fr. Rome: Moyen Âge* 107 (1995): 207–31.

thius,[49] the beneficial effects of music on mood are frequently quoted among the recommendations. Further proof of the diffusion of this integrated vision of man and his
affective life is that this therapy of, and by, emotions, became one of the major fields
in the growing competition between theologians and medical doctors. Reacting in
some way to the prescription of the Fourth Lateran Council (1215), which enjoined
doctors to encourage their patients to first consult a doctor of the soul, both theologians and doctors refused to divide their patients according to their fields of competence, each asserting their claim to understand the whole person. Still, no physician of
the period presumed to replace a priest—and this they stated very clearly.[50]

Thus from the eleventh century onward, as the medical discourse began to reflect
on both the physiological causes and psychological symptoms of emotions, a new
approach to emotions emerged that considered them to be naturalized and embodied.
Along with the (limited) popularization of medical texts, this change in the understanding of emotions reveals the growing impact of medical knowledge on medieval
society. Not only did expanding numbers of elites turn to medical practitioners with
their problems, and not only were problems being increasingly treated (at least par-
tially) using a physiological approach, many texts of literature, political science, theology, and education show the capillary diffusion of medical theories and knowledge
in the upper strata of society.[51] Physicians appear as a recognized profession, and recent studies show well the presence of medicine in the everyday lore of urban society,
mainly, but not only, around the Mediterranean.[52] Yet medical doctors proposed remedies for the emotional disorders that were aligned with the class of their wealthy patients: listening to delicate music, reading poetry, or drinking good wine would have
been available to or suitable for only the most affluent individuals. Stereotypes about
the emotional agitation and instability of common people were thus strengthened as a
result, while other emotions, like melancholy, became a mark of the elites. In fact, one
might say, strangely enough, that the medical consideration and treatment of emotions reinforced social discrimination and caused emotions to be further used as a tool
of social distinction.[53]

TWELFTH-CENTURY MONASTIC ANTHROPOLOGY:
A SPIRITUALIST PSYCHOLOGY LINKING BODY AND SOUL

As seen above, psychological reflection was not the exclusive purview of urban
schools in the twelfth century, a period characterized by a great fluidity in the circu-

---

[49] See Anicius Manlius Severinus Boethius, *Fundamentals of Music*, ed. Claude V. Palisca, trans.
Calvin M. Bower (New Haven, Conn., 1989).

[50] Ziegler, *Medicine and Religion* (cit. n. 43), 252–4; Ziegler, "*Ut dicunt medici*: Medical Knowledge and Theological Debates in the Second Half of the Thirteenth Century," *Bull. Hist. Med.* 73
(1999): 208–37; Luc Berlivet, Sara Cabibbo, Maria Pia Donato, Raimondo Michetti, Marilyn Nicoud,
eds., *Médecine et religion: compétitions, collaborations, conflits (XIIᵉ–XXᵉ siècles)* (Rome, 2013).

[51] Cohen-Hanegbi, "A Moving Soul" (cit. n. 17).

[52] On this subject, in addition to the ongoing research of Cohen-Hanegbi, see Florence Eliza Glaze
and Brian K. Nance, eds., *Between Text and Patient: The Medical Enterprise in Medieval and Early
Modern Europe* (Florence, 2011), esp. the essay by Joseph Ziegler, "Medicine and the Body at the
Table in Fourteenth-Century Italy: Book One of Philip of Ferrara's *Liber de introductione Loquendi*,"
in ibid., 121–36.

[53] See, e.g., Paul Freedman, "Peasant Anger in the Late Middle Ages," in *Anger's Past: The Social
Uses of an Emotion in the Middle Ages*, ed. Barbara H. Rosenwein (Ithaca, N.Y., 1998), 171–9; C.
Stephen Jaeger, *Ennobling Love: In Search of a Lost Sensibility* (Philadelphia, 1999).

lation of knowledge. Cistercian monks and the regular canons of Saint-Victor of Paris collectively produced a human science that created a fertile dialogue between affective spirituality and speculative thought. As proof of this, one of the most popular thirteenth-century sources concerning psychology was an anonymous treatise, most likely written by a Cistercian in the 1170s, entitled *De spiritu et anima*.[54] Weaving together both ancient and contemporary writings, this text offered an impressive synthesis and was quickly attributed to Saint Augustine. Furthermore, Cistercians started to master the art of an ancient but long-forgotten genre: the *De anima*, or treatise on the soul. William of Saint-Thierry, Aelred of Rievaulx, and Isaac of Stella (d. ca. 1170) all wrote such treatises.[55] These men, warning against the arrogance of philosophers and the materialism of the disciples of Hippocrates, embarked upon an ambitious project—designing the cartography of the human soul, with all its powers, qualities, and links to the body—hence taking up the challenge of combining monastic introspection with a new kind of knowledge privileging observation and logic. The purpose of this monastic reflection on the nature of the soul remained aligned with the project of Augustine as corrected by Anselm of Canterbury at the end of the eleventh century—that is, to create a synergy between the science of man and his spiritual destiny by combining the philosophical reflection on the powers of the soul with a spiritual perspective.

The most important contribution of the Cistercians to medieval anthropology was the integration of affectivity into the classification of the powers of the soul.[56] Cistercian authors started by taking up elements of Platonic and Stoic philosophy transmitted by the Fathers, first in the division of the soul into three powers, as Isaac of Stella described: "The soul then is rational, positively appetitive (concupiscible), and negatively appetitive (irascible)—a kind of trinity of its own, as it were."[57] Next, they attributed the four affects (Stoic in origin) to two of these powers: joy and hope, belonging to the concupiscible part, were linked to the faculty to desire; and pain and fear, being associated with the irascible faculty, were connected to hate, which acts by rejection.[58] The lists of affects established by the authors on this basis could differ in their details. With the irascible faculty, Isaac associated a series of other affects, such as compunction, boredom, fear, sadness, worry, *acedia*, jealousy, anger, indignation, and so forth; with the concupiscible part, he associated joy, hope, delectation, celebration, elation, and charity, among others.[59] All these inclinations, which radiated out from the heart (and were thus called *affectus cordis*), guaranteed the mobility of the soul.[60] Isaac maintained that the affective capacity of the soul, *affectus*, paired with its

[54] *De spiritu et anima*, trans. Bernard McGinn, in *Three Treatises on Man: A Cistercian Anthropology*, ed. Bernard McGinn (Kalamazoo, Mich., 1977), 181–288.

[55] William of Saint-Thierry, *De natura corporis et animae*, in *La Nature du corps et de l'âme*, ed. and trans. Michel Lemoine (Paris, 1988). For the English translation, see McGinn, *Three Treatises* (cit. n. 54), 103–52. Aelred of Rievaulx, *Dialogue on the Soul*, trans. Charles H. Talbot (Kalamazoo, Mich., 1981). Isaac of Stella, *Epistola de anima*, *Patrologia Latina* (hereafter *PL*) 202, 1877 B. For the English translation, see McGinn, *Three Treatises* (cit. n. 54), 157.

[56] This process is highlighted by Boureau, *De vagues individus* (cit. n. 6), 45–54.

[57] Isaac of Stella, *Epistola de anima*, *PL* 202, 1877 B (cit. n. 55); translated by McGinn in *Three Treatises* (cit. n. 54), 157.

[58] Isaac of Stella, *Epistola de anima* (cit. n. 55), 1879 B.

[59] Isaac of Stella, "Sermo 17," in *Sermons* I, ed. Anselm Hoste (Paris, 1967), Sources chrétiennes 130, sec. 14–26, 320–8; Isaac of Stella, *Epistola de anima* (cit. n. 55), 1879 B.

[60] William of Saint-Thierry, *De natura corporis* (cit. n. 55), 105–52.

ability to know, *sensus*, constituted the two poles of the dynamic of being.[61] Bernard of Clairvaux's list was slightly different and was adopted by the author of the anonymous *De spiritu et anima*.[62]

Aelred of Rievaulx agreed with Isaac concerning the participation of affectivity in the dynamic of being. He believed that an affect (*affectus*) was a "spontaneous and pleasant inclination of the soul towards someone"[63]—a definition that was convenient only for the first kind of emotions, those of the concupiscible faculty. While Isaac simply omitted the link between the rational faculty and affects by insisting on the spontaneity of affective movements, even qualifying them as impulses (*impetus*), Aelred recognized the possibility of a concurrence, even of an open conflict, between affect and will, as well as between affect and reason.[64] Reasonable will was long-lasting and created the possibility of controlling the action; affect was unpredictable and fleeting. Claiming this, Aelred slipped from the three faculties of the soul inherited by monastic anthropology from Augustine (memory/reason/will) toward a new tripartition (affect/will/reason). He considered, however, that a fruitful cooperation was possible between them, when affect, reason, and will combined into an ordered love:

> Love arises from attachment when the spirit gives its consent to this attachment, and from reason when the will joins itself to reason. A third love can also be brought about from these two when the three—reason, attachment, and will—are fused into one.[65]

For Aelred, there was a love according to reason alone, and a love according to emotion alone, but they were both incomplete as long as they did not combine their forces in the same direction. Affect, which was subordinated to the will in the anthropology of Augustine, came to be liberated by Aelred, who made it one of the faculties of the soul, thereby conferring autonomy to the appetitive sensibility of the soul. Still, distinguishing and opposing affect and reasonable will, he returned from a three-part to a two-part, dual organization scheme that reproduced the dynamic duality of soul.

In twelfth-century anthropological discussions, emotions took a central role in the issue of the relation between body and soul, the two distinct substances that make up the human being. While their difference was irreducible, according to Saint Bernard, they constituted a natural unity [*unitas naturalis*], essential to life and eternal beatitude.[66] In order to explain this improbable combination, the Cistercians explored several paths, attributing unity to the formal analogy of body and soul, or to the existence of mediating element(s), or modes of presence, of the soul in the body. In the discussions of the association between body and soul, emotion—considered as occupying an intermediary position between them, and dynamic in its principle—was frequently

---

[61] Isaac of Stella, "Sermo 19," in *Sermons*, II, ed. Anselme Hoste and Gaetano Raciti (Paris, 1974), Sources chrétiennes 130, sec. 13, 176–7. The vision of the canons of Saint-Victor is very near to this sight. See Hugh of Saint Victor, *De sacramentis christianae fidei*, PL 176, 331B.

[62] For an inventory, see Damien Boquet, *L'Ordre de l'affect au Moyen Âge: Autour de l'anthropologie affective d'Aelred de Rievaulx* (Caen, 2005), 169–70n111; *De spiritu et anima*, PL 40, 829–30.

[63] Aelred of Rievaulx, *De speculo caritatis* (cit. n. 26), 3.31.119.

[64] See, e.g., ibid., 3.54.131, 1.105.60.

[65] Ibid., 3.48.128. For the English translation, see Aelred of Rievaulx, *The Mirror of Charity*, trans. Elizabeth Connor, with introduction and notes by Charles Dumont (Kalamazoo, Mich., 1999), 3, c. 20, 253.

[66] Bernard of Clairvaux, *Liber de diligendo Deo*, ed. and trans. Françoise Callerot, Sources chrétiennes 393 (Paris, 2010), sec. 30, 136.

cited as a proof or as a form of their unification. While all the quoted authors proposed their solutions to the problem, William of Saint-Thierry went the furthest in attempting to match the new knowledge on natural phenomena with the Fathers' spiritualist doctrine. His ideas on this issue appeared in a text written around 1140, entitled *De natura corporis et animae*.[67] This unique treatise is composed of two parts. The first offered a physical approach to the body—a kind of synthesis of the medical knowledge of his time, based primarily on the translations of Constantine the African and Alfano of Salerno[68]—while the second part was devoted to the soul. The two books of the treatise do not speak to one another. According to Michel Lemoine, William probably wrote his treatise on the soul first, to which he felt the need to add another part on the body, in order to reply to the philosophers of his time, whom he considered materialist.[69] We find here the same divide, then, as we find in the two treatments Aristotle proposed to the question of anger.[70] While passions are discussed in the second part, dealing with the soul, William attributes them alternatively to body and soul, according to the context and the author that he uses as his source.[71] In fact, he distinguishes different kinds or levels of passions that effect a shift from body to soul. He describes a physiological passion, which is a kind of illness or lesion of an organ,[72] but there is also a passion understood as a capacity to suffer, a *passibilitas*, a feature of the soul that distinguishes it from any other spiritual substance and links body and soul, as the instance by which the body acts on the soul and the soul is present in the body.[73] In this way, two types of passions exist for William: those of the flesh and those of the spirit, the first being irrational and complicit with the body, while the second, being rational, participates in the elevation of the spirit. In this way, when "the spirit follows the affects of the flesh and the senses, the 'animal passions' become vices of the soul."[74] William describes a passion-vice that goes up to the spirit from the bodily impulse, ultimately contaminating the thoughts themselves and mutating into vices. This pessimistic vision, by placing the origin of vices in the impulses of the "animal soul" [*anima bestialis*] recalls the anthropology of the Fathers, especially that of Gregory the Great. Still, speaking of spirits, virtues, and "animal passions," as well as establishing analogies between the qualities and functions of the body and soul,[75]

[67] William of Saint-Thierry, *De natura corporis* (cit. n. 55). For an analysis, see Damien Boquet, "Un nouvel ordre anthropologique au XIIᵉ siècle: réflexions autour de la physique du corps de Guillaume de Saint-Thierry," *Cîteaux* 55 (2004): 5–20; McGinn, *Three Treatises* (cit. n. 54), 27–47; Svenja Gröne, "Le premier écrit scientifique cistercien: le *De natura corporis* de Guillaume de Saint-Thierry (†1148)," *Rives nord-méditerranéennes* 31 (2008): 115–30.

[68] On William's sources, see Michel Lemoine, "Les ambiguïtés de l'héritage médiéval: Guillaume de Saint-Thierry," in Besnier, Moreau, and Renault, *Les Passions* (cit. n. 6), 297–308.

[69] William of Saint-Thierry, *Epistula de erroribus Guillelmi de Conchis*, edited by Paul Verdeyen (Turnhout, 2007), 61–71; William of Saint-Thierry, *Lettre sur les erreurs de Guillaume de Conches*, in Michel Lemoine and Claude Picard-Parra, *Théologie et cosmologie au XIIᵉ siècle* (Paris, 2004), 183–97; Michel Lemoine, "Guillaume de Saint-Thierry et Guillaume de Conches," in *Signy l'abbaye et Guillaume de Saint-Thierry*, Actes du colloque international d'études cisterciennes, 9–11 septembre 1998, ed. Nicole Boucher (Signy-l'Abbaye, 2000), 527–39; Danielle Jacquart, *La Médecine médiévale dans le cadre parisien* (Paris, 1998), 34–6.

[70] Cohen-Hanegbi, "A Moving Soul" (cit. n. 17).

[71] See the synthesis on the use of the term *passio* by William in Lemoine, *La Nature du corps* (cit. n. 55), 195n91; Boquet, *L'Ordre de l'affect* (cit. n. 62), 146–7; Alain Boureau, "Le statut nouveau des passions de l'âme au XIIIᵉ siècle," in Boquet and Nagy, *Le Sujet des émotions* (cit. n. 14), 190–1.

[72] William of Saint-Thierry, *De natura corporis* (cit. n. 55), II, sec. 54, 133, and sec. 68, 151.

[73] Ibid., sec. 51, 128–31.

[74] Ibid., secs. 75–6, 158–61.

[75] Ibid., secs. 88–91, 175–9.

William adopts the vocabulary of medicine and natural philosophy, so that the frontiers of body and soul become entirely blurred. His description of the passions and their activity produces the same effect: trying to set up a spiritualist psychology, William succeeds in convincing his readers of the close union of body and soul.

With the integration of new questions from natural philosophy and medicine, monastic anthropology became emancipated from the pessimistic view of the Church Fathers, for whom human beings, by their very nature, were oriented toward sin. Without severing this heritage, twelfth-century monks redesigned the links between body and soul and the questions of freedom and of moral responsibility. By doing this, they created an original path, just before the explosion provoked by the advent of Aristotelian hylomorphism. The result was not a promotion of the body but an understanding of the "inferior" faculties of the soul, those in contact with the body and those on which spiritual activity was based: sensitivity, imagination, and affectivity. Between spiritual anthropology, theology, natural philosophy, and medicine, the classic conception of passion as a perturbation of reason and of will started to be inverted at the beginning of the twelfth century, making the affective faculty a constitutive force of human nature.

TOWARD AN ACADEMIC SCIENCE OF THE PASSIONS OF THE SOUL

It was during the eleventh and twelfth centuries that the most important questions leading to the renewal of the anthropology of emotions arose. Was affectivity a power of the soul? What was the role of the passions in the union of body and soul? Was one responsible for one's spontaneous emotional surge? How could one order emotions to become virtues? These questions, engaging spiritual issues with scientific exploration, called for the constitution of an integrated intellectual approach, which, as we have seen, succeeded slowly in the middle of the twelfth century, with the strange coexistence of body and soul in the spiritual anthropology of William of Saint-Thierry. The anthropological treatises of Hugh and Richard of Saint Victor, as well as the *De anima* of Cistercian monks, represent a moment of equilibrium between "before" and "after," and between the two kinds of approaches they fused. At the end of the twelfth century, Western thought on the passions of the soul was in a process of renewal because of the opening of the West to the *logica nova* and the ethics of Aristotle, as well as to the medical and philosophical writings of Avicenna. The social and institutional context of the production of knowledge also changed for both clerics and laymen with the development of universities and of lay courts, particularly in the urban sphere. Within two generations, between the end of the twelfth and the middle of the thirteenth century, a scientific thought of emotions took shape, equidistant from theology and spirituality on one side and from medicine and natural science on the other.

This scholastic science of emotions in the thirteenth century fell within the settings of what historians of philosophy call the "psychology of faculties," which considered psychic phenomena to come under the faculties of the soul.[76] Its elaboration took advantage of readings of the treatises on the soul by Aristotle and Avicenna, as well as of their commentaries, and integrated both the Augustinian heritage and the contributions of medicine. In the first decades of the thirteenth century, a series of psy-

---

[76] King, "Emotions" (cit. n. 11).

chological treatises, written by David of Dinant (d. ca. 1217), Alexander Neckham (d. 1217), and John Blund (d. 1248), gave particular importance to the classification of emotions, understood by Avicenna as the actions of the estimative faculty. John Blund, the English theologian, in his *Treatise on the Soul*, written around 1210, elaborated a system in which the emotions of the concupiscible and irascible faculties—the faculties of what is desirable and what is to be avoided—were placed into binary opposition: love/hate, joy/pain, desire/disgust. In so doing, he, like several of his contemporaries, perfected a scheme coming from the Fathers, according to which the emotions of the concupiscible aspired toward virtue, while the emotions of the irascible kept away the vices.[77]

The great synthetic treatises of moral philosophy written during the same period, such as the *Summa aurea* of William of Auxerre or the *Summa de bono* of Philip the Chancellor, contained original propositions on the psychology of affects in relation to virtues. This is the case, for instance, with the anonymous treatise *De origine virtutum et vitiorum*, written in the first decades of the thirteenth century, which proposed a particular dynamics of emotions.[78] Its author, perhaps an English monk, conceived the passions—joy, hope, sadness, and fear—as movements of one of the four spiritual powers, the "faculty of alacrity," which was a kind of vigorous desire. These passions of alacrity echoed the actions of will, considered a power of the spirit: love, desire, hate, and distrust.[79] In this way, passions of alacrity and actions of will conditioned the movements of the soul driving toward virtue, and affects became one of the four forces of the flesh, always ready to drag one toward vice.[80] Emotions, divided between passions and affects, could equally be oriented toward good and evil.

While the end of the twelfth century and the first decades of the thirteenth represented a period of great creativity, the process of academic normalization that characterized the next decades put an end to the proliferation of different classifications. This proves the fluidity of ideas and of vocabulary during this time but also testifies to the pastoral concerns of those who formulated them. One of the authors whose moral writings best reveal the imbrication of pastoral and scientific issues in the first half of the thirteenth century was William of Auvergne, who was a canon of Notre Dame before serving as bishop of Paris (1228–49). William never lost sight of his pastoral mission. He described emotions as the manifestations of the faculty of movement, and the will as what controlled and steered all movement of the soul.[81] Emotion contained a passive dimension, as it reacted to the stimulation of the senses, but an active emotion, such as desire, was carried by the will. In this way, to William, passions

---

[77] John Blund, *Tractatus de anima*, ed. Daniel Angelo Callus and Richard William Hunt (London, 1970), secs. 62–73, 18–21; discussed by Knuuttila, *Émotions* (cit. n. 6), 226–30, and by King, "Emotions" (cit. n. 11), 175.

[78] István Péter Bejczy, "*De origine virtutum et vitiorum*: An Anonymous Treatise of Moral Psychology (c. 1200–1230)," *Arch. Hist. Doct. Litt. Moyen Âge* 72 (2005): 105–45.

[79] Ibid., 127–8, 107.

[80] Ibid., 140.

[81] On the anthropology of William of Auvergne, see Édouard-Henri Wéber, *La Personne humaine au XIIIᵉ siècle: L'avènement chez les maîtres parisiens de l'acception moderne de l'homme* (Paris, 1991), 76–84; Franco Morenzoni and Jean-Yves Tilliette, eds., *Autour de Guillaume d'Auvergne (†1249)* (Turnhout, 2005). In the same volume, see the essential studies on the passions by Silvana Vecchio, "*Passio, affectus, virtus*: il sistema delle passioni nei trattati morali di Guglielmo d'Alvernia," 173–87, and Carla Casagrande, "Guglielmo d'Auvergne e il buon uso delle passioni nella penitenza," 189–201.

were not *of us* [*a nobis*], as if provoked by an external element, but *in us* [*in nobis*], as movements of the cognitive faculty adapted by will. This process of appropriation transformed passion [*passio*] into affect [*affectus*].[82] For William, virtues were *affectus* made durable: they were considered as *habitus*, ways of being. In this way, love and hate, pain and joy, anger and kindness [*mansuetudo*] formed the material of vices and virtues, as they became lasting dispositions.[83] William made explicit a fundamental idea of the thirteenth century, according to which virtue was a passionate *habitus* or way of being, motivated and morally connoted by emotion. In this conception, affective intensity was no longer regarded as harmful, and it survived in the lasting disposition of virtue.

This complex vision of emotions and virtues had particular importance in the penitential process of contrition, confession, and satisfaction.[84] While generally, the first phase of penance—contrition—was considered the most emotional by theologians, for William, contrition, which was associated with the pain of repentance, elicited many different emotions (fear, shame, indignation, hate, etc.) and became a virtuous action marked by a certain stability.[85] The difficulty then became finding the right balance of different emotions, unstable by nature, during this process. This explains why William discusses shame extensively in the framework of the sacrament of penance and distinguishes arrogant shame, which is harmful, even diabolic, as it prevents confession, from virtuous shame, which is a gift of God, as it enhances contrition.[86] The issue of this passage from *affectus* as a movement to *habitus* as a way of being was particularly challenging for preachers and confessors. Not only did they have to encourage their flock to evaluate their own emotions according to complex criteria; they also needed to help them feel some of these emotions in a quasi-permanent way because they were considered the foundations of virtue.

In the same period, the work of the Franciscan master of theology, John of Rupella/La Rochelle (d. 1245) brought about an important change in the scholastic anthropology of emotions. John elaborated a systematic psychology of emotions and took an essential step in the difficult process of naturalizing emotion, that is, in its liberation from sin. John, who belonged to the first generation of Franciscans in the university, wrote two treatises on the soul in the 1230s. His *Summa de anima* had a direct influence on some of the greatest theologians of the thirteenth century, among whom were the Franciscan Alexander of Hales (his master and colleague at the University of Paris); Bonaventure, a student of Alexander; and the great Dominicans, Albert the Great and Thomas Aquinas.[87] John's reading of the passions of the soul in the *Summa*

---

[82] William of Auvergne, *De virtutibus*, in *Opera omnia* (Orléans-Paris, 1674), 1:119. We follow here the commentary of Vecchio, "*Passio, affectus, virtus*" (cit. n. 81), 178.

[83] William of Auvergne, *De virtutibus* (cit. n. 82), 1:122.

[84] Paul Anciaux, "Le sacrement de pénitence chez Guillaume d'Auvergne," *Ephemerides theologicae lovanienses* 24 (1948): 95–118.

[85] William of Auvergne, *De Sacramentis*, in *Opera omnia*, (cit. n. 82), 1:465; Casagrande, "Guglielmo d'Auvergne" (cit. n. 81), 194, 196–7.

[86] Franco Morenzoni, "La bonne et la mauvaise honte dans la littérature pénitentielle et la prédication (fin XIIᵉ–début XIIIᵉ siècle)," in *Shame between Punishment and Penance: The Social Uses of Shame in the Middle Ages and Early Modern Times*, ed. Bénédicte Sère and Jörg Wettlaufer (Florence, 2013), 177–93.

[87] John of Rupella, *Summa de anima*, ed. Jacques-Guy Bougerol (Paris, 1995). On the anthropology of John of Rupella/la Rochelle, see Boureau, *De vagues individus* (cit. n. 6), 93–129. See also Odon Lottin, "Les traités sur l'âme et les vertus de Jean de la Rochelle," *Rev. Néo-Scolastique Phil.* 32 (1930): 5–32; Knuuttila, *Emotions* (cit. n. 6), 230–6.

de anima offered a synthesis of knowledge, ancient and new. In *De spiritu et anima*, he combined the Augustinian tradition with the teachings of John of Damascus and of the *De anima* of Avicenna. John, who deliberately showed the compilation of his sources in the structure of his *summa*, exposed in its second part the theory of the soul, first according to the pseudo-Augustine, then to John of Damascus, and finally and at length to Avicenna, who most interested him. He integrated elements of medical knowledge into his system, describing, for instance, the affective dispositions of people as, among other things, dependent on the complexion of the body, that is, on the variation of humors.[88] In this way, John characterized the scholastic conception of emotions, understood as acts of the appetitive power of the soul and accompanied by bodily modifications, and distributed them around the powers of the soul in a complex system.[89]

In the first part of his *Summa*, John dealt with the union of body and soul and raised the question of the reason for their union. In his reply, he maintained that the natural unification of soul and body took place before original sin. He assumed that, in the human condition consecutive to the Fall, this union had reciprocal advantages for the body and soul, even if this could not happen without reciprocal tribulations. It was in this context that he asked whether the passions originated from nature or from sin. His answer distinguished natural passion, which depended on human nature, from nonnatural passion, which resulted from the Fall.[90] His notion of natural passion included the instability and the "immoderate character" of movement. In other terms, the assaults of passion on the soul were not, for John, an effect of sin: they came from its very nature, a feature that John considered an imperfection, a weakness or defect [*defectus*]. However, as a consequence of original sin, this weakness had become a component constitutive of all earthly existence. Before the Fall, the instability of passions was a possibility; after the Fall, it had become a necessity.

By all appearances, John of La Rochelle's treatment of emotions did not challenge the scheme of Augustine, which linked emotional perturbation and sin. But, in reality, John introduced an important shift by refusing to consider the instability of passionate movements as the symptoms of sin. He distinguished a natural emotional vigor, common to all human beings before and after original sin, and a *passibilitas* of misfortune, resulting from the Fall.[91] From then on, Franciscan scholastic anthropology considered emotions as expressions of both fragility and freedom, characterizing human existence. Understood in this way, passion became proper to humankind.

As a consequence of the growth of reflection concerning emotions in different fields and directions over the course of an entire century, in the middle of the thirteenth century, masters of theology started to reserve a specific section for the passions of the soul in their theological *summae*, many developing what we can call a "treatise on passions." The Dominican Albert the Great (d. 1280) dealt with the question in his different writings, integrating passions into various aspects of the frame-

---

[88] John of Rupella, *Summa de anima* (cit. n. 87), II, IV, c. 108, 263.

[89] Ibid., II, IV, c. 105, 254; II, IV, c. 107, 256–62. For the scheme of the passions as depending on the faculties of the soul, see Silvana Vecchio, "Passions de l'âme et péchés capitaux: les ambiguïtés de la culture médiévale," in *Laster im Mittelalter/Vices in the Middle Ages*, ed. Christoph Flüeler and Martin Rohde (New York, 2009), 58n35.

[90] John of Rupella, *Summa de anima* (cit. n. 87), I, VIII, c. 46, 147–50; Boureau, *De vagues individus* (cit. n. 6), 128–9.

[91] John of Rupella, *Summa de anima* (cit. n. 87), I, VIII, c. 46, 150.

work of the Aristotelian worldview.[92] In his treatise *On the Movement of Animals*, a commentary on Aristotle's treatise of the same name, among his "books on nature," or *libri physici*, passions were considered solely in their physiological dimension, while in other texts they were regarded only from the moral perspective: "Passion is a movement of the sensitive appetitive power which is born of the representation of a good or an evil."[93] Albert, who in the 1240s utilized the complete translation of the *Nicomachean Ethics*, pioneered the embrace of different aspects of the Aristotelian science of passions. Nevertheless, he still seemed nearer to the moralist theologians of the beginning of the century than John of La Rochelle. In his *De bono*, Albert inserted a long classification of passions into his discussion of the virtues, between "temperance" and "prudence." As if the psychology of passions could not be addressed in an autonomous way, his study of passions remained subordinated to a discourse on virtue.[94]

The last step toward the autonomy of passions was taken a few years later, by Thomas Aquinas (d. 1274), the most famous student of Albert. Reserving twenty-six questions for the passions of the soul in his *Summa theologiae*, Aquinas deserves to be presented by historians of theology as the author of the first "treatise of passions."[95] Written around 1270, the treatise is both substantial and coherent; it precedes the part of Thomas's *Summa* devoted to *habitus* and virtues. Thomas defines passions as movements of the sensitive appetite of the soul that answer to exterior stresses and are accompanied by bodily transformations. He identifies eleven main passions, sorted between those of the concupiscible (love, hate, desire, aversion, joy, and sadness) and those of the irascible (hope, despair, fear, audacity, and anger).[96] He also establishes hierarchies and synergies between them, following the dynamic of contrary emotions structured in pairs. Presented as both psychology and physiology, his treatment of the passions proves that an autonomous "science of passions" is henceforth conceivable inside the framework of a theological discourse. In this regard, the *Summa theologiae* marks the completion of a process that started more than a century before.

While we have to recognize that Thomas introduced only a few new ideas—his sources are similar to those of his predecessors and his taxonomy is less complex than

---

[92] Pierre Michaud-Quentin, "Le traité des passions chez Albert le Grand," *Rech. Théol. Ancien. Médiév.* 17 (1950): 90–120; Michaud-Quentin, *La Psychologie de l'activité chez Albert le Grand* (Paris, 1966), 91–113.

[93] Albert the Great, *Super Ethica: Commentum et quaestiones*, in *Opera omnia*, vol. 14, ed. Wilhelm Kübel (Münster, 1987), 2.5.113–4. Albert gives other definitions of passions in *De Bono*, *Opera Omnia*, 3.5.1; see Michaud-Quentin, *La Psychologie* (cit. n. 92), 92–3.

[94] Silvana Vecchio, "Il discorso sulle passioni nei commenti all'*Etica Nicomachea*: da Alberto Magno a Tommaso d'Aquino," *Doc. Stud. Trad. Filos. Mediev.* 17 (2006): 93–119.

[95] Thomas Aquinas, *Summa theologiae*, in *Sancti Thomae Aquinatis Opera omnia iussu Leonis XIII edita*, vols. 4–12 (Rome, 1888–1906), Ia–IIae, q. 22–48. Thomas deals with passions elsewhere too, in his commentary on the *Sentences* of Peter Lombard and in that on the *Nicomachean Ethics*. Finally, concerning the passions of the soul in Aquinas, among the numerous publications, see the introduction of Silvana Vecchio to the translation of the treaty of passions: Thomas Aquinas, *Le Passioni dell'anima* (cit. n. 15), 5–18; Peter King, "Aquinas on the Emotions," in *The Oxford Handbook to Aquinas*, ed. Brian Davies and Eleonore Stump (Oxford, 2012), 209–26; Paul Gondreau, *The Passions of Christ's Soul in the Theology of Saint Thomas Aquinas* (Münster, 2002); and Ruedi Imbach, "Physique ou métaphysique des passions?" *Critique* 716–7 (2007): 23–35.

[96] In fact, the list of passions in Aquinas cannot be limited to the eleven basic passions. Barbara H. Rosenwein found several dozen derived passions in the *Summa theologiae* alone. See Rosenwein, "Emotion Words," in Boquet and Nagy, *Le Sujet des émotions* (cit. n. 14), 104–5.

the one elaborated by John of La Rochelle—his presentation was remarkable in its systematization and synthesis, revealing the principal achievements of scholastic thought regarding the bodily implications of passions. Convinced of Aristotelian hylomorphism, Thomas gave equal attention to the double nature of the passions, both psychic and somatic. Bodily changes accompanying the passions were not only effects of the movements of the soul but whole participants in the emotional process. Thus, passion was presented as a psychosomatic phenomenon, associating a formal element (in the sense of Aristotle, for whom the soul was the form of the body) with a material one (physiological modification).[97] An immediate consequence of this conception was the complete integration of medical theory into the presentation of passions. Another original feature concerned the cognitive dimension of passions: as movements of the sensitive appetite, they escaped reason, at least at the start. Still, they were not considered sensations but rather acts of sensitive cognition, which came, as Avicenna first theorized, from the evaluation of the object perceived. An emotional movement supposed the appreciation of the "tendency" or direction contained in the cause of the emotion in an objective way.

Thomas's third original contribution concerned the ethical implications of the passions. Separating his "treatise of passions" from the "treatise of virtues," Thomas broke with the traditional way of considering emotions in Christianity as a symptom of vice or virtue. Undoubtedly, this shift was already in progress, but it was Aquinas who finalized it, making the passions natural movements of the soul, which expressed no moral value in and of themselves. As a consequence, the value of passions came from the way reason and will took charge of them, as the rational powers had the function of commanding passions. Hence, Thomas could claim that all passions served virtue as far as they subordinated themselves to reason: "The passions of the soul, in so far as they are contrary to the order of reason, incline us to sin: but in so far as they are controlled by reason, they pertain to virtue."[98] Thus, while rehabilitating the passions, Aquinas manifested an absolute trust in the human ability to act according to reason.

Last but not least, Thomas showed a real taste for psychological inquiry, reflecting the new consideration of emotions in the last centuries of the Middle Ages. Actually, his treatise on passions in the *Summa* is twice as long as the treatise on virtues, and the twenty-six questions are composed with tremendous detail, nuance, and extremely concrete considerations of the feeling and the complexity of emotions. The greatest part of the treatise is devoted to particular emotions: how they occur, by which physical symptoms they are made manifest, what their effects are, and how to cultivate their benefits and fight against their hindrances.

At the end of the thirteenth century, the debates about the psychology and ethics of the passions continued in the academic context and deepened the trend toward the autonomy of affective studies, especially following Augustine and Bonaventure, who had granted a prominent place to the will as the summit of the soul.[99] This position was also held by the Franciscan Duns Scotus (d. 1308), who believed that there were

---

[97] See esp. Aquinas, *Summa theologiae* (cit. n. 95), Ia–IIae, q. 37, a. 4, along with the comment of Knuuttila, *Emotions* (cit. n. 6), 241.

[98] Aquinas, *Summa theologiae* (cit. n. 95), Ia–IIae, q. 24, a. 2.

[99] On the theories of passions in scholasticism after Thomas Aquinas, see Dominik Perler, *Transformationen der Gefühle: Philosophische Emotionstheorien, 1270–1670* (Frankfurt am Main, 2011).

passions that did not act against the will but served as a "contributing force," as Olivier Boulnois has put it.[100] While the metaphysics of Duns Scotus was opposed to that of Aquinas, on the question of the moral value of emotions, his position was not very far from that of Thomas. With regard to the regulation of passions by the spirit, they both believed that passions participated in good will and, ultimately, in virtue. Duns Scotus's statement, "Where passions are, there is virtue" [*Ubi sunt passiones, ibi est virtus*],[101] might be considered a vindication of this unpopular part of human nature that had suffered so much degradation at the hands of medieval thinkers.

CONCLUSION

The elaboration of a scholastic discipline of emotions in the twelfth and thirteenth centuries, considered here from the point of view of physiology, psychology, and other realms of knowledge,[102] illustrates the slow but definitive transformation of the Christian way of understanding the human as an emotional being. It demonstrates how a pacified anthropology took shape, at least in academic culture, achieved through the naturalization of the body and of emotional life. Over the course of two centuries, schoolmasters strove to collect the pieces of humanity shattered by the consequences of original sin, in order to consider the wholeness of humanity (i.e., humans made of body and soul), so that, with these faculties that move the human being restored, they could engage humankind on a new path toward salvation.[103]

Without losing its moral dimension, emotion became part of human nature. Over time, this emotion-passion shifted from the Stoic view that it would come from outside the self, as an illness, to a new view that passion was part of human nature and natural to man. Emotion was no longer considered a legacy of original sin or a vehicle for redemption but instead an ontological part of the created being. The way to salvation proposed by late medieval thinkers was no longer bound in an austere dialectic opposing spiritual and carnal man but instead proceeded from the new understanding of the divine project as expressed by the physical world. The work of Thomas Aquinas exemplifies the argument that the normalization of theological discourse on the passions simultaneously reflected an acceptance of emotions in human life and a commitment to discipline emotions by reason. The rational spirit, the noble part of the soul that Augustine called *mens*, had to come to terms with its auxiliary forces: sensation, imagination, emotionality, all linked to otherness, to what was outside control. If the realm of the spirit was not openly questioned, a real paradigm shift occurred in the twelfth and thirteenth centuries, leading to what Alain Boureau has called a "fed-

---

[100] Olivier Boulnois, "Duns Scot: existe-t-il des passions de la volonté?" in Besnier, Moreau, and Renault, *Les Passions*, 285; Knuuttila, *Emotions*, 265–72 (both cit. n. 6).

[101] John Duns Scotus, *Opus Oxoniense*, in *Opera Omnia*, vols. 5–10, ed. Lucas Wadding (Lyon, 1639), III, d. 33, q. 1, sec. 3. Photoreprint by Georg Olms (Hildesheim, 1968), 7, 696; quoted by Boulnois, "Duns Scot" (cit. n. 100), 282.

[102] See, e.g., the debate on interjections in the expression of emotions by language: Irène Rosier-Catach, "Discussions médiévales sur l'expression des affects," in Boquet and Nagy, *Le Sujet des émotions* (cit. n. 14), 201–23; Christopher Lucken, "Éclats de la voix, langage des affects et séduction du chant: Cris et interjections à travers la philosophie, la grammaire et la littérature médiévales," in *Haro! Noël! Oyé! Pratiques du cri au Moyen Âge*, ed. Didier Lett and Nicolas Offenstadt (Paris, 2003), 179–201.

[103] Wéber, *La Personne humaine* (cit. n. 81).

eral conception" of the human subject, made of several united pieces.[104] The human subject had become more efficient, but also more erratic, as he/she became exposed to internal rebellions and foreign incursions, whether they were divine or diabolic.

In terms of the history of emotions, we could go still further. The fundamental transformations of the conception of affectivity that we have tried to describe here led to the birth of a science of the soul where emotions are central to the definition of humanity. This new intellectual attention paid to affectivity is only one component of a larger shift in the heart of medieval Christendom: the emergence of emotions as a decisive factor in Christian anthropology, occurring in synergy in different segments of society and visible in different fields of medieval culture between the eleventh and thirteenth centuries. As we have analyzed it elsewhere, this "medieval turn to passions" in anthropology is organically embedded in a theological turn to Passion— and, simultaneously, to the Incarnation, that is, to the embodied Christ, promoted as an "emotionology" by the Roman Church.[105] The joint effects of this large cultural movement can be read in lay, vernacular literature as well as in the decoration of churches and cathedrals or in spiritual and mystic writings. Rather than an inverse relationship between emotions and science, as in the great narrative of modernity,[106] in the Middle Ages one can observe instead a convergence, a mutual strengthening between the history of emotions and that of the science that theorized them. The period of the second part of the Middle Ages shows emotions being increasingly honored in both fields, as all the relevant sources from the period attest.

[104] Boureau, *De vagues individus* (cit. n. 6), 26.
[105] See Damien Boquet and Piroska Nagy, *Sensible Moyen Âge: Une histoire des émotions dans l'Occident médiéval* (Paris, 2015), esp. chaps. 4 and 8; for the notion of "emotionology," coined by Peter and Carol Stearns, see Dror, Hitzer, Laukötter, and León-Sanz, "An Introduction" (cit. n. 27).
[106] See Dror, Hitzer, Laukötter, and León-Sanz, "An Introduction" (cit. n. 27).

A Moving Soul:
Emotions in Late Medieval Medicine

by Naama Cohen-Hanegbi*

ABSTRACT

Theories of the soul and its faculties, including emotions, are recognized to have evolved significantly from the twelfth to the fifteenth century. While these concepts were widely researched, they have been to a large extent isolated to their theoretical realm with little attention given to their practical application. This essay begins with a question asked by natural philosophers, theologians, and physicians throughout the thirteenth century: "Can the soul be moved by the body?" While the proposed answers to this question had substantial implications for understanding the nature of living creatures, I argue that they had very practical ramifications for formulating and treating emotions within medical practice.

In his immense achievement in translating (and possibly commenting on) a large body of Aristotle's work from Greek to Latin, James of Venice (d. after 1147) made a major contribution to the intellectual life of the Middle Ages.[1] Aristotle's *De anima* was one of James's translations, and he sought not only to translate as much Aristotelian material as possible but also to answer to the growing curiosity among Latin scholars. As Dag Hasse noted, a budding interest in natural philosophy was already evident in the late eleventh century, with the dissemination of the translations of medical texts by Constantine the African, which proposed an original method for inquiry into human nature.[2] Together with James's translation, and the significant contributions of Dominicus Gundissalinus (fl. 1160–80), which made Avicenna's *Kitāb an-nafs* accessible, psychology, or *scientia de anima*, the study of the human soul, was to become a central discipline of investigation over the next four centuries. This investigation certainly had precursors among Christian thinkers, as noted by Damien

---

* Department of History, Gilman Building #375, Tel Aviv University, P.O. Box 29040, Tel Aviv, 6997801, Israel; naamaco@post.tau.ac.il.

[1] Unfortunately, the personal and professional life of James remains almost completely unknown; see Steven J. Livesey, "James of Venice," in *Medieval Science, Technology, and Medicine: An Encyclopedia*, ed. Thomas Glick, Steven J. Livesey, and Faith Wallis (New York, 2005), 282. On James's possible authorship of commentaries on Aristotle, see Sten Ebbesen, "Jacobus Veneticus on the Posterior Analytics and Some Early 13th-Century Oxford Masters on the Elenchi," *Cah. Inst. Moyen-Âge Grec Latin* 21 (1977): 1–9; David Kristian Bloch, "James of Venice and the Posterior Analytics," *Cah. Inst. Moyen-Âge Grec Latin* 78 (2008): 37–50.

[2] Dag Nikolaus Hasse, *Avicenna's "De anima" in the Latin West: The Formation of a Peripatetic Philosophy of the Soul 1160–1300* (London, 2000), 29–30.

Boquet and Piroska Nagy.[3] Nevertheless, from the twelfth century on, it was through this natural philosophy science of the soul that medieval scolars put forth their theories about the nature and function of the occurrences of the soul. These occurrences, which appear in various terms in Latin, including *passiones anime*, *affectiones*, *motus anime*, or *accidentes anime*, can be quite safely translated into modern definitions of emotions. Though there are certain distinctions and even arguments about the features of each category, particularly with respect to how each represents the relationship between body and soul, they all correlate well enough with contemporary assumptions about emotions as situations involving both body and mind. Because the focus of my analysis is precisely how this association between body and mind/ soul was understood within medieval medicine, the commonly used but rather amorphous term "emotions" is sufficiently accurate here.[4]

Much has been written about this twelfth-century development and the evolution of the analysis of the soul in late medieval universities. As we learn from Boquet and Nagy, interest in the workings of the soul within the body was already apparent in the works of scholars in the cathedral schools and among the Cistercians, but a more systematic approach was developed toward the end of the twelfth century.[5] With the rise of universities, Aristotle's *De anima* became a required textbook in the curricula of bachelor of arts programs; it emerged as the source for many commentaries and *quaestiones* by anonymous students as well as the great and renowned minds of the age. As Carlos Bernard Bazán noted, commentaries were a means not only for clarifying difficult passages but also for constructing the authors' own doctrines. The commentators offered various ways of explaining the character of the link between the body and the intellect, thereby proposing diverging theories on the topic.[6] Thus, the language and frame of thought of this examination of the soul influenced dramatically much of the period's central learned discourse. For example, it contributed to and shaped theological discussion on beatific visions, sin, and the operation of human will and discussions on the various faculties of the soul (among which were included the estimative and sensitive powers and the appetites from which the *passiones anime* originated) were often at the basis of explorations of human nature.[7]

---

[3] Boquet and Nagy, "Medieval Sciences of Emotions during the Eleventh to Thirteenth Centuries: An Intellectual History," in this volume.

[4] See also Boquet and Nagy's argument for accepting the use of this term (ibid.).

[5] Ibid.

[6] See B. Carlos Bazán, "13th Century Commentaries on *De anima*: from Petrus Hispanus to Thomas Aquinas," in *Il Commento filosofico nell'Occidente latino (secoli XIII–XV)*, ed. G. Fioravanti, C. Leonardi, and S. Perfetti (Turnhout, 2002), 119–84, on 123; L. Minio-Paluello, "Le texte du *De anima* d'Aristote: La tradition latine avant 1500," in *Autour d'Aristote, Recueil d'études de philosophie ancienne et médiévale offert à monseigneur A. Mansion* (Leuven, 1955), 217–43; Gérard Verbeke, "Les progrès de l'Aristote Latin: le cas du *De anima*," in *Rencontres de cultures dans la philosophie médiévale: traductions et traducteurs de l'antiquité tardive au XIVe siècle*, ed. Jacqueline Hamesse and Marta Fattori (New Leuven, 1990), 195–201. The ongoing influence and importance of *De anima* has been made clear through research into commentators on the work from the Renaissance through the seventeenth century. See F. E. Cranz, "Perspectives de la Renaissance sur le *De anima*," in *Platon et Aristote à la Renaissance* (Paris, 1976), 359–76; and also the essays by Olaf Pluta, Lorenzo Casini, and Paul J. J. M. Bakker in *Mind, Cognition and Representation: The Tradition of Commentaries on Aristotle's "De anima,"* ed. Paul J. J. M. Bakker and Johannes M. M. H. Thijssen (Burlington, Vt., 2007), 109–78.

[7] The literature on this topic is vast and includes monographs focusing on particular thinkers as well as thematic and comparative studies. Key studies in the field include Richard Dales, *The Problem of the Rational Soul in the Thirteenth Century* (Leiden, 1995); Katharine Park and Eckhard Kessler's chapters on psychology in *The Cambridge History of Renaissance Philosophy*, ed. Charles B. Schmitt

It was through this occurrence of the passions that the relationship between body and soul, as well as reason and instinct, could be negotiated. This negotiation of ideas transpired not only between scholars and schools of thought but also between and across disciplines, such as theology and medicine. Medical authors, whose primary focus was the functioning of the body, emphasized particular elements pertaining to their concerns. Thus, we find in the medical analysis of the passions of the soul terms familiar to natural philosophy and theology, in addition to notions peculiar to medicine.

Contemporary scholarship, inspired by the growing interest in emotions, has produced several studies on the ways medieval thinkers conceptualized the appearance and manifestation of *affectiones* and *passiones anime*.[8] These studies uncovered the resonance of both Aristotle's and Avicenna's scientific investigations of the soul within the medieval perception of emotions and also established certain developments within scholastic thought. Nonetheless, even with the growth of literature dedicated to exploring the conceptualization of emotions within the particular culture and language of late medieval scholasticism, the focus has often been too limited. Some fundamental aspects of this field—the body being first and foremost among them— have been cast aside as less formative to the understanding of the soul and the passions. This neglect imposed, in turn, a very partial view of medieval emotions that failed to acknowledge the interaction between ideas on emotions and the culture of emotions. This artificial division derives, to a large extent, from disciplinary interests that go well beyond medieval studies. As this volume endeavors to emphasize, the important contribution of the sciences to the ways societies throughout history defined and experienced emotions has too often been overlooked. Whether as a provider of explanatory models or as a facet of culture, scientific and medical understanding and the treatment of emotions have not been sufficiently studied as part of the evolution of emotions in history. In medieval histories of emotions, the interrelationship between the intellectual production and its social and cultural resonance remains underexplored, often producing overly general comments on humoral theory or the appetites. Histories of science and medicine, for their part, habitually remain within the scholastic world of the universities and its learned production with a certain disinterest in their impact on broader communities. This narrow view is particularly remiss in the case of the medical profession, which, insofar as it was a discipline of knowledge, was directed to providing care and transmitting its knowledge-based practices to patients.

By taking into account the cultural role of medicine and its theoretical underpinnings in treating emotions, this article charts the contribution of scientific and medical thought to the formation of the culture of emotions in the late medieval period. The

---

et al. (Cambridge, 1988), 453–534; Dominik Perler, "Introduction: Transformations of the Soul: Aristotelian Psychology 1250-1650," *Vivarium* 46 (2008): 223–31.

    [8] A selective list includes Damien Boquet, *L'ordre de l'affect au Moyen Âge: Autour de l'anthropologie affective d'Aelred de Rievaulx* (Caen, 2005); Simo Knuuttila, *Emotions in Ancient and Medieval Philosophy* (Oxford, 2004); Dominik Perler, "Emotions and Cognitions: Fourteenth-Century Discussions on the Passions of the Soul," *Vivarium* 43 (2005): 250–74; Henrik Lagerlund and Mikko Yrjönsuuri, eds., *Emotions and Choice from Boethius to Descartes* (Dordrecht, 2002); Carla Casagrande and Silvana Vecchio, *Passioni dell'anima: Teorie e usi degli affetti nella cultura medievale* (Florence, 2015); Barbara H. Rosenwein, *Generations of Feeling: A History of Emotions (600–1700)* (Cambridge, 2016).

essay will inquire into three distinct issues. The first will focus on medical theories of emotions as presented in the commentaries of core medical texts used for medical education by examining various authors' engagement with a question of both philosophical and theological import: "Can the soul be moved by the body?" The answers proposed to this question had substantial implications for understanding the nature of living creatures and the manner in which the body and soul interacted. The discussions provoked by this significant issue illustrate how physicians construed the role of medicine in caring for the soul. Their disagreements and dialogue highlight the formulation of a disciplinary method of thought that converses with kindred disciplines in natural philosophy and even theology and yet distinguishes itself from them by adopting a specific medical approach. This professional position, which medical authors sought to establish, particularly with regard to the soul, will be at the center of the second concern of this paper. As part of the process of early professionalization of learned medicine in the late medieval period, the identification of boundaries was a necessary consideration. Late medieval physicians seem to have been eager to define the nature of medical intervention, though various definitions necessarily arose. Their inquiries were clearly linked to a broader debate on the definition of medicine as *scientia* or *ars*; while learned medicine could claim theoretical knowledge and was therefore considered by some as *scientia*, its practical end and its technical, sometimes manual, methods, led others to consider it *ars* and as such a less valued vocation.[9] The medical concern with the form of medical intervention shaped the discussion on the relevance of the accidents of the soul to medical treatment and fashioned its boundaries. Finally, the essay will conduct a brief inquiry into the practical end: through the analysis of records of medical practice that include therapeutic advice concerning emotional states, the degree of transmission of theory into practice will be questioned. Medical case studies provide an illuminating resource for inquiring into the position physicians chose to assume in the formulation or maintenance of the emotional mores of their patients. Such evidence enables us to assess the social role physicians devised for themselves in addressing the public's emotional life.

MEDICAL THEORIES OF EMOTIONS

In a passage in *De anima* regarding the relationship of the body and soul, Aristotle offered two answers to the question, What is anger? "For the [dialectician it is] an appetite for vindication, or something of this sort, while the [naturalist defines it as] the raging of blood or heat around the heart."[10] Aristotle's remark implies two assumptions. The first is that scholars from different disciplines will employ disparate terms and emphasize diverse aspects of the same reality—in this case, the philosopher will treat the form and the natural philosopher the matter. The second assumption is that, while anger's matter is found in the body and its form in the soul, both are in fact a single occurrence that belongs to the body and soul. As the study of *De anima* penetrated university learning, Aristotle's definition of anger was to become

[9] Nancy G. Siraisi, "Taddeo Alderotti and Bartolomeo da Varignana on the Nature of Medical Learning," *Isis* 68 (1977): 27–39; Siraisi, *Medicine in the Italian Universities* (Leiden, 2001).

[10] I quote Aristotle, here and elsewhere, from Robert Pasnau's translation of Moerbeke's version of *De anima*, in Thomas Aquinas, *A Commentary on Aristotle's "De anima,"* trans. Robert Pasnau (New Haven, Conn., 1999), 12.

widespread, to be quoted by medical authors, philosophers, and theologians as well as authors of manuals of confession.[11] Although an analytic approach to the passage may at first glance seem straightforward in establishing the joint involvement of the body and soul in emotional occurrence, questions about the process through which emotions come to be remained unanswered. Physicians were particularly keen on grasping how the body functions in emotional occurrences. As such, they inquired specifically: (1) to what extent emotions were to be considered an integral part of the human being and (2) in what order the body and soul would respond to the onset of emotional states. Both questions inspired discussions within medical textbook commentaries and *quaestiones* that elucidated medicine's approach to the physicality of emotions and to the ability of the body to influence the soul.

Whether the emotions were external to the body or an integral part of the human being was one of the questions that invited physicians to engage with these issues. Commentators of medical textbooks between the late twelfth and fifteenth centuries grappled with the discrepancy between the theories of Galen and Avicenna on this matter and tried to reconcile their differences. According to Galen, emotions were occurrences external to the person, imprinting first the soul and then the body, while Avicenna suggested seeing the emotions as part of the *virtutes animalis*—that is, a natural state of the body. Galen's view predominated in most commentaries on the basic medical textbooks, such as Galen's *Tegni* and Avicenna's *Canon*.[12] Hence, emotions were most often included among the *res non naturales*, as implied by Galen, yet certain elements of Avicenna's theory were also incorporated. One solution, provided by several physicians, including such leading figures as the Bolognese physician Taddeo Alderotti (1210–95) and the Sienese Ugo Benzi (1376–1439), drew a temporal sequence between the parts of the soul and body in emotional occurrences. According to this view, two movements were distinct within the operation of the accidents: a movement of powers (*virtutes*) of the body and a movement of something external to the person that imprints the soul. This external movement then affects the animal power, stirring the heat and the spirit of the body.[13] Taking this position, physicians proposed an intricate argument that can be said to serve both general scholastic culture and their own professional concerns. At face value, the distinction between the affairs of the body and the affairs of the soul implied an inherent hierarchical position: it was the soul that instigated bodily movement, and its reaction to external stimuli guided the movement of the physical body. As the prolific medical

---

[11] A few examples include: "quoniam ira supercalefacit omnia membra et propter fervorem cordis omnes actus racionis confundit" (Arnaldus de Villanova, *Regimen sanitatis ad regem Aragonum*, in *Arnaldi de Villanova Opera Medica Omnia*, vol. X.1, ed. L. García-Ballester, J. A. Paniagua, and M. R. McVaugh [Barcelona, 1996], 436); "Ira est motus sanguinis circa cor propter appetitum vindicte" (Guido de Monte Rocherii, *Manipulus curatorum* [London, 1508], unpaginated); "Nam ire stumulus accensi cor palpitat corpus tremit" (Antonius de Butrio, *Speculum de confessione* [Leuven, 1483], unpaginated); "Ira est accensio sanguinis circa cor propter appetitum vindicte" (Domingo de Valtanas Mejía, *Summa Confessorum* [Seville, 1526], fol. 112v).

[12] Both were elementary medical texts taught and commented upon in the schools of medicine. See Nancy G. Siraisi, *Medieval and Early Renaissance Medicine: An Introduction to Knowledge and Practice* (Chicago, 1990), 71.

[13] Taddeo Alderotti, *In C. Gal. Micratechnen commentarii Secunde Editionis Emaculati . . .* (Naples, 1522), fol. 162r. Ugo Benzi expressed a similar position in his commentary on Avicenna's passage 1.1.1.6.4 of the *Canon*. Benzi, *Super primo canon: Avicenne preclara expositio* (Pavia, 1518), fol. 59r.

commentator Jacopo da Forlì (d. 1414) noted clearly in his *questiones* on the *Canon*, "the natural and animal vital power is the body of the soul, naturally subordinated to it (the soul) in its movements."[14] These observations affirm fully the accepted Christian view of the soul as a prime mover, as the "cause" of the body. I will briefly turn to describe the main lines of argument postulating this opinion.

The importance of maintaining this orientation can be seen in the way Thomas Aquinas chose to expound on Aristotle's formulation of the relationship between the soul and the body in emotions in his commentary on *De anima* 408b1-30. Aristotle argued that the soul is unable to move without the body and that the soul in its nature has no mobility and can only be moved by an accident.[15] Thomas chose to defend the higher power of the soul. He explained that Aristotle rejected the power of the passions of the soul to move the soul in itself. And thus, it would be misleading to define emotions in relation to the soul (e.g., saying that the soul is sad or joyful). A more accurate description would refer to the conjoined movement of soul and body together. Thomas's interpretation, however, veers slightly from the literal reading of the passage. Rather than relating to the unification of soul and body, his emphasis turns to the position of the soul as the source of all passions as well as the movement of the body. Reaffirming this focus on the soul, the next passages, and indeed most of Thomas's commentary, are dedicated to describing the powers of estimation and judgment (which are functions of the soul), through which the soul stimulates the passions:

> And it is likewise clearly apparent in other cases that these operations are movements not of soul but of the compound, nevertheless they come from soul, as, for instance, in the case of getting angry. For soul judges something to be worth getting angry about, and as a result an animal's heart is moved and blood rages around it.[16]

Here Thomas emphasizes not only the sequence of events but also the superiority of the soul by which the body is moved. In this depiction of the emotional occurrence, the alterations that the body undergoes are determined by the reactions of the soul. Such an approach implies that altering emotional reactions may only be possible if the appetitive responses could be altered.

To be sure, Thomas, who had a great interest in and appreciation for Aristotelian natural philosophy, did not discard the claim that there were bodily manifestations correlating with the workings of the soul. Yet his attention was focused on the latter.

---

[14] Jacobus da Forlivio, *Expositio et quaesiones in primum canonem Avicennae* (Venice, 1547), Questio XLV, 217r. This can also be seen in Jacopo da Forlì's argument against Avicenna's theory regarding the power of the imagination to move all matters. Jacopo criticized this theory, stating that the imagination is always united with only one *proprium* body and not with all bodies. This is again an affirmation of the union of the soul with the body, but at the same time it affirms the powerful hold of the imagination as a mover of the body. On the commentators' struggles with Avicenna's position, see Per-Gunnar Ottosson, *Scholastic Medicine and Philosophy: A Study of Commentaries on Galen's Tegni (ca. 1300–1450)* (Naples, 1984), 264. Jacopo da Forlì argued that emotions require cognition and therefore could not be considered solely natural operations, but in order to accommodate his opinion with the authority of the *Canon* he defined two aspects of the emotions, one that appears in the body and the other in cognition. Ibid., fols. 198v–199r.

[15] "If, above all, being in pain or being joyful or intellectively cognizing are movements, and if each one is some-thing's being moved, still being moved comes from the soul—being angry or fearful, for instance, in that the heart is in a certain way moved" (Thomas Aquinas, *Commentary* [cit. n. 10], 80).

[16] Ibid., 83.

Considering that one of the central applications of the theory of the appetite, both in the works of Thomas and certainly in scholastic culture more broadly, was the formulation of the doctrine of sin and virtue and the function of the will to deter one from doing evil, the secondary role attributed to the body supported disinterest (or at least minor interest) in the physicality of emotions. This can be seen, for example, in the development of the concept of intellectual passions, in addition to sensitive passions. Again, here is Thomas on this notion:

> It is important to know as well that just as we find both an appetitive and an apprehensive power in sense, so too do we find both an appetitive and apprehensive power in intellect. Thus these—love, hate, joy and things of this sort—can be understood both as they exist in sensory appetite, in which case they have a conjoined bodily movement, and also as they exist in intellect and will alone, without any sensory affect. And in this case they cannot be said to be movements, because they do not have a conjoined bodily movement.[17]

This parallel system of appetites of the intellect explains its uniqueness and independence, as well as proclaiming the exceptionality of human nature, arguing that some emotions, such as love, hate, and happiness, appear only in humans and are not found in animals.[18] Such an argument discloses an important underlying motivation for defining some passions as nonphysical. Physicality undermined the difference between man and beast and devalued affects as love, which in its pure form was considered to be divine. Downplaying the animal nature of emotions could assist in employing the passions in devotional life as a utilization of humane sensitivity toward religious piety and, in another way, could substantiate the role of the will in the theology of sin in which emotions such as anger and envy had a significant place. Thus in the years following Thomas's interpretation, various thinkers developed more thorough suggestions regarding the relationship between the intellect and emotions. Such was the case with Duns Scotus (1265/6–1308), whose theory of the passions stressed the existence of two kinds of passions, those of the sensitive soul and those originating in volition, which are thereby essentially linked to reason and the rational soul. As Simo Knuuttila noted in his survey of the development of medieval philosophy of emotions, Sco-

[17] Ibid., 86. Thomas Aquinas offered a problematic construction when suggesting the term "passions of the intellect": strictly speaking, the passions of the intellect cannot be named passions because the intellect does not have movement (the rational soul is perfect and as such it cannot undergo a process of change), similar to that found in the sensitive soul; although these passions originate from appetitive powers of the rational soul, they are not accompanied by any sensory activity. The choice of improper terminology only emphasizes that Thomas saw a need to define a parallel emotional construction that is detached from the body. In his commentary on *De anima*, Albertus Magnus (1206–80) also referred to the existence of "delights" in the rational soul. Such delights could not be regarded as passions per se because of the impassibility of the soul, yet they are similar in their formal aspect to passions. See Albertus Magnus, *De anima*, in *Opera omnia*, ed. Wilhelm Kübel (Westfal Monastery, 1968), 7.1:43.

[18] The operations of the sensitive soul were considered to be shared by all living animals; thus animals were thought to have emotional experiences. The discussion on animals' cognitive capacities appears in various places in Aristotle's *De anima*. Of particular interest to the topic of emotions is passage 432b20–30. Medieval commentators repeatedly discussed the issue, following Avicenna's example of the sheep's fear of the wolf found in Avicenna, *Liber de anima seu Sextus de naturalibus*, ed. S. van Riet (Leuven, 1968–72), 38. Hasse describes the development and use of this example by medieval philosophers; see Hasse, *Avicenna's "De anima"* (cit. n. 2), 127–53. See also Jack Zupko's essay, "Self-Knowledge and Self-Representation in Later Medieval Psychology," in Bakker and Thijssen, *Mind* (cit. n. 6), 87–106.

tus's differentiation of the passions was echoed strongly in the conceptualization of emotions by later authors, such as William of Ockham (1288–1348) and Jean Buridan (1300–58).[19] These new postulations exhibited a shift from the emphasis in the two preceding centuries in which a greater focus was placed on the intellect and the emotional experience of the separate soul, as opposed to the soul-body compound. This shift may have its roots in the waning interest of commentators on *De anima* in aspects relating to the workings of the body, and the more narrow focus, instead, on aspects of the soul and its immaterial entities. This becomes much more apparent in Thomas de Vio Caietanus's (1469–1534) early sixteenth-century *Commentaria in de Anima Aristotelis*. Caietanus opened his commentary with a general exposition on the first part of book I, singling out as its main problem whether there are passions that belong to the soul alone in addition to those of the body-soul union.[20] Further proof of the growing emphasis on the intellectual soul appears outside the commentary tradition: the development of the doctrine of the sacrament of penance, with its articulation of the notion of full contrition as an intellectual pain of remorse that in its purest form has no sensitive manifestation. In addition, this can be seen in the development of the notion that souls experienced suffering in hell and pleasure in heaven.[21] These religious positions were taught in the schools and were circulated to broader audiences in sermons and lay treatises, in both Latin and the vernaculars.[22] The primacy of the soul was therefore widely entrenched in the period, and its subordination of the body was well accepted. And yet the medicine of the body could not sustain this view without a challenge.

The difficulty that arose in the practice of physicians from this "safe" position can be identified in Gentile da Foligno's (d. 1348) less conventional opinion. Gentile was a student of Taddeo at Bologna who later taught at Padua and Perugia and became recognized as an influential and thorough commentator on Avicenna. Unlike his colleagues, Gentile proposed a staunch materialistic view of emotions. In a commentary on a passage in the *Canon* mentioned above, he rejected the idea that the physical occurrence of emotions is a by-product of the affairs of the soul on the grounds that this position implied that the physical movement was not part of the emotion itself. He suggested instead that there were two alternative ways of depicting emotions that were not linked in sequential order. Gentile argued that it is possible to consider emotions as an outcome of the movement of the appetites, which would locate emotions within the animal powers of sense and movement, or one could refer to emotions as a movement of the heat and spirits, which belong to the operation of the vital virtues.[23] Although all scholastic physicians agreed that a physical component to emotions existed, only Gentile maintained that the movement of the body should be seen as equal to the movement of the soul and not only as a result of it. This "strong hylomorphism," which urged one to see the body as fully attached to the soul in the living body, is, one might say, a sensible position for a physician to hold. It maintains the body's

---

[19] Knuuttila, *Emotions* (cit. n. 8), 256–82; see also the discussion of "Volitional Passions" in the philosophy of William Ockham and Adam Wodeham by Perler, "Emotions and Cognitions" (cit. n. 8).

[20] Thomas De Vio Cardinalis Caietanus, *Commentaria in de Anima Aristotelis*, ed. P. I. Coquelle, OP (Rome, 1938), 14–5.

[21] Esther Cohen, *The Modulated Scream: Pain in Late Medieval Culture* (Chicago, 2010), 182.

[22] Alan E. Bernstein, "Heaven, Hell and Purgatory: 1100–1500," in *The Cambridge History of Christianity*, vol. 4, *Christianity in Western Europe, c. 1100–c. 1500*, ed. Miri Rubin and Walter Simons (Cambridge, 2009), 200–16.

[23] Gentile da Foligno, *Primus (et secundus) Avicennae* (Venice, 1520), 85v.

significance for the living being. Such a position had much importance to medicine's view of emotions because it affirmed that emotions do manifest, if not occur, in the body. Thus, though often presenting a less materialist position, physicians widely claimed that emotions are essentially a physical phenomenon. While not positing that emotions are simultaneously mental and physical, the notion that the body is the locus of emotions was pivotal to medical consideration of the "accidents of the soul." Medical texts of different genres (commentaries, regimens of health, etc.), and of both theoretical and popular orientations, repeatedly attempted to substantiate this view by ascribing the physical activity of emotions to a specific part of the body—the brain and/or the heart—or to change in the bodily equilibrium of the qualities (heat, cold, dry, and wet).[24] These explanations depended on medical authorities, but they relied no less on the way practical medicine interpreted bodily reactions. In this spirit, Tommaso del Garbo (1305–70), who taught in both Perugia and Bologna, argued that the ability of emotions to cause transmutations in matter is evident from experience [*quod sit patet per experientiam*].[25] The experience of physicians with human bodies was thus held as a source of knowledge that carried some authority.

In Aristotle's passage on anger and the fragmented representation of anger by each discipline, it appears that with regard to the physicality of emotions, disciplinary interests were defended by physicians and others. Nevertheless, in turning now to the debate that developed on the ability of the body to move the soul, we will see how the *experientia* argument brought new challenges to the appetitive theory of emotions.

CAN THE BODY MOVE THE SOUL?

In the popular medical poem *Regimen sanitatis salernitanum* (probably composed in the thirteenth century), several verses were dedicated to defining the four temperaments. Each of the temperaments is said to produce personal characteristics with dominant behavioral and emotional traits. The sanguine person is easily angered but also joyful and loving; the choleric is known to be angry; the phlegmatic, slow and sleepy; the melancholic, sad and envious.[26] The complexional theory was widely accepted and used during the late medieval period, even beyond the medical profession, and its value for describing and judging character was recognized by artists and literary authors.[27] Complexional theory was even referenced in pastoral manuals in which the priest was instructed to be aware of a penitent's complexion in discerning their sins and in assigning penance.[28] Humoral change was, therefore, understood (at least by some) to be able to change one's mental state. But it was not only the body's matter that was thought to be able to influence the soul. The six nonnaturals in general and particularly material substances, such as wine, were perceived to influence bodily complexion and through it to alter cognitive and mental states. Medical treatises on wine

---

[24] Naama Cohen-Hanegbi, "The Matter of Emotions: Priests and Physicians on the Movements of the Soul," in "Convergence/Divergence: The Politics of Late Medieval English Devotional and Medical Discourses," ed. Denis Renevey and Naoë Kukita Yoshikawa, special issue, *Poetica* 72 (2009): 21–42.

[25] Tommaso del Garbo, *Summa medicinalis* (Venice, 1531), fol. 114v.

[26] Salvatore de Renzi, ed., *Collectio Salernitana* (Naples, 1859; repr., Naples, 2001), 1:48–9.

[27] See, e.g., Piers D. Britton, "Humoral Exemplars Type and Temperament in Cinquecento Painting," in *Visualizing Medieval Medicine and Natural History, 1200–1550*, ed. Jean Ann Givens, Keren Reeds, and Alain Touwaide (Aldershot, 2006), 177–203.

[28] Alain de Lille, *Liber poenitentialis*, ed. Jean Longère (Leuven, 1965), 2:31.

and the *regimen sanitatis* literature often noted that wine was a potential cure for timidity and cowardice.[29] Wine's ability to alter cognitive states and to harm one's judgment was also widely recognized.[30] The biblical verse "wine gladdens the heart of
man" was mentioned repeatedly, and it became a common proverb, one that is still
heard today.[31] Nevertheless, although all these examples point to an acceptance of the
possible impact of the body and other kinds of matter on the soul and the mind as
its faculty, such views were challenged by two prominent modes of thought of the
period. The first appeared in the spiritual discourse, which by its very nature revealed
a tendency to dissociate emotions from the body. This tendency may be seen in the common interpretation of the above quoted psalm verse, to which theologians referred as
mystical joy experienced upon imbibing the blood of Jesus in the sacrament of the
Eucharist. Because this verse was used in Eucharist sermons, devotional treatises, and
handbooks for priests for performing the mass, this nonphysical reading of it reached
wide audiences, and it had significant influence in disseminating this particular notion
of the emotions.[32]

The second challenge to the body's influence on the soul came from the discourse
on the nature of the soul, which has been addressed above. The possibility that physical complexion and humoral change can alter and move the soul demanded reformu-
lation of the relationship between the body and the soul in emotions. The scholastic
tradition expressed in the commentaries and independent treatises repeatedly argued
for seeing the emotions and related mental states as products of the soul alone. Thomas,
in his *De motu cordis*, stated clearly that despite attempts by physicians to identify
the vital movements of the body, its movements are always governed by the soul and
its appetites.[33] Whether explicitly or *ex silencio*, Thomas's opinion became the most
common interpretation of the hylomorphic nature of the emotions. The notion that
the movement of emotions originated only in the soul safeguarded the supremacy
of the soul over the body.

Some philosophers mitigated this sharp view by submitting some agency to the
body as initiating movement. The commentary ascribed to the thirteenth-century logi-
cian Peter of Spain, for example, discussed certain movements that follow an action
in the soul and thereby, accidentally, move the body; other movements involve the
body and soul hybrid—as in the case of matter affecting form (such as in sleep),
or form that moves matter (as in imagination and sensation).[34] Nevertheless, despite

---

[29] Bartholomaeus of Salerno, "Practica," in de Renzi, *Collectio Salernitana* (cit. n. 26), 4:376;
Stiftsbibliothek Melk, MS 728 (967), fol. 56v. See also L. Elaut, "The Walcourt Manuscript: A Hygienic Vade-Mecum for Monks," *Osiris* 13 (1958): 184–209. For the association between wine and
courage, see Matthew Klemm, "Medicine and Moral Virtue in the *Expositio Problematum Aristotelis*
of Peter of Abano," *Early Sci. & Med.* 11 (2006): 302–35.

[30] See the exempla in Thomas de Cantimpré, *Bonum universale de apibus* (Douai, 1627), 251, 536.

[31] Ps. 104:15.

[32] Guido Monte Rocherii, *Manipulus curatorum* (London, 1508), unpaginated; "Life of Soul,"
trans. Paul F. Schaffner, in *Cultures of Piety: Medieval English Devotional Literature in Translation*,
ed. Anne Clark Bartlett and Thomas Howard Bestul (Ithaca, N.Y., 1999), 134; Manuel Ambrosio Sánchez Sánchez, *Un sermonario Castellano Medieval: El ms. 1854 de la Biblioteca universitaria de Salamanca* (Salamanca, 1999), 295.

[33] In *De motu cordi*, Thomas presented a very clear argument for the soul's dominance to refute the
conjecture that complexion may dictate emotional states. See Thomas Aquinas, *Opera omnia*, vol. 43
(Rome, 1976), 127–30.

[34] The identity of the philosopher Peter of Spain is still unknown after much scholarly debate. He is
assumed to be a Dominican who studied in Paris for a time. See *The Stanford Encyclopedia of Phi-*

Peter's allowing a material influence on the soul, emotions belonged to the latter kind of movement—that of form influencing matter.[35] Almost a century later, Nicholas Oresme (ca. 1320–82) offered a more complex account of this relationship between the body and the soul. Oresme suggested two kinds of movements of passions: a movement of the soul that moves the body, as in the case of the psychological phenomenon of anger that changes the state of the body, and a bodily state that generates psychological occurrences, as in a choleric complexion that makes a person prone to anger.[36] Though a number of scholars mentioned the movement of the humors and their imbalance with regard to the emotions, Oresme was exceptional in arguing that it can be the cause of emotional change rather than its outcome.[37] However, his interests lying elsewhere, he did not expand on the means of altering emotional states; thus the value of medicine, or material means for moving the soul, was ignored. Nonetheless, one cannot deny that his position clearly assumes a strong mutual influence of body and soul.

What emerges from this brief survey is that during the thirteenth and fourteenth centuries a guiding theory on the way the body and the soul interacted was established. This theory was nevertheless accompanied by an array of alternative, sometimes contesting, notions. Considering Thomas's reference to physicians and the mention of humoral theory, it seems reasonable to argue that the spread of medical knowledge, both learned and popular, figured prominently in articulating varied forms of the relationship between body and soul. As Damien Boquet and Piroska Nagy suggest in another essay in this volume, the science of emotions developed in tandem with the medical theory of the soul. But while scholars who were not physicians introduced medical concepts without commitment to the medical discipline, professors of medicine sought to accommodate medical theory and practice within the structure of the period's philosophical conventions in a much more careful way.

A significant stumbling block for medical authors was the claim that matter can influence cognitive and mental activities, if not the soul as a whole. It was a fundamental argument as it substantiated the efficacy of medicine and its ability to alter the state of the soul by medicinal means. Engaging with this problem, however, posed a problem for defending the philosophical grounds of medicine. Thus, theoretical works mostly skirt around the theoretical implications of the movement of the soul when writing about the relationship between emotions and humors. This allowed the authors of these works to express unconventional views and, at the same time,

---

losophy, s.v. "Peter of Spain," by Joke Spruyt, http://plato.stanford.edu/archives/win2012/entries /peter-spain/ (accessed 10 February 2016).

[35] Pedro Hispano, *Obras filosóficas: Comentario al "De anima" de Aristóteles*, ed. S. I. P. Manuel Alonso (Madrid, 1944), 294–9. A somewhat more radical contemporary opinion, that the movement of emotions occurred in the sensitive soul and the body at the same time, was suggested by an anonymous teacher in Paris (ca. 1273–7); see "Un commentaire semi-Averroïste du traité de l'âme," in *Trois commentaires anonymes sur le traité de l'âme d'Aristote*, ed. Maurice Giele, Fernand Van Steenberghen, and Bernard Bazán (Leuven, 1971), 184. This position echoes Gentile da Foligno's argument in his medical commentary on the *Canon* mentioned above.

[36] Nicolaus Oresme, *Nicolai Oresme Expositio et quaestiones in Aristotelis "De anima,"* ed. Benoît Patar and Claude Gagnon (New Leuven, 1995), 11–2.

[37] Albertus Magnus noted that changes in the humoral balance occur in emotions and that every person might feel emotions and bodily expressions differently because of their individual nature. Albertus Magnus, *De anima* (cit. n. 17), 12. A similar explanation can be found in the commentary of the fourteenth-century philosopher Jean Buridan. See Buridan, *Expositio in Aristotelis De anima: Le traité de l'âme de Jean Buridan*, ed. Benoît Patar (New Leuven, 1991), 12–4.

seemingly to agree with the accepted views of philosophers. The attempts to discuss this issue directly were few and far between. Two appear in commentaries on Hippocrates's twenty-third aphorism of his sixth book: "Patients with fear or depression of long standing are subject to melancholia."[38] Taddeo Alderotti emphasized the physical nature of the emotions but refrained from claiming that complexion, a material substance, could be the cause of an emotion. Instead he added cautiously that the complexion might determine a mental state but not the emotion itself.[39] Thus, while hinting at the importance of well-tempered complexions, he did not contend that it was possible for emotions to be changed through matter (e.g., through the use of medical substances). About a hundred and fifty years later, Ugo Benzi was not as hesitant when he suggested that the influence could be mutual and that just as fear and pusillanimity may cause melancholy, melancholic humor can cause fear because of its black and tenebrous characteristics.[40] A certain correlation can be detected between Ugo Benzi's opinion and the ideas expressed in the commentary on the *De anima* by Nicholas Oresme discussed above. Their similar opinions may indicate a more open acceptance of the impact of the body on the soul from the second half of the fourteenth century on, though this hypothesis calls for further verification. The two medical commentaries, by Ugo and Taddeo, that related to the impact of complexion on the soul share another interesting aspect—both treated a text that had practice as its focus. Perhaps due to the nature of the genre, commentaries of a more theoretical bent (on the *Tegni* and *Canon*) related to the topic in a rather implicit and less committed manner, preferring to avoid the hazardous problem of the influence of matter on the soul.

Peter of Abano (1257–1316) was another author who elaborated on the relationship of the body and soul. In his *Conciliator*, he argued that the soul is dependent on the body and that complexion may alter the state of the soul. His work, which preceded Ugo Benzi's commentary by over a century, thoroughly investigated the nature of complexion. Peter, of Italian origin, was a generation younger than Taddeo; he taught at both Paris and Padua and wrote treatises on medicine and astrology. His *Conciliator* was devoted to accommodating natural philosophical and medical approaches, designed to establish a more solid and codified theory of medicine grounded in philosophy. Though he argued against the proposition that the soul was complexion, he nevertheless claimed that in the living body the balance of complexion is influential on the state of the soul.[41] Preferring this "medical perspective on the soul," as Matthew Klemm defined it, Peter endorsed the notion that medical intervention is bound by a moral obligation because it may influence a patient's behavior. As Peter emphasized clearly, the soul is medicine's most important subject, notwithstanding the fact that it was not treated directly. This is a significant statement because of its rejection of the dichotomous split between the care of the body and the care of the soul. Here we may see the professional implications of the definitions of the relationship between the soul and the body. By defining the soul through the means of the body, Peter

---

[38] Hippocrates, *Hippocratic Writings*, ed. G. E. R. Lloyd, trans. J. Chadwick and W. N. Mann et al. (London, 1978; repr., New York, 1983), 229; Ugo Benzi, *Expositio super aphorismos Hippocratis et super Galeni commentum* (Venice, 1498), fol. 148v.

[39] Taddeo Alderotti, *Expositiones in arduum aphorimos Ipocratis . . .* (Venice, 1527), fol. 175v.

[40] Ugo Benzi, *Expositio super aphorismos* (cit. n. 38), fol. 148v.

[41] Matthew Klemm, "A Medical Perspective on the Soul," in *Psychology and the Other Disciplines*, ed. Paul J. J. M. Bakker and Sander W. de Boer (Leiden, 2012), 275–95.

made a strong case for the necessity of knowledge of the body and medical intervention when the balance of complexion is harmed. Thus, the engagement with the accidents of the soul permitted an avenue for claiming the justification of the medical profession and particularly medicine's moral and even spiritual function.

MEDICAL INTERVENTION: THE BOUNDARIES OF TREATING EMOTIONS

Peter of Abano's claim for medicine's moral role touched on a topic of much contention. Discussions of the relevance of emotions to medicine were not easily defended because a primary assumption was that the physician should treat the physical body. Nonmedical texts that directly criticized university physicians for attempting to assume authority and knowledge that did not rightfully belong to them (Petrarch's invectives being the most famous of them) are scant, yet much evidence for treating patients who would nowadays be considered mentally ill derives from saints' miracle collections and folk exorcism remedies.[42] We can assume, therefore, that in terms of practice, people of the thirteenth and fourteenth centuries would not turn to physicians for emotional care. This may have changed in the late fourteenth century as physicians were called upon more often to examine cases of suspected mental disturbance, such as demonic possession.[43] Nevertheless, internal professional debates on the topic of physicians' authority over the soul surface much earlier. Most medical texts that discussed the rudiments of the medical profession carefully touched on the limits of the medical treatment of the emotions and the soul more generally, but they did not articulate the implications of this aspect on medical practice.

It became, for example, a recurrent topic in the commentaries of Galen's *Tegni*, a basic medical text taught in every European medical school between the late twelfth and fifteenth centuries. The third book of the *Tegni* opens with an outline of various causes that alter the humors and the complexions, among which emotions are mentioned:

> It is indeed required to abstain from intemperance of the passions of the soul, namely, from anger, sorrow, and joy and fury and fear and envy and worries, since these expel and move the body from that which is its balance according to nature.[44]

This brief work leaves out any explanation as to why such an alteration of the body occurs with emotions. A more intricate account, which acknowledges the exceptionality of emotions as materia medica, appears in the first chapter of his *De regimine sanitatis*. Describing the proper regimen for youth, Galen devoted a rather large sec-

---

[42] Francesco Petrarca, *Invectives* (Cambridge, Mass., 2003), 72. For further discussion of Petrarch's views on medicine, see George W. McClure, *The Culture of Profession in Late Renaissance Italy* (Toronto, 2004), 4–10. See the various essays in Sari Katajala-Peltomaa and Susanna Niranen, *Mental (Dis)Order in Later Medieval Europe* (Leiden, 2014); Catherine Rider's essay, which particularly deals with medicine, considers only theoretical materials with no case studies: Rider, "Demons and Mental Disorder in Late Medieval Medicine," 47–69.

[43] Judith Bonzol, "The Medical Diagnosis of Demonic Possession in an Early Modern English Community," *Parergon* 26 (2009): 115–40.

[44] Two translations of the Greek original circulated between the thirteenth and fifteenth centuries, and each used a different term to refer to the emotions: *Accidentia anime* in the translation from Arabic and *Passiones anime* in the translation derived directly from the Greek. Galen, "Liber Tegni Galieni," in *Articella* (Venice, 1483), fol. 187v.

tion to the education of suitable habits and behavior, closely linking behavior and health. Although peripheral to Galen's discussion, it is in this chapter that we find a distinction that would become pivotal to the debates of later physicians in their attempts to clarify the place of emotions within medical practice. Stressing the need for physicians' involvement in the care of emotions, Galen differentiated between addressing emotional behavior for the sake of health and the prevention of sickness and doing so in order to instill moral propriety.[45] Ethics and morals are the business of philosophers, not physicians, wrote Galen. His words and the distinction he made resonate more loudly than intended in the writings of many of his successors, who often introduced them in order to expound or modify them. This is particularly true in the case of the deeply influential commentary of the eleventh-century physician Alī ibn Ridwān (or, as the Latin authors referred to him, Haly Rodoan [d. ca. 1061]), whose commentary served as a trigger for the debate within late medieval Christian commentators.[46]

In contrast to Galen, Haly pointed out distinctly that the accidents of the soul are an unusual case among the nonnaturals and should perhaps not be included among them. He did acknowledge the various manners in which emotions may affect the body, yet he questioned their actual relevance to medical practice. Haly then quoted Galen's above-mentioned reservation but turned it into a central part of his argument by referring to it before any other consideration of the topic and developing it further. He claimed that not only the aims of philosophy and medicine, but also the methods they may apply in practice, differ. This limitation was further strengthened by the adamant statement that an existent medical method for dealing with emotions should be followed.[47] Significantly, while Haly restricted the involvement of physicians with emotions, he did not prohibit it completely. His position, that physicians should regard emotions only with respect to illness, was later described by commentators as extremely antagonistic and categorical, which seems to have served their own rhetorical needs more than it reflected Haly's position. Haly came to represent, for most Latin commentators, the restrictive perimeters of the professional boundaries of medicine. His comparison of medicine with philosophy and law singled out medicine's specific realm—the realm of the body. By identifying philosophy and law as the disciplines responsible for dictating correct emotional behavior, medicine's task was merely to monitor the fluctuation of the humors.[48] Though dealing with the seemingly peripheral issue of emotions, this definition of the boundary between medicine and philosophy captured an essential difficulty university physicians encountered with regard to their formulation of the medical profession.

As the research of Nancy Siraisi, Danielle Jacquart, and Michael McVaugh has shown, university physicians struggled to clarify and place their position somewhere between the abstract field of philosophy (in particular natural philosophy, but also with regard to the philosophy of ethics) and the mechanics of medical care.[49] These

[45] Galen, "De regimine sanitatis," in *Opera omnia*, vol. 2 (Venice, 1490), fol. 144r.

[46] On the *Commentum Haly*, see Cornelius O'Boyle, *The Art of Medicine: Medical Teaching at the University of Paris, 1250–1400* (Leiden, 1998), 93–4.

[47] "Commentum Haly," in *Articella* (cit. n. 44), fol. 187v.

[48] A brief account of Haly's position on emotions in medicine as well as those of some of his commentators appears in Per-Gunnar Ottosson, *Scholastic Medicine and Philosophy: A Study of Commentaries on Galen's Tegni, ca. 1300–1450* (Naples, 1984), 259–64.

[49] Siraisi, "Taddeo Alderotti" (cit. n. 9); Michael R. McVaugh, *Tractatus de intentione medicorum*, in *Arnaldi de Villanova Opera Medica Omnia*, vol. V.1, ed. Michael R. McVaugh (Barcelona, 2000),

discussions revolved around a number of issues, including the necessity of theoretical knowledge, the manner in which it should be accommodated with medical knowledge, and the particular benefits of practical experience. The affinity between the philosopher's role and the physician's trade appeared already in one of the earliest Latin commentaries on the *Tegni*, that of Bartholomaeus of Salerno (fl. 1150–80), in the passage on sexual intercourse. According to the historian Faith Wallis, Bartholomaeus's notion of equilibrium and the necessity of balance to health owes a significant debt to the writings of Aristotle. This shared use of terminology and concepts assisted him in establishing the semblance between philosophy and medical knowledge; both disciplines offered guidance for those who sought to restore their equilibrium.[50] The physician, as the philosopher, was responsible for wisely offering this counsel. This was not only a technical parallel. As Wallis shows, Bartholomaeus's formulation of the content of medical guidance was infused by knowledge of Aristotelian ethics and the mores of Christian life.[51] Thus, sexual intercourse is to be suggested in accordance with Christian lore and social mores. Although the preceding paragraph on the accidents of the soul ignored moral concerns, relating instead only to the material influence, the introduction of an ethical point of view to the commentary is significant. It would become an ongoing part of commentaries over the next three centuries, specifically in the chapters dealing with emotions.[52] So much so, in fact, that it may be said that emotions were often discussed in medical literature not only per se but also with respect to a set of more general and wide-ranging issues concerning the relationship of philosophy and medicine and the value of medicine as a *scientia*.

Grappling with the (somewhat anachronistic) image of the philosopher/lawyer whose concern is the care of the soul set by Haly, Taddeo Alderotti distinguished between two objectives that the treatment, or the consideration, of emotions may try to achieve: the preservation of the soul and the preservation of the body.[53] The first aims to prevent sin, the second to prevent illness. Medical treatment of emotions, Taddeo concurred, should relate exclusively to physical health. However, in contrast to Haly's prohibiting tone, which discouraged the consideration of emotions by physicians, Taddeo undermined the prohibition by emphasizing the necessary role of medicine in the care of emotions. In addition, Taddeo argued that the care of emotions should be not only considered in view of the disposition of illness, as suggested by Haly, but also recognized as fundamental to the conservation of health.[54] This line of argument was then followed and developed by the next generations of physicians. For example, Taddeo's student, Petrus Tusiggnano, emphasized the superior abilities of philosophy

---

143–5; Pedro Gil-Sotres, *Regimen sanitatis ad regem Aragonum*, in García-Ballester, Paniagua, and McVaugh, *Arnaldi de Villanova Opera Medica Omnia*, vol. X.1 (cit. n. 11), 805.

[50] Faith Wallis, "12th Century Commentaries in the *Tegni*: Bartholomaeus of Salerno and Others," in *L'Ars Medica (Tegni) de Galien: Lectures antiques et médiévales*, ed. Nicoletta Palmieri (Saint-Etienne, 2008), 140–1. On Bartholomaeus's acquaintance with Aristotle's Ethics, see Charles de Miramon, "Réception et oubli de l'Ethica Vetus. Salerne et Bologne (1150–1180)," in *Mélanges en l'honneur d'Anne Lefebvre-Teillard*, ed. Bernard d'Alteroche et al. (Paris, 2009), 734–8.

[51] Wallis, "12th Century Commentaries" (cit. n. 50), 147–8.

[52] Stadt- und Regionalbibliothek, Erfurt, CA 4o 294, fol. 68v.

[53] See Gil-Sotres, *Regimen sanitatis* (cit. n. 49), 805–7.

[54] Taddeo Alderotti, *Commentum in microtegni Galieni* (Naples, 1522), fol. 162r. Petrus Hispanus, a physician (1205–77?), offered a similar position in his commentary, but there, in addition to the debate with Haly, the author also delved into the discussion on the nature of the passions and whether they are part of the vital powers or an external force. By weaving together this theoretical issue with

and law to influence and enforce appropriate emotional habits, yet, at the same time, stressed that the body and soul are united in living creatures and maintain a mutual relationship of influence. This being the case, the implication was that it is impossible to actually differentiate between the emotions of the body and the soul. This claim was further substantiated by a phrase he quoted and attributed to Galen: "evil deeds follow the *species* of distempered complexion." This phrase, which can be found with slight variations (and various attributions), challenges the disciplinary divide by closely attaching the preservation of bodily complexions with the prevention of "bad behavior."[55] This ethical implication was further expanded upon by authors of popular regimens of health, notably Bernard de Gordon (1258–1311) and Maino de Maineri (d. 1368); their interpretation of the need to care for emotions went beyond the physical affairs of the body to include health's influence on a person's moral behavior. This is particularly apparent in Maino's version of the phrase: "bad men follow humoral intemperance."[56] Physicians, it is implied, may cure men of their evil habits. Another influential *regimen* author, Arnau de Vilanova (1240–1311), exemplifies how the ethical and the physical intermingled in his advice to King James II. He advised James to avoid sorrow and anger because of the imbalance they cause, supporting his opinion with a passage on the impropriety of a ruler who expresses them without moderation.[57]

Advice such as that found in the *regimina* literature appears to be more permissive in referring to morality. This tendency of practical writings was perhaps shaped by the personal ties physicians had with their patients, often kings and nobility, to whom they dedicated their works. When writing the theoretical treatise *Speculum medicine*, Arnau was adamant about the need to refrain from overstepping the boundaries of medical intervention by addressing morality. A large part of the chapter on the accidents was devoted to specifying the topics relating to the emotions that a physician should refrain from discussing with his patients.[58] Yet when counseling his patient, the king, he decidedly pointed out not only the morality of health preservation but a specific emotional regimen appropriate for his complexion and royal office. This dual voice, here condemning moral advice, there allowing, can therefore be ascribed to the difference between theory and practice and the more casual approach of texts written for popular audiences. A number of popular fifteenth-century medical treatises corroborate this assumption.[59] But this generic distinction does not give full justice to the problem at hand. Close examination of medical literature on the topic reveals that this dual voice is to be found repeatedly within theoretical works as well. Whether uttered quietly in Taddeo's and Petrus Tusiggnano's texts, or more openly in

---

the professional one, Petrus exemplified well how the theoretical queries on the nature of emotions carried strong implications for the articulation of medicine's professional authority. MS 1877, fol. 42v, Biblioteca Nacional, Madrid.

[55] Petrus Turisanus, *Plusquam commentum in microtegni Galeni* (Venice, 1498), fol. 87v. Although this sentence captures the spirit of Galen's ideas in the first book of *De regimine sanitatis*, I have not found this exact quote. This is noteworthy because variations of similar sentences are found in other texts (e.g., the health regimens of Bernard de Gordon and Maino de Maineri), and they seem to represent an ongoing oral tradition regarding the mutual influence of behavior and health.

[56] Maino de Maineri, *Regimen sanitatis magnini mediolanensis* (Paris, 1483), fol. 143r.

[57] Arnaldus de Villanova, *Regimen sanitatis* (cit. n. 11), 436.

[58] Arnaldus de Villanova, "Speculum medicine," in *Hec sunt opera Arnaldi de Villanova que in hoc volumine continentur . . .* (Lyon, 1504), fol. 28v.

[59] Chiara Crisciani, "Éthique des *consilia* et de la consultation: à propos de la cohésion morale de la profession médicale (XIIIe-XIVe siècles)," trans. Marilyn Nicoud, *Médiévales* 46 (2004): 23–44.

those of later physicians, both the criticism against aspiring to lead patients to virtuous ways (caring for the *homo virtuosus*) and the endorsement of emotional health (which subscribes to Christian culture's understanding of virtue) as a moral pursuit find expression. Thus, despite stating that medicine should treat only the body, Niccolo Falcucci (1350–1412) chose to expand dramatically the emotions he considered to influence the body (he included hatred among others); Ugo Benzi noted that if physicians ignored emotional states, the soul and the intellect may incur great damage; and the Florentine Michele Savonarola (1384–1466) noted that knowledge of the patient's habits and customs is essential to offering emotional advice.[60] These statements disclose an ongoing concern for clearly setting the professional parameters of doctors' knowledge and practice. Yet they also reveal a growing willingness for physicians to take part in the design and regulation of emotional codes.

Though relatively peripheral, it is nevertheless evident that emotions figured in the formulation of the image of the physician during this period. Acknowledging the importance of the treatment of the accidents of the soul implied that medicine dealt with more than "just" the body. Although anxious to assert their mandate over the body and its physical state, medical practitioners and theorists were also interested in the soul, its faculties, and their correspondence with the living body and were necessarily prone to consider whether habits and feelings were part of their domain. Addressing such manifestations of the soul enabled them to extend their reliance on philosophy beyond the system of natural philosophy (as seen in the debates on the nature of the relationship between the body and the soul) to the realm of ethics and morality.

EMOTIONS IN PRACTICE

The encounter with patients' emotional lives, whether in person or on paper, offered an important challenge to medicine's professional definitions beyond the matters of the body. The evidence discussed above largely derives from medical works written for physicians' professional use. Turning to case studies produced by physicians of the same milieu, we can see how the developments in the conceptualization of the profession translated into practice. *Consilia* and *Regimina* are sources that need to be treated with caution. Written by practicing physicians, the reports provide engaging insights into a prescribed course of treatment of patients, who at times are mentioned by name.[61] However, it is necessary to keep in mind that these cases were written mostly in Latin and often were directed to the elite of society and to fellow physicians as illustrative theoretical texts and therefore would likely be reformulated and refashioned accordingly. And yet these texts, written and published, certainly add to

---

[60] Niccolo Falcucci, *Sermones medicinales septem* (Venice, 1490), fol. 41v; Ugo Benzi, *Expositi Ugonis Senensis super libros Tegni* (Venice, 1498), fols. 73r–v; *Practica Iohannis Michaelis Savonarole* (Venice, 1497), fol. 19r.

[61] On the history of late medieval medical consultations see Jole Agrimi and Chiara Crisciani, *Les Consilia medicaux*, trans. Caroline Viola, Typologie des sources du moyen âge occidental, vol. 69 (Turnhout, 1994); Gianna Pomata, "Sharing Cases: The *Observationes* in Early Modern Medicine," *Early Sci. & Med.* 15 (2010): 193–236; Pomata, "Observatio ovvero Historia: Note su empirismo e storia in età moderna," *Quad. Stor.* 91 (1996): 173–98; Marylin Nicoud, "Expérience de la maladie et échange épistolaire: Les derniers moments de Bianca Maria Visconti (mai–octobre 1468)," *Melanges l'Ecole Franç. Rome: Moyen Âge* 112 (2000): 311–458; see also Chiara Crisciani, "L'*Individuale* nella medicina tra medioevo e umanesimo: I *Consilia*," and Nancy G. Siraisi, "L'*individuale*' nella medicina tra medioevo e umanesimo: I 'casi clinici,'" both in *Umanesimo e medicina*, ed. Roberto Cardini and Mariangela Regoliosi (Rome, 1996), 1–32, 33–62.

our knowledge of how physicians sought to represent their craft, as they offer a fair account of what physicians thought would be the best, if not the ideal, practice. The few cases described briefly below show the ways practitioners understood the influence of the accidents of the soul on health, the suitable manner of treatment, and their own role in restoring or setting emotional balance to their patients. The three physicians to be referenced were well educated in the theoretical discussions and debates on the emotions and the body, and emotions in medicine. Whether their approach was formed by what they read as theory or by practice is difficult to determine, though a certain interplay between the two is most likely. These three cases are in general much richer than many of the cases on this topic, and although identifying differences among cases and authors would be worthwhile, this desideratum lies beyond the scope of this essay.

Bartolomeo Montagnana (1380–1452), a professor of medicine at Padua, presented in great detail the case of a young man suffering from melancholy with manic episodes. The disease is described in complexional terms—a general warm complexion but a cold and humid liver, heart, and brain. According to Bartolomeo's analysis, this situation affected the patient's cognition, debilitating his reason and leading him to imagine violent situations of quarrels, trickery, and insult. The gravity of the situation seems to have raised Bartolomeo's doubt whether it was at all possible to heal the patient. Yet his advice was decidedly directed to comfort the patient and reduce, as much as possible, the melancholic imbalance. When relating the appropriate regimen for the accidents of the soul, his suggestions centered, therefore, on avoiding any state that might cause a rapid and vehement movement in the body. Producing a calm, soft, and unexciting environment emotionally, but also with respect to other nonnatural factors such as exercise and sleep, is the repeating refrain in Bartolomeo's instructions.[62]

Pleasure and comforting words were recommended to appease the patient's mind and temper the vehemence of the movement of his spirits. At the same time, stern reprehension was voiced against excessive sexual activity that might further debilitate the body. Reiterating an ongoing medical discussion on the influence of sex, Bartolomeo argued that some activity might be useful to discharge the young man's abundant timidity, but only as long as it was carried out in moderation and *moraliter regulatum*.[63] Nevertheless, the text continues beyond a customary admonishment against immoral behavior. Bartolomeo continues with a rebuke: engaging in the "shameful act" with cheap and hateful morals (*vilius et destabilius*), with fleeting lasciviousness and delectation (*per lasciviam et delectacionem*) like a *muliercula*—that is, in a weak and womanlike manner—was severely condemned. Sexual activity should be moderate and followed by sleep so that the corporal and imaginative faculties would be restored.[64] Pleasure was thus endorsed to ease the melancholic emotional symptoms, but desire, regarded as a detestable inclination, was not. Bartolomeo's instructions were concerned with health and the preservation of the balance of the humors,

---

[62] MS 1239 (0.4.8), fol. 93v, Trinity College Library, Cambridge (hereafter cited as TCL).

[63] The call to connect sexual activity to Christian morality appears, e.g., in Bartholomaeus of Salerno's commentary on the *Tegni* and in Peter of Abano's commentary on the *Problemata*, discussed above. See Wallis, "12th Century Commentaries" (cit. n. 50), 125–64; Gijs Couke, "*Non adeo est honesta ut delectabilis. Sexual Pleasure in Medieval Medicine: The Case of Pietro d'Abano*," in *Piacere e dolore: Materiali per una storia delle passioni nel Medioevo*, ed. Carla Casagrande and Silvana Vecchio (Florence, 2009), 117–48.

[64] MS 1239 (0.4.8), fol. 95v, TCL.

but health was governed no less by moral concepts. Similar warnings against effeminacy due to excess sexual activity appeared repeatedly in Bartolomeo's *consilia*, emphasizing the association of the physical occurrence of the release of too much heat and humidity with immoderate and unbecoming delights. The nature of the male physical constitution was thus threatened by inordinate sexual delights. At the same time, it was the patient's imbalanced humors that encouraged the physician to warn against excessive pleasures, perhaps because he suspected the patient to be more prone to them. Bartolomeo's advice, therefore, did not distinguish between the patient's evil habits and the threat of health to both body and soul.

Sexual pleasures often drew a "Christian" reaction from physicians. Frequency of sexual activity, time of day, and modes and manners were all primary issues pertaining to health and open to the regulation of physicians, but as health and moral behavior were deeply intertwined in these matters, physicians claimed authority in this regard as well, not distinguishing the ethical from the medical. But other kinds of emotions also drew their attention to moral implications. Anger (*ira*) was one of the dominant emotions rebuked by physicians, primarily because of its heating qualities. As Aristotle observed, it boils the blood around the heart. But anger was also considered a vice according to the Church's list of the seven deadly sins, and its harm to the souls of men was a common topic in sermons, devotional treatises and art, confession inquiries, and the like. Several medical authors also mentioned the sinful nature of anger in discussing its harm.

In a personal regimen to Archbishop Pedro Gómez Barroso (1331–90), the physician Estéfano of Seville (fl. 1380) charted the regimen of passions beneficial for the bishop. Anger, wrote Estéfano, should be avoided as it does not comply with the archbishop's "good complexion"; he then distinguished between *yra* and *saña*—the latter, which is anger without offence (*enojo*), is not a sin and more favorable to the body because the fervor of the natural heat is tempered without descending into vengeance. It also vivifies and draws the blood to all the members in order to increase and preserve their natural heat. *Yra*, in contrast, tends to escalate, as in the case of the cholerics, and consequently destroys natural heat, often resulting in death and "sin of the soul," caused by the desire to do evil unto one's neighbor.[65] Clearly this kind of vengeful anger does not befit an archbishop, but Estéfano reassured him by adding, "The senior Archbishop removes from himself the desire for vengeance with great patience of the virtue of God." This sort of flattery to the powerful patron is apparent throughout the chapter, with Estéfano repeatedly referring to the archbishop's "angelic person" and his good complexion. The reference to sin was thus clearly chosen to accommodate the patient. And still, the convergence of medical and religious senses of health is noteworthy, as Estéfano allowed himself to comment on matters of faith and interpret them according to his medical knowledge.

The last example to be mentioned here is the medical correspondence of the humanist physician Giovanni Manardi (1462–1536). The letter offers advice to an unnamed friend suffering from black bile. Despite its humanist flourish, it is still a technical medical regimen containing advice about a helpful potion and the proper diet for his complexion. But the letter begins with a warning against ill health deteriorating into sin and is followed by extensive advice about the comforts of the soul, which

---

[65] MS 18052, fol. 13r, Biblioteca Nacional, Madrid.

in turn is followed by Christian-humanist consolatory words.[66] Although relying heavily on the fashion of the age in the structure and content of the advice, it is not incidental that the letter was published in a collection of medical epistles and that the title, opening, and ending are dedicated to the disease and its cure. Moreover, the physician's authoritative tone reinforces the notion that this case study is about a doctor bestowing counsel on his patient.

All three examples deserve further scrutiny. The circumstances of the cases and the lives of the physicians certainly contributed to their decision to express their advice in such an explicitly moral tone. But in their vivid remarks, the process in which consideration of the soul is incorporated into their medicine is exposed. The soul and the emotions they treated were certainly manifestations of imbalanced humors, but they were not merely bodily circumstances. The care of the soul in the spiritual sense is engulfed in their medical care. Because all three were learned physicians, writing in the accepted genres of the period, we might safely assume that they were well versed in the arguments concerning the relationship between body and soul and the appropriate role of the physicians. Recalling Peter of Abano's argument, it may be said that Bartolomeo, Estéfano, and Giovanni were ultimately caring for their patients' souls, feeling perhaps less inhibited in doing so.

CONCLUSION

The category of the accidents of the soul posed a challenge for late medieval physicians. In the process of developing a systematic discipline and asserting the profession of the learned physician, emotions unsettled the clear-cut division of body and soul. This challenge was simultaneously presented to the theory of medicine and its practice. It was thus necessary to articulate the manner in which the body and the soul interact in emotional states, as well as to devise the correct and appropriate method for treating them. But it was precisely this ambiguous nature of emotions that allowed the reframing of medical treatment of emotions and would also alter and refigure the professional role of the physician on a broader scale. Thus the developments in the ways emotions were perceived lent force to, but perhaps also depended upon, transformations within the discipline. Framing this shift within the historiography of emotions, we may find it useful to consider the new discourse of emotions as a shift in the emotional style of medical literature. The growing focus within medicine on the agency of the body alongside the role of emotions as moral states brought about a new style of thinking about emotions in medical thought that fostered a change in medicine as a profession.[67]

Finally, the evolution of the treatment of emotions in medieval medicine sheds light on the importance of addressing the history of emotions from the perspective of the history of medicine. Although it is by now accepted wisdom that the theory

---

[66] Giovanni Manardi, *Epistolae medicinales in quibus multa recentiorum errata, & antiquorum decreta reserantur . . .* (Paris, 1528), fol. 39r–40r. For more about consolation literature, see George W. McClure, *Sorrow and Consolation in Italian Humanism* (Princeton, N.J., 1991). For more about consolation in medical literature, see Naama Cohen-Hanegbi, "Mourning under Medical Care: A Study of a *consilium* by Bartolomeo Montagnana," *Parergon* 31 (2014): 35–54.

[67] A helpful account of the term and its use in historiography appears in Otniel E. Dror, Bettina Hitzer, Anja Laukötter, and Pilar León-Sanz, "An Introduction to *History of Science and the Emotions*," in this volume.

of humors incorporated emotions and that emotions were explained with regard to the complexions and temperaments, much research needs to be done on the medical understanding of emotions and on how they were treated in practice. As medicine developed in the late medieval period, it became a very active and influential part of society and culture. If we want to understand fully the function of emotions in this society, we cannot ignore this substantial discipline of knowledge, the social role of the physician, or this "cultural event" of the encounter with the physician. The history of the development of the medical profession similarly lacks an important dimension when the treatment of the emotions is overlooked.

The Feeling Body and Its Diseases:

How Cancer Went Psychosomatic in Twentieth-Century Germany

by Bettina Hitzer* and Pilar León-Sanz‡

ABSTRACT

This essay examines how psychosomatic medicine, as it emerged between 1920 and 1960, introduced new ideas about the emotional body and the emotional self. Focusing on cancer, a shift can be mapped over the course of the twentieth century. While cancer was regarded at the beginning of the century as the organic disease par excellence, traceable to malignant cells and thus not caused or influenced by emotions, in later decades it would come to be thoroughly investigated within the field of psychosomatic medicine. This essay illuminates why and how this shift occurred in Germany and how it was affected by the earlier turn toward a psychosomatic understanding of cancer in the United States.

INTRODUCTION

Shortly before his death in the summer of 1934, Georg Groddeck, the eminent but controversial pioneer of German psychosomatic medicine, thoughtfully wrote in a text that would not be published for more than thirty years: "Physicians are becoming more and more interested in the psychosomatics of sickness and health. However, it is remarkable that almost no one has tried to discover the psychic causes of the most significant modern ailment, that is, cancer."[1] Groddeck's statement was certainly reflective of the 1920s and 1930s, yet this situation would dramatically change in the postwar era.[2]

---

* Max Planck Institute for Human Development, Berlin, Germany; hitzer@mpib-berlin.mpg.de.

‡ School of Medicine, University of Navarra, Pamplona, Spain; mpleon@unav.es.

Earlier versions of this text have been discussed with Uffa Jensen, Alexa Geisthövel, and Pedro Gil-Sotres as well as with our two coeditors, Otniel Dror and Anja Laukötter. We are very grateful for their comments and suggestions as well as for the comments and suggestions of the anonymous reviewers and the editor of *Osiris*. We also wish to thank Kate Davison for her help with the copyediting of this article.

[1] Georg Groddeck, "Von der psychischen Bedingtheit der Krebskrankheit," in *Psychoanalytische Schriften zur Psychosomatik*, ed. Günter Clauser (Wiesbaden, 1966), 380–5 (trans. Bettina Hitzer).

[2] Not a single article from *Der Nervenarzt*, one of the most important German journals for psychosomatic medicine from 1928 onward, was dedicated to a psychosomatic understanding of cancer. Similarly, there were no issues of the journals *Allgemeine Ärztliche Zeitschrift für Psychotherapie und Psychohygiene* (first published 1928, and after 1930 as *Zentralblatt für Psychotherapie*) or *Zentral-*

The historiography of cancer usually considers the first half of the twentieth century as a time when Western societies discovered that cancer was a widespread disease, though physicians and researchers had a feeling they had not yet fully come to understand what caused it. In general, medical experts agreed that chronic inflammation—be it due to mechanical, biochemical, or infectious agents—played a role in generating cancer. Some research was also geared toward identifying parasites or hereditary factors in carcinogenesis; however, the possibility of direct infection by a cancer germ was refuted by the majority of researchers. Because there was more or less general consensus that cancer began as a local disease, early detection was promoted in order to treat cancer with surgery, X-rays, or radium therapy—the three main therapeutic methods in use at the time.[3] In laboratories in Berlin, Frankfurt, London, Paris, and New York, physiologists, hematologists, and others experimented with cells, tissues, and animals with the aim of understanding carcinogenesis, while surgeons and radiologists tried to cure already diagnosed cancer patients—mostly with no lasting effect. At the same time, a growing number of medical practitioners in Europe and the United States pleaded for a holistic understanding of the relationship between humans and medicine, an appeal that was part of a more general shift toward holism promoted by an influential group of intellectuals and natural scientists during the 1920s and 1930s.[4] Cancer researchers, by contrast, seem to have been an exception to this trend.

One could indeed wonder why, in their efforts to tackle the unsolved problem of carcinogenesis and the deficiencies of treatment, cancer researchers and physicians remained more or less unaffected by the coeval holistic reasoning of their colleagues in philosophy, the natural sciences, and other branches of medicine, especially considering that older "holistic" models did integrate cancer, most notably humoral pathology.[5] In fact, up until the 1880s, melancholy, depression, and grief figured prominently in explanations of cancer's onset, both in medical textbooks and in encyclopedias intended for the lay public.[6]

---

blatt für Psychoanalyse und Psychotherapie: medizinische Zeitschrift für Seelenkunde (1911–4) dedicated to cancer.

[3] See, e.g., the 1929 Encyclopedia Britannica entries on "cancer" and "cancer research," which were in part based on a resolution stipulated by an international meeting organized by the American Society for the Control of Cancer in 1926. See Encyclopedia Britannica, vol. 4, 14th ed. (New York, 1929), s.v. "cancer," by George A. Soper, 731–4, and s.v. "cancer research," by Walter Sydney Lazarus-Barlow, 734–8. For an overview of cancer research in Europe and the United States, see Patrice Pinell, The Fight against Cancer: France 1890–1940 (New York, 2002); James T. Patterson, The Dread Disease: Cancer and Modern American Culture (Cambridge, Mass., 1987); Wolfgang U. Eckart, 100 Years of Organized Cancer Research—100 Jahre organisierte Krebsforschung (Stuttgart, 2000); Gabriele Moser, Deutsche Forschungsgemeinschaft und Krebsforschung 1920–1970 (Stuttgart, 2011).

[4] Christopher Lawrence and George Weisz, eds., Greater than the Parts: Holism in Biomedicine, 1920–1950 (Oxford, 1998); Anne Harrington, Reenchanted Science: Holism in German Culture from Wilhelm II to Hitler (Princeton, N.J., 1999).

[5] Until the nineteenth century, cancer was understood as an inflammation process. Hippocrates described the cause of cancer as an excess of black bile, a view further elaborated by Galen (AD 129–ca. 200). Within Galenism, a flux of black bile could give rise to scirrhus, one form of which was related to or capable of changing into cancer. Other explanations suggested that a flux of black bile unmixed with blood gave rise to cancer forthwith, most often in female breast tissue. Over the centuries, physicians added complementary theories, arriving at the conclusion that cancer was a corrupt form, but the same process that made normal tissues also made abnormal tissues. David Cantor, "Cancer," in Companion Encyclopedia of the History of Medicine, ed. William F. Bynum and Roy Porter (London, 1997), 537–61, on 540.

[6] Bettina Hitzer, "Healing Emotions," in Emotional Lexicons: Continuity and Change in the Vocabulary of Feeling 1700–2000, ed. Ute Frevert, Monique Scheer, Anne Schmidt, Pascal Eitler, Bettina

In the modern move toward holism, psychosomatic models also played a major role, indicating the growing importance of emotions.[7] Emotions do not necessarily have to figure into psychosomatic models, since these models might—and some certainly did—refer solely to psychic influences and personality features without taking emotions into consideration. However, based on booming research on physiology, sensory perception, psychiatry, and psychoanalysis in the late nineteenth and early twentieth centuries, emotions were usually central to these models either as part of the psyche or as a faculty that linked body and psyche.

Against this background, one major aim of this essay is to solve the riddle that Groddeck posed: Why is it that only a handful of the rising number of physicians and researchers who gravitated toward psychosomatic issues in the first half of the twentieth century took a direct interest in cancer? Conversely, why did so few oncologists use psychosomatic ideas to explain and treat cancer at that time? And, looking ahead, why and how did this state of things change? In grappling with these questions, a second major aim of this investigation is to reveal underlying assumptions both about the interrelation of the emotions-psyche-body triad and about the possible imperviousness of the material body to cancer, where the body has sometimes been seen as a kind of basic organic structure that is not affected by the psyche or the emotions. In this context, conceptions of cancer function as a kind of litmus test for analyzing how far assumptions about the interrelation between body and emotions reach. Third, this study on emerging psychosomatic understandings of cancer aims to unpack the multifaceted twentieth-century trajectories and trends in conceptions of the body and its diseases as being subject to emotions. Far beyond the narrow understanding of a psychoanalytical psychosomatic medicine, these trajectories have experienced multiple peak periods during the past hundred years and are today fueled by a tendency within the life sciences to question the established dichotomy of body and mind by emphasizing the role of emotions in cognition.

The study focuses on Germany, since holism, especially medical holism, had become more prevalent and culturally influential in post–World War I and early Nazi Germany. To explain this change and to qualify its very nature, one has to take into account the transatlantic influences and developments that would prove to be of crucial importance in post-1945 West Germany.[8] Moreover, some of these influences and developments proceeded from the work of German-speaking doctors who had be-

---

Hitzer, Nina Verheyen, Benno Gammerl, Christian Bailey, and Margrit Pernau (Oxford, 2014), 118–50, on 131–4; Patricia Jasen, "Malignant Histories: Psychosomatic Medicine and the Female Cancer Patient in the Postwar Era," *Can. Bull. Med. Hist.* 20 (2003): 265–97.

[7] The following is based on a broad understanding of psychosomatics encompassing all models that link body and psyche. It works on the assumption that the psyche affects the functions of the body, its health, and its diseases in various ways. This differs from a more narrow understanding of psychosomatics as psychoanalytically informed that is often used in the historiography of twentieth-century psychosomatics. For an overview of the history of the psychosomatic movement in America, see Edward Shorter, *From Paralysis to Fatigue: A History of Psychosomatic Medicine in the Modern Era* (New York, 1992); Theodore M. Brown, "The Rise and Fall of American Psychosomatic Medicine," paper presented to the New York Academy of Medicine, New York, 29 November 2000, http://human -nature.com/free-associations/riseandfall.html (accessed 26 October 2015).

[8] In the introduction to their history of emotions and their relation to science, editors Frank Biess and Daniel M. Gross argue that even from a broader scientific perspective, the transatlantic context is a "crucial factor in explaining the shifting status of emotions as an object of scientific inquiry." Biess and Gross, "Emotional Returns," in *Science and Emotions after 1945: A Transatlantic Perspective*, ed. Frank Biess and Daniel M. Gross (Chicago, 2014), 1–38, on 14.

come established in the United States prior to World War II.[9] For this reason, one part of this essay discusses those features of American psychosomatic cancer medicine from the 1930s to the 1950s that influenced West German thinking about emotions and cancer and contributed to the establishment of West German psychosomatic cancer medicine in the late 1950s.

AT THE MARGINS: CANCER AND PSYCHOSOMATIC MEDICINE IN EARLY TWENTIETH-CENTURY GERMANY

As Groddeck had indicated, a tiny minority of German physicians were interested in revealing the psychic causes of cancer. Apart from Groddeck himself, the only major medical "school" to take an interest in the psychic aspect was anthroposophic medicine, which was a form of holistic thinking based in German-speaking countries at that time that is still current today. Why did Groddeck and other holistic thinkers believe emotions played a part in causing cancer when the overwhelming majority of German physicians and researchers did not? Why were their efforts not recognized more broadly? To answer these questions, this section will take a closer look, first, at anthroposophic medicine and, second, at Groddeck himself.[10]

The Invention of the "Cancer Psyche" by Anthroposophic Medicine

Rudolf Steiner, founder of the anthroposophic movement, held a holistic view of the human being that distinguished four dimensions of the human body. Steiner called these dimensions the physical body, the etheric body, the astral body, and the ego, arguing that they were inextricably linked to one another and had to be maintained in a balance. The link between the astral and physical body was established by the faculties of representation, feeling, and will. The corresponding elements of the physical body were the nerves, the respiratory system, and the metabolism.[11]

Two younger anthroposophic physicians, Werner Kaelin and Gerhard Suchantke, closely cooperated on the development of an early detection test for what they understood to be precancerous states at the world's first anthroposophic clinic, the privately run Clinical-Therapeutical Institute in Arlesheim, Switzerland.[12] The test had the pur-

---

[9] Some prominent physicians, including Franz Alexander, emigrated to the United States before 1933, as did a number of Jewish scientists and physicists in subsequent years. See, e.g., Mitchell G. Ash and Alfons Sollner, *Forced Migration and Scientific Change: Emigré German-Speaking Scientists and Scholars after 1933* (Cambridge, 1996); Volker Roelcke, Paul Weindling, and Louise Westwood, eds., *International Relations in Psychiatry: Britain, Germany, and the United States to World War II* (Rochester, N.Y., 2010).

[10] One could also name the work of Wilhelm Reich, the Austrian-born psychiatrist and psychoanalyst. However, because his work on the assumed relationship between carcinogenesis and the inability to fully give oneself away in orgasm began primarily after he left Germany in 1933, it is not investigated here. Moreover, his studies on cancer were only marginally acknowledged by medical science in the 1940s and 1950s before gaining widespread attention within the context of the student movements of the 1960s and 1970s. Reich, *The Cancer Biopathy*, vol. 2 of *The Discovery of the Orgone* (1948; repr., New York, 1973). See also James E. Strick, *Wilhelm Reich, Biologist* (Cambridge, Mass., 2015), 186–217.

[11] Werner Kaelin, *Krebsfrühdiagnose—Krebsvorbeugung: Krebsdisposition und Krebs als Zeitkrankheit* (Frankfurt am Main, 1956), 16–22. For an overview of anthroposophic medicine, see Gunver S. Kienle, Hans-Ulrich Albonico, Erik Baars, Harald J. Hamre, Peter Zimmermann, and Helmut Kiene, "Anthroposophic Medicine: An Integrative Medical System Originating in Europe," *Glob. Advances Health Med.* 2 (2013): 20–31.

[12] The clinic was founded in 1921 by the physician Ita Wegman. Now named the Ita-Wegman-Klinik after its founder, it was the only clinic for anthroposophic medicine in the world until 1960 and is still one of the major centers for it in German-speaking countries.

pose of enhancing early treatment options for cancer patients. Kaelin argued that the blood of cancer patients had characteristic properties. In a series of laboratory experiments, using blood samples from cancer patients and "normal" ones, Kaelin examined the forms or shapes the blood took when it was transferred from a petri dish onto a piece of paper following a specific procedure, as well as the time it took for the blood to develop these forms. He compared the results from both groups and found that the forms produced by the cancer patients' blood deviated from those produced by the blood of "normal" patients.[13]

Kaelin and Suchantke were convinced that cancer was a disease that encompassed the whole body, the tumor being only the last and latest stage of an illness that had originated years or even decades before. Even in its early stages, though, cancer could have repercussions that could be identified by analyzing the physiology of the patient's proteins (on which Kaelin's test was based). These repercussions could also be identified by analyzing certain moods and feelings, all of which were considered to contribute to as well as to indicate carcinogenesis.[14] Kaelin and Suchantke held that the foundation for what they called the "cancer psyche" was laid in early childhood. At risk were children who were subjected to coldness, lack of understanding, suffering, pain, or shock, for whom it would subconsciously enter the body via the emotions, thus inhibiting the body's capacity to develop in inner harmony.[15] If the child suffered difficult or unhappy experiences or emotional shocks, the resulting disposition could then lay the groundwork for a fully developed "cancer psyche," an emotional state of depression and detachment that constricted the creative forces of human life and tended to "swallow" negative experiences. The "cancer psyche" could then develop independently and unnoticed until triggered by some other external factor.[16] The image of "swallowing" was deliberately reminiscent of the psychoanalytic concept of repression. And like this concept, it was thought to inhibit an active, healthier mode of processing difficult experiences.[17] Consequently, both Kaelin and Suchantke argued that the creative forces of such individuals were stunted so as to leave them with an inner emptiness and muteness that they themselves might even be unaware of or might (perhaps even intentionally) hide behind a facade of pleasantness.[18] As a result, their bodily awareness and self-perception were presented as defective, which explained why cancer patients would usually only recognize their symptoms very late or tend to minimize or ignore them.[19]

---

[13] See Werner Kaelin, *Die prophylaktische Therapie der Krebskrankheit* (Stuttgart, 1930). He also published a condensed version of his findings in a highly respected oncological journal: Kaelin, "Versuche zu einer Frühdiagnose des Krebses aus dem Blut nach capillar-dynamischer Methode," *Z. Krebsforsch.* 34 (1931): 457–72. For a revised and extended version, see Kaelin, *Krebsfrühdiagnose* (cit. n. 11).

[14] Gerhard Suchantke, "Zum Problem der Krebspsyche," *Natura* 4 (1929–30): 365–8.

[15] Kaelin, *Krebsfrühdiagnose* (cit. n. 11), 24–5, 32–3.

[16] Suchantke, "Zum Problem" (cit. n. 14), 365, 368; Kaelin, *Krebsfrühdiagnose* (cit. n. 11), 32, 36, and 38–44. External factors could be certain ailments or their components, alcohol, tobacco, X-rays, or chemicals like aniline. In this respect, Kaelin drew on discussions about carcinogenic agents in food, stimulants, and the working environment that were especially prevalent during National Socialism. See Robert N. Proctor, *The Nazi War on Cancer* (Princeton, N.J., 1999).

[17] In a later article, Suchantke thus explicitly referred to Freud and his concepts, which he interpreted in light of Rudolf Steiner's anthropology. See G. Suchantke, "Wert und Unwert des Begriffs der Krebspsyche," *Beiträge zu einer Erweiterung der Heilkunst nach geisteswissenschaftlichen Erkenntnissen* 4 (1951): 145–56, on 150.

[18] Kaelin, *Krebsfrühdiagnose* (cit. n. 11), 30–1; Suchantke, "Zum Problem" (cit. n. 14), 365.

[19] Suchantke, "Wert und Unwert" (cit. n. 17), 150.

To cure not only the tumor, a symptom that appeared only in the later stages of the disease, but also the "cancer psyche" itself, practitioners of anthroposophic medicine had little interest in psychoanalysis and instead looked to drug therapy in order to re-balance the interplay of the four main dimensions of the human body. The drug of choice was to be found in the mistletoe plant, which had previously been recommended as a treatment for cancer by Steiner and Ita Wegman, who together founded anthroposophic medicine. Steiner and Wegman were inspired by the parasitic character of the plant to claim that cancer could be cured using drugs that shared fundamental characteristics with the disease itself. Along with the idea that health was a state of balance produced between the four dimensions of the body, both bodily symptoms and the emotional states of cancer patients were seen to be accessible through direct intervention into the physical body.

This approach differed starkly from the psychoanalytically oriented psychosomatic medicine that emerged and gained momentum between 1900 and the 1930s. Yet both were part of the shift toward holism that was symptomatic of the intellectual critique of the perceived dominant attitude in the natural sciences and especially in medicine, namely, a materialistic and mechanistic worldview devoid of any deeper meaning. On the one hand, anthroposophic medicine aligned with the better part of the *Neue Deutsche Heilkunde*, the doctrine developed in the early years of National Socialism, the aim of which was to implement "biological medicine." The *Neue Deutsche Heilkunde* advocated naturopathic and homeopathic methods similar to those used in anthroposophic medicine.[20] On the other hand, anthroposophic institutions faced serious problems because they got in the way of the National Socialist *Gleichschaltung*, the process by which the state sought to bring the whole of society under uniform control.[21] Thus, in 1935, the Anthroposophic Society was forced to disband. Kaelin and Suchantke nevertheless continued to practice and publish both during National Socialism and afterward.[22]

Over the course of the 1930s, as National Socialist medicine became more radicalized and turned increasingly toward military medicine, all strands of "biological" medicine, including *Neue Deutsche Heilkunde* and anthroposophic medicine, gradually lost their support. Kaelin and Suchantke's studies into the "cancer psyche" therefore went more or less unnoticed.

There are, however, some indications that Kaelin and Suchantke's work was influential even if it was not always explicitly referenced. For example, Johannes Kretz, director of the General Hospital in Linz, Austria, argued in 1941 that psychic behavior played a role in causing cancer.[23]

Georg Groddeck: Cancer as a Symbol of Unfulfilled Emotional Needs

While anthroposophic medicine had a marginalized position within academic medicine from the start, the psychosomatic pioneers in Germany were all more or less dis-

---

[20] Detlef Bothe, *Neue Deutsche Heilkunde 1933–1945*. Abhandlungen zur Geschichte der Medizin und Naturwissenschaften, vol. 62 (Husum, 1991), 270–99.

[21] Uwe Werner, *Anthroposophen in der Zeit des Nationalsozialismus 1933–1945* (Munich, 1999).

[22] While Kaelin stayed in Arlesheim until his death in 1973, Suchantke moved to Berlin in 1934, where he was employed at the university clinic for naturopathy, then to Tyrol in 1939, and shortly afterward to Bavaria. In 1954, he returned to Arlesheim, where he died in 1956. Kaelin published articles on cancer in the flagship journal of the *Neue Deutsche Heilkunde*, entitled *Hippokrates*. See Werner Kaelin, "Frühdiagnose mittels der kapillar-dynamischen (K.D.) Reaktion," *Hippokrates* 2 (1934): 48–58.

[23] Johannes Kretz, "Das Krebsleiden als Allgemeinerkrankung," *Z. Krebsforsch.* 51 (1941): 6–35, on 26.

tinguished physicians, some of them even heading clinical departments.[24] Most of them were specialists in either internal medicine or neurology. During the 1920s, this first generation of psychosomatic physicians were very much concerned with those physical ailments that had—in their view—something to do with the autonomic nervous system. In studying paralytic symptoms, allergies, cardiac problems, and peptic ulcers, they were careful to indicate that physicians should first exclude organic causes before diagnosing a psychosomatic disease.[25] To explain how emotions could work on the body, they pointed either to a rather vague concept of psychogenesis or to psychoanalytic conversion theory. Cancer did not figure within this category of psychosomatic ailment because there was no clear link to the autonomic nervous system, and in any case, it seemed obvious to most of them that cancer had a very distinct organic cause. In addition, most psychosomatic physicians were clinicians concerned with treatment and, as far as possible, curing people. Whether emotions were involved in carcinogenesis or not, it was probably inconceivable for these physicians, some of whom had witnessed the inexorable decline and death of cancer patients, that psychotherapeutic treatment could do anything to hold back or reverse this process.

There are two factors that likely explain Groddeck's conception of cancer as psychosomatic disease, which was unique compared to those of other German psychosomatic pioneers, and his concomitant decision to treat some cancer patients using exactly the same approaches he used with his other patients. First, Groddeck did not work at a clinical department of internal medicine but had founded his own sanatorium in 1900 in the Black Forest city of Baden-Baden. There he experimented with treating chronically or severely ill patients, who were often considered to be incurable. Even though the bulk of these patients did not suffer from hysteric or other neurological conditions, Groddeck's therapeutic approach combined the techniques of massage and spa treatment directed toward the body with "older" mental techniques like hypnosis and suggestion, as well as the "new" psychoanalytical form of treatment.[26]

This rather idiosyncratic mixture of techniques was acceptable for treating those who were considered within standard medical frameworks to be incurable, but it also reflected Groddeck's particular understanding of analytical psychology. And this was the second aspect that led Groddeck to conceive of cancer as a psychosomatic disease. In 1917, he first summarized his long-standing personal and medical experiences in a small but widely noticed book, claiming that all organic ailments were mentally (co)determined and could thus be treated with psychoanalysis.[27] Drawing on a notion of the "unconscious" that was close to but nevertheless different from the Freudian concept, he rejected all forms of mind-body dualism, even in terms of a psychophysical reciprocity fervently discussed by his contemporaries. Unlike other influential

[24] A prominent example is Gustav von Bergmann (1878–1955), who held a professorship in internal medicine first in Marburg, then in Frankfurt am Main, and eventually in Berlin.

[25] Felix Deutsch, "Das Anwendungsgebiet der Psychotherapie in der inneren Medizin," *Wiener Med. Wochenschr.* 72 (1922): 809–16, on 815.

[26] For a biography and general introduction to Groddeck's achievements, see Herbert Will, *Die Geburt der Psychosomatik: Georg Groddeck, der Mensch und Wissenschaftler* (Munich, 1984).

[27] Georg Groddeck, *Die psychische Bedingtheit und psychoanalytische Behandlung organischer Leiden* (Berlin, 1917), 11–3. This book was reviewed enthusiastically by Sandor Ferenczi, a close collaborator of Freud who later held Groddeck in high esteem. Ferenczi, review of *Die psychische Bedingtheit und psychoanalytische Behandlung organischer Leiden*, by Georg Groddeck, *Int. Z. Ärztl. Psychoanal.* 4 (1917): 346–7. See also Will, *Psychosomatik* (cit. n. 26), 48.

psychosomatic thinkers of the 1920s like Ernst Simmel or Felix Deutsch, Groddeck did not subscribe to the idea that unconscious conflicts could be "converted" into organic ailments, and thus his main focus was not on revealing the psychogenesis of organic diseases.[28] On the contrary, he insisted on his reading of the unconscious as neither psychic nor somatic, defying all attempts to define the exact nature of its mechanisms.[29]

In his most significant work, *Das Buch vom Es* ("Book of the It"), published in 1923, Groddeck further developed his understanding of the unconscious—what he called the "It"—as the governing force or essence of human life, a force that develops psychic or organic ailments in order to safeguard itself from the impositions of modern public morality.[30] Groddeck therefore considered diseases to be symbols, purposeful expressions of the It that the physician (and the patient) had to uncover by asking, why? Disclosing the meaning of the symptoms—be they mental or somatic—could free the It from its constraints, thus releasing the inner powers of healing, without which all medical efforts would be in vain.[31]

Groddeck had first mentioned cancer in his 1917 study, interpreting the disease as a means for the female It to counteract strong sexual impulses, the enjoyment of which was not permitted by modern girls' or women's education, which imposed sexual frigidity as a feminine ideal or even predisposition.[32] In his 1934 investigation into the psychic determination of cancer, the link to repressed sexual impulses had weakened in favor of a different but nevertheless similarly sexually determined association. In the later study, Groddeck understood cancer as a symbol of a pregnancy impeded by various social factors, that is, as the "seed of an abhorrent changeling."[33] Drawing on the findings of modern laboratory medicine, he argued that this correlation could be intensified by the increase of female hormones in cancer patients, the similarity of some elements in tar to female hormones, and the relation of neoplastic cells to embryonic cells. The correlation was further supported by the etymological proximity of the words *Neubildung*, *Krebs*, and *Geschwulst* (respectively, "neoplasm," "cancer," and "lump"), which were also used to describe the embryo in the womb. Whereas in 1917 Groddeck had conceived of the disease as a means used by the It to protect itself from destruction, he later interpreted cancer as something that resulted from feelings of guilt and a longing for punishment arising out of socially coerced or volitional sterility, a phenomenon that Groddeck understood as typical for modern European societies.[34] This explained the rise in cancer morbidity that contemporary epidemiologists had observed, the causes of which were never-

---

[28] The German neologism *psychophysisch* (literally "psycho-physical") was one of the key terms of early German psychosomatic discussion—even before today's term "psychosomatic," which was easier to translate into English, gathered momentum. See Will, *Psychosomatik* (cit. n. 26), 4.

[29] Groddeck, *Psychische Bedingtheit* (cit. n. 27), 27–9.

[30] Georg Groddeck, *Das Buch vom Es: Psychoanalytische Briefe an eine Freundin* (Leipzig, 1923). It appeared later in English as *The Book of the It* (London, 1935). Freud borrowed Groddeck's term for his seminal "Das Ich und das Es," published in the same year, and later translated into English as *The Ego and the Id* (London, 1927).

[31] Groddeck, *Psychische Bedingtheit* (cit. n. 27), 29.

[32] Ibid., 15.

[33] Georg Groddeck, "Von der psychischen Bedingtheit der Krebserkrankung," in *Psychoanalytische Schriften zur Psychosomatik*, ed. Günter Clauser (Wiesbaden, 1966), 380–5, on 382.

[34] Ibid., 381.

theless hotly debated, with some attributing the rise simply to better techniques of observation.[35]

Curiously, even though he thought that cancer was associated with female hormones and an unfulfilled or (self-)denied longing for a child, Groddeck did not conceive of cancer as a woman's disease. Because he was convinced that the male unconscious could and did indeed imagine itself as pregnant—expressed in the everyday locution *Geisteskind* ("brainchild")—he had no difficulty in explaining cancer in men as a symbol of guilt felt due to the absence of a (brain) child. In his view, the primary sites of cancer in men—mouth, stomach, and rectum—confirmed this interpretation because they were the bodily sites where men receive, retain, and excrete.[36]

Why a person's unconscious would choose a substitute child—that is, cancer—that would eventually kill the person if he or she failed to discover his or her true ailment was a conundrum that Groddeck did not explain but that might well have appalled contemporary clinicians trying to treat patients dying of cancer. Even though Groddeck conceived of cancer (and disease in general) as a kind of circumvention chosen by the It to avert more serious harm, somewhat akin to a wake-up call, he did not attempt to conceal the possibility that psychoanalysis might only improve the fate of cancer patients without actually curing them.[37] Nor did he elaborate on precisely how the It develops the disease as a symbol of the more serious problems it faces.[38] Because he placed so much emphasis on emotions—feelings of guilt and the longing for a child—one can assume that in his view, emotions were the seminal agents of this mechanism that affected both body and psyche.

THE "GRANITE OF THE MATERIAL PROCESS": CANCER IN 1930s GERMANY

Despite Groddeck's efforts to treat and conceive of cancer as though emotions were involved in a coeval psychic-somatic process, other psychosomatically oriented physicians in Germany remained silent with regard to an emotionally predisposed "cancer psyche." Several things were responsible for that reluctance. First, in the early 1930s some important Central European psychosomatic theorists relocated to the United States, notably Franz Alexander, who emigrated to Chicago in 1930, and Felix Deutsch, who went to Washington in 1936.[39] Those who stayed in Germany continued working in internal medicine or clinical neurology departments as physicians. Restoring the patient's productivity had already been an important feature of their practice during the Weimar Republic but was now described as a fundamental imperative. Second, most psychosomatic theorists continued to think and practice within the framework of the

---

[35] Epidemiologists have highlighted the growing mortality associated with cancer since the nineteenth century. During the twentieth century, the epidemiology of cancer shifted in light of research into its etiology, whereby two polarities emerged: those favoring endogenous factors (such as genetic mutation) and those favoring exogenous factors (such as viruses, parasites, environmental chemicals, or physical agents such as radiation). In addition, increased publicity surrounding various carcinogens meant that the disease became more visible. Cantor, "Cancer" (cit. n. 5), 537, 556.

[36] Groddeck, "Bedingtheit der Krebserkrankung" (cit. n. 33), 382–3.

[37] Ibid., 385.

[38] As a physician treating patients on a day-to-day basis, Groddeck was interested not so much in explaining how it worked but in proving that a psychoanalytical co-treatment was effective. See Groddeck, *Psychische Bedingtheit* (cit. n. 27), 29.

[39] Hungarian-born Sandor Ferenczi, another important figure and a close friend of Groddeck, died in 1933. Groddeck himself died in 1934; he had not left Germany after 1933, although he had an ambivalent relationship with National Socialism. See Will, *Psychosomatik* (cit. n. 26), 104–6.

concept of neurosis, investing their research efforts into experiments exploring the psychology of perception as well as Gestalt psychology. There was a deep rift between this strand of research and what was going on in cancer research proper, which was preoccupied with investigating the role of chemical agents, vitamins, and hormones in relation to carcinogenesis.[40]

Initially, those practitioners who stayed in Germany found themselves in a somewhat unclear situation. Under National Socialism, psychoanalysis had been declared a "Jewish" discipline, and the bulk of Freud's writings burned in 1933. However, this official denunciation was a strategic maneuver to enable its appropriation into National Socialist health policy. Thus, many psychoanalysts and psychosomatic researchers were able to continue practicing as before. Foremost among them were the proponents of what Pedro Laín Entralgo has called the Heidelberg School of Psychosomatics, as well as those who, after the beginning of National Socialism, reassembled under the umbrella of the German Institute for Psychological Research and Psychotherapy, directed by the Adlerian psychotherapist Matthias Göring, a cousin of *Reichsmarschall* Hermann Göring.[41] And it was Göring who very actively tried to emphasize the importance of a *Deutsche Seelenheilkunde*—the psychological corollary to the *Neue Deutsche Heilkunde*—to help forge and educate a healthy *Volk* for the National Socialist state. He resolutely advocated a holistic medical approach by stating that "mental factors played a big, quite often decisive role in causing numerous diseases."[42] But he distinguished the "new" psychotherapeutic and psychosomatic approach from other, especially Freudian, ones by underscoring that the *Deutsche Seelenheilkunde* was not concerned with the individual mind insofar as it involved the well-being of the individual, but rather with its value for the whole people, the *Volksganze*.[43] Psychotherapeutic and psychosomatic interventions were therefore directed toward restoring the productivity of the *Volksgenosse*, or National Socialist citizen, by ensuring their mental and physical health and hence their usefulness to society and the state, as Göring's colleague Harald Schultz-Hencke rushed to point out.[44]

[40] Moser, *Deutsche Forschungsgemeinschaft* (cit. n. 3), 55–239.

[41] Pedro Laín Entralgo, "Viktor von Weizsäcker und die ärztliche Praxis," in *Viktor von Weizsäcker zum 100. Geburtstag*, Schriften zur anthropologischen und interdisziplinären Forschung in der Medizin, vol. 1, ed. Peter Hahn and Wolfgang Jacob (Berlin and Heidelberg, 1987), 23–44, on 24; Anne Harrington, *The Cure Within: A History of Mind-Body Medicine* (New York, 2008), 86; Geoffrey Cocks, *Psychotherapy in the Third Reich: The Göring Institute*, 2nd rev. and exp. ed. (Piscataway, N.J., 1997); Ulrich Schultz-Venrath and Ludger M. Hermanns, "Gleichschaltung zur Ganzheit: Gab es eine Psychosomatik im Nationalsozialismus?" in *Neues Denken in der Psychosomatik*, ed. Horst-Eberhard Richter and Michael Wirsching (Frankfurt am Main, 1992), 83–103. For an overview of the Heidelberg school's reestablishment after 1945, see Thomas Henkelmann, "Zur Geschichte der Psychosomatik in Heidelberg: V. v. Weizsäcker und A. Mitscherlich als Klinikgründer," *Psychotherap. Psychosomat. Med. Psychol.* 42 (1992): 175–86.

[42] Matthias H. Göring, *Über seelisch bedingte echte Organerkrankungen* (Stuttgart, 1937), 13–4.

[43] Matthias H. Göring, "Die nationalsozialistische Idee in der Psychotherapie," in *Deutsche Seelenheilkunde: Zehn Aufsätze zu den seelenärztlichen Aufgaben unserer Zeit*, ed. Matthias H. Göring (Leipzig, 1934): 11–6, on 15. On the renaming of psychosomatic terms during National Socialism, see Mechthilde Kütemeyer, "Die Sprache der Psychosomatik im Nationalsozialismus," in *Gift, das du unbewußt eintrinkst: Der Nationalsozialismus und die deutsche Sprache*, ed. Werner Bohleber and Jörg Drews (Bielefeld, 1994), 61–82.

[44] Harald Schultz-Hencke, "Die Tüchtigkeit als psychotherapeutisches Ziel," in *Deutsche Seelenheilkunde*, ed. Matthias H. Göring (Leipzig, 1934), 84–97. Although Göring—having in mind the burning of Freud's books—ostensibly tried to emphasize the new character of psychotherapy under

Viktor von Weizsäcker and Richard Siebeck, the leading figures of the Heidelberg clinic's psychosomatic approach during the 1930s and 1940s, did not dissociate themselves from this National Socialist dictum of productivity as the therapeutic aim of medical interventions. Weizsäcker in particular, whose role during national socialism is the subject of heated debate even today, repeatedly discussed whether and how physicians should carry out their duties toward both society and the patient when making decisions about life, death, and capacity for work. For Weizsäcker, this was not only a theoretical issue but part of his medical practice: from assessing a patient's ability to work in view of a claim of incapacity due to wartime trauma ("Rentenneurose") in the 1920s, to delivering a series of lectures in the summer of 1933 pondering a "medical doctrine of extermination," to recommending the actual sterilization of epileptic patients during National Socialism.[45]

In terms of productivity, cancer was increasingly perceived as a serious threat since epidemiologists had proven that it was on the rise. However, within the localist paradigm that was predominant in medical research and practice at that time, it was thought that the best option to cure cancer was to catch it early and then to "combat" it radically with surgery or radiation. Thus, most efforts were directed toward early detection and prevention—efforts that were actively promoted by the German Cancer Society (*Reichsausschuß für Krebsbekämpfung*). The society did not tire of pointing out that cancer was not a disease of the elderly but of people in the most productive age category—middle-aged persons—whether in the form of cervical cancer, which predominantly affected women between the ages of thirty and fifty, or stomach or lung

---

National Socialism, one could argue that restoring people's productivity in view of the nation's good was nothing new to psychotherapeutic thinking. Michael Hagner, "Naturphilosophie, Sinnesphysiologie, Allgemeine Medizin," in *Der Hochsitz des Wissens*, ed. Michael Hagner and Manfred D. Laubichler (Zurich, 2006), 315–36, on 329–35.

[45] In 1920s Germany, there had been much discussion about "Rentenneurose," usually related to war experiences. Men claimed to be unable to work, and they requested a pension even though physicians could not diagnose any corporeal ailment. The difficulties they had were attributed to neuroses, and physicians had to judge whether these neuroses were "real" or "faked" in order to determine whether the patients would receive money from the state. Weizsäcker examined many patients who supposedly suffered from "Rentenneurose." In most cases, he recommended occupational therapy for these patients and strongly advised that they not be given a pension. In analyzing Weizsäcker's writings from 1933 and 1934, Udo Benzenhöfer reasoned that these texts oscillated between (linguistic) adaptation and approval of certain National Socialist ideas about productivity, "Volksgemeinschaft," and even sterilization on the one hand, and a cautious criticism insisting on the physician's autonomy and authority on the other. Benzenhöfer, *Der Arztphilosoph Viktor von Weizsäcker: Leben und Werk im Überblick* (Göttingen, 2007), 106–31. On the series of lectures Weizsäcker held at Heidelberg University in the summer of 1933 (published in 1934), see Michael Hagner, "Values and the Body: Sketches for a History of Psychosomatics in Germany," in *Sternwarten-Buch: Jahrbuch des Collegium Helveticum*, ed. Gerd Folkers and Helga Nowotny (Zurich, 1997), 68–80, on 75–6. Equally unresolved is Weizsäcker's role as head of neurology in Wrocław from 1941 onward, where one of his assistants studied the brains of children murdered within the euthanasia program. It remains unclear whether Weizsäcker knew that the children were murdered. See Karl Heinz Roth, "Psychosomatische Medizin und 'Euthanasie': Der Fall Viktor von Weizsäcker," *1999: Z. Sozialgesch.* 1 (1986): 65–99, and Benzenhöfer, *Arztphilosoph*, 152–60. The Heidelberg archives of the Neurology Department contain a number of expert witness statements signed by Weizsäcker that recommended the sterilization of epileptic patients, e.g., the recommendation provided to the Erbgesundheitsgericht Darmstadt (Hereditary Health Court, Darmstadt), 26 September 1938, signed by Dr. Ansorge and Prof. Dr. v. Weizsäcker, in Krankenakte D. Sch., *6 February 1902, Heidelberg University Archives (hereafter cited as UAH), preliminary signature: Acc. 30/01 (Neuro), neu: L-V, 1938, no. 898.

cancer, which most often affected men in their prime.[46] Moreover, it was not only the German Research Council (*Deutsche Forschungsgemeinschaft*) that attributed greater importance to cancer during the 1930s. It was also a political requirement driven by Hitler's personal interest in cancer and was thus promoted in special exhibitions and in local early detection initiatives.[47] Psychotherapy as a means to treat cancer seemed to offer few possibilities for early detection and prevention and was more or less inconceivable within the localist paradigm.

This split between cancer as pure somatic disease and other organic diseases that matched with the concept of neurosis is clear when we turn to Weizsäcker's clinical practice in the 1930s and 1940s. Originally trained as a specialist in internal medicine, he later turned toward neurology, directing the Department of Neurology at the Heidelberg-based Ludolf Krehl Clinic from 1920 onward.[48] There he encountered cancer not only when treating patients with brain tumors, which differed from other tumors as a result of the neurological and psychic changes they caused, but also when treating patients with other forms of cancer.[49] His colleague Richard Siebeck, also a specialist in internal medicine, was head of the clinic from 1931 to 1934 and again from 1941 until his retirement in 1951. Siebeck dealt with patients who had cancer on a regular basis, and thus he devoted an entire chapter to cancer in his seminal 1949 work, *Medizin in Bewegung*. While he thought extensively about the interrelation of personality traits and gastric ulcers, he limited himself in the book to the problem of how to talk to and take care of patients with stomach cancer.[50]

Weizsäcker and Siebeck together developed a new concept of psychosomatic medicine that they characterized as medical anthropology. In doing so, they completely abandoned the idea that disease was an "objective," well-defined event that could be diagnosed by examining the texture of the body and its parts, external and internal,

---

[46] A 1931 health exhibition *Kampf dem Krebs* ("Fight Cancer"), organized by the German Hygiene Museum Dresden and the German Cancer Society, showed a number of posters that advised middle-aged parents to check their bodies regularly in order to avoid dying of cancer. See, e.g., *Exhibition: Kampf dem Krebs*, ca. 1931, Deutsches Hygiene Museum DHMD 2001/247.1. For a closer look at the link between emotions and health education, see Anja Laukötter, "How Films Entered the Classroom: The Sciences and the Emotional Education of Youth through Health Education Films in the United States and Germany, 1910–30," in this volume.

[47] Hitler, whose mother died of cancer, took a great interest in cancer research and treatment, so much so that the last thing he asked his *Propagandaminister* Joseph Goebbels to report on, before taking leave on the eve of attacking the Soviet Union on 22 June 1941, were recent advances in cancer research; see Elke Fröhlich, *Die Tagebücher von Joseph Goebbels*, vol. 4 (Munich, 1987), 711. On cancer exhibitions during National Socialism, see Lilo Berg, Katharina Klotz, and Susanne Roeßiger, *Rechtzeitig erkannt—heilbar: Krebsaufklärung im 20. Jahrhundert*, ed. Deutsches Hygiene Museum (Dresden, 2002); Bettina Hitzer, "Körper-Sorge(n): Gesundheitspolitik mit Gefühl," in *Performing Emotions: Interdisziplinäre Perspektiven auf das Verhältnis von Politik und Emotion in der Frühen Neuzeit und in der Moderne*, ed. Claudia Jarzebowski and Anne Kwaschik (Göttingen, 2013), 43–68, on 57–63. On early detection measures, see Proctor, *War on Cancer* (cit. n. 16), 27–34.

[48] Weizsäcker held the position until 1941, when he took up the chair of neurology at the University of Breslau/Wrocław. After 1945, he returned to Heidelberg and took up the chair for general clinical medicine, which would later be renamed the chair of psychosomatic medicine. He retired in 1952.

[49] A case in point is the patient "E. N.," who had been admitted to the Neurology Department in 1935. She was diagnosed with metastases after having had a mastectomy in 1934 but was apparently treated within the Neurology Department because of emotional troubles, which were detailed at length in her medical record. Patient file "E. N.," *1 June 1877, UAH, preliminary signature: Acc. 30/01, neu: L-V, (Neuro), 1935, no. 635.

[50] Richard Siebeck, *Medizin in Bewegung: Klinische Erkenntnisse und ärztliche Aufgabe* (Stuttgart, 1949), 18–37, 397–408.

or by measuring bodily parameters. Based on his *Gestaltkreis* theory, which postulated a circular unity of external stimulation, perception, and movement, Weizsäcker defined disease as a subjective phenomenon that took place between the subject, his or her environment, and the doctor.[51] The way the diseased body was felt and discussed when talking to the doctor was crucial, because it conveyed the true essence of the disease, which had to be understood as something embedded in the biography of the feeling subject.[52] Weizsäcker argued that while psychoanalysis interpreted what the patient said in order to understand the psyche, medical anthropology had to take these statements seriously as a self-perception of bodily and mental processes. This idea was based on the assumption that one could perceive the inner processes, functions, and their respective interplay through bodily sensations and fantasies alike.[53]

Shortly after the end of World War II, Weizsäcker tried to clarify the understanding of psychosomatics that he had elaborated during the 1930s while at Heidelberg.[54] He then explicitly rejected the idea that organic disorders or diseases might be caused by psychic factors—a concept that had been intensely discussed under the heading "psychogenesis" by psychoanalytically oriented physicians like Felix Deutsch, Franz Alexander, and others who had worked and published in the United States during the 1930s. For Weizsäcker, there could not be a causal connection leading from the psyche to the body, since he insisted on the parallel structure of both, without claiming that either one took primary position in terms of time or relevance.[55] Every disease—be it organic or psychic—was, in this view, the materialization of an unsolved conflict. Thus, every disease had a hidden "aim" that the clinician had to decipher through psychotherapy, the goal of which was to lead the suffering subject to discover and accept the true meaning of his or her life, including death as an integral part of human experience.[56]

---

[51] While Weizsäcker first mentioned elements of his *Gestaltkreis* theory in the 1920s, he continued working on it during the 1930s and published a synthesis of his findings in 1940. Viktor von Weizsäcker, *Der Gestaltkreis: Theorie der Einheit von Wahrnehmen und Bewegen* (Leipzig, 1940).

[52] Reviewing all writings published by Weizsäcker during these years, one will find that his conception of the body-psyche relation was not always consistent. Whether this inconsistency should be viewed as a fruitful "structural instability" or as a product of theoretical indecision is open to debate. See Gerlof Verwey, "Medicine, Anthropology, and the Human Body," *Growth Med. Knowl. Phil. Med.* 36 (1990): 133–62, on 147–54.

[53] Weizsäcker first read Freud's writings during the 1920s. In 1926, he paid Freud a visit in Vienna. They remained in contact, especially after Weizsäcker sent him his *Körpergeschehen und Neurose* in 1932, discussing each other's respective ideas and concepts. Weizsäcker did not fully embrace Freud's psychoanalytical concept but used it more as an inspiration to develop his own ideas about the interrelation between bodily, psychic, and social life fundamental for his medical anthropology. See Viktor von Weizsäcker, *Gesammelte Schriften*, ed. Peter Achilles, Dieter Janz, Martin Schrenk, and Carl Friedrich von Weizsäcker (Frankfurt am Main, 1986), 1:154; Benzenhöfer, *Arztphilosoph* (cit. n. 45), 69.

[54] Of course, Weizsäcker's concept developed and changed slightly over time. In his published works, he elaborated his theory by presenting individual case studies, which were apparently based on his psychotherapeutic conversations with patients but were not reported in the patient files one can find in the Heidelberg University Archives. He did not use any psychological tests or statistics to prove his assumptions. See, most notably, Viktor von Weizsäcker, "Körpergeschehen und Neurose: Analytische Studien über somatische Symptombildungen," *Int. Z. Psychoanal.* 19 (1933): 16–116; Weizsäcker, *Studien zur Pathogenese*, Schriftenreihe zur Deutschen medizinischen Wochenschrift, vol. 2 (Leipzig, 1935).

[55] Viktor von Weizsäcker, "Psychosomatische Medizin," in *Gesammelte Schriften* (cit. n. 53), 6:459–60. Originally published in *Verhandlungen Deutsch. Gesell. Innere Med.* 55 (1949): 13–24.

[56] Ibid., 461–4.

Yet even though Weizsäcker considered his theory to be a concept of general medicine, his case studies were mainly conducted in the context of neurological illnesses, which excluded what were considered to be fatal organic diseases. In December 1943, in a letter to his disciple and Heidelberg colleague, Wilhelm Kütemeyer, Weizsäcker explained his reticence:

> There are many questions yet to be answered, particularly if we leave behind the neuroses and turn toward organic diseases or even to psychosis. I tried hard but mostly in vain to solve this problem. . . . I soon also rejected the model of psychogenesis and confined myself to proclaiming a formal analogy of the psychic and the organic drama. The contents, the motifs of the psychic processes are not able to elucidate why first this organ, then that one was affected and why it was affected in this or that way.[57]

Kütemeyer, however, was not put off by this admission and went on to apply the concept of medical anthropology both to psychoses and to fatal diseases.[58]

HOW THE "GRANITE OF THE MATERIAL PROCESS" BECAME EMOTIONALLY ACCESSIBLE

In a Festschrift published on the occasion of Weizsäcker's seventieth birthday, Kütemeyer purposefully chose cancer as a topic in order to exemplify the idea that the different spheres (somatic, psychic, and metaphysical) provided "mutual elucidation" (*gegenseitige Erläuterung*) as the disease ran its course.[59]

What Kütemeyer meant by "mutual elucidation" became clear when he presented his case study of a thirty-four-year-old Hodgkin's lymphoma patient. The man had once been an open-minded, sentient little boy coddled by his mother, who was repeatedly beaten by his brutal father. Despite his love for his mother, he identified with his father, who then became the prototype of authority for him. This identification entailed an inner split that deepened when his only friend moved away. As a corollary, his inner life was numbed, while he nevertheless was able to continue adapting to his environment and especially to authority without any apparent effort. During World War II, he became the perfect example of a soldier who had no qualms about killing or being killed.

Kütemeyer concluded from this biographical sketch that the patient's psychic dimension was marked by emptiness and numbness, and his metaphysical dimension by an oblivious desperation, and the malignant process, which Kütemeyer described as a "monstrous fertility of the soul," corresponded to the somatic dimension.[60] Even more than Weizsäcker, who was also in favor of including the social dimension, and who thus highlighted the need for the physician to embrace his political mission, Kütemeyer emphasized the significance of the sociopolitical dimension of the disease

---

[57] Viktor von Weizsäcker to Wilhelm Kütemeyer, 22 December 1943, Deutsches Literaturarchiv Marbach, A: Sternberger, 89.10.6952/9 (trans. Bettina Hitzer). That Weizsäcker did not answer the question how psyche and soma were interrelated was a criticism that was also voiced by other physicians. See Suchantke, "Wert und Unwert" (cit. n. 17), 145.

[58] In reference to Weizsäcker he chose a variation of Weizsäcker's 1933 title: Wilhelm Kütemeyer, *Körpergeschehen und Psychose*, Beiträge aus der Allgemeinen Medizin 9 (Stuttgart, 1953).

[59] Wilhelm Kütemeyer, "Anthropologische Medizin in der inneren Klinik," in *Arzt im Irrsal der Zeit: Eine Freundesgabe zum siebzigsten Geburtstag am 21.4.1956, Viktor von Weizsäcker*, ed. Paul Vogel (Göttingen, 1956), 243–65, on 246.

[60] Ibid., 255.

by viewing personal and political history as parallel. Kütemeyer applied medical anthropology to both social and individual pathology and viewed cancer as a kind of prototypical "German" or even "European" disease and as a means to come to terms with the National Socialist past.[61] This was even more obvious in a paper he presented in 1965 at the Fourth International Conference on Psychosomatic Aspects of Neoplastic Disease. There he argued that

> these [cancer] patients adopt in their dependency on the dominating figures and all their representatives in the service of the taboos, which have been erected by them and which throttle their lives and the destination of their life, an attitude like that of the "liberated" concentration camp convict towards his torturer: he remains at his master's feet, singing his favorite melodies with a feeble voice, in order to amuse and appease him. Thus the patient tries to exist in two worlds, which contradict and negate each other fundamentally. The reservoir of accumulated hatred and the destructive force equivalent to it, has, with the exception of insignificant outlets onto the surface, only the possibility of expressing itself in a psycho-spiritual-social invisible way, that is in the malignant process.[62]

Kütemeyer thus not only held a particular psychic structure (submissiveness to authority leading to irresolvable inner conflict) to be responsible for cancer but moreover claimed that it was precisely the powerful feelings that resulted from this psychic structure that generated cancer, feelings that could not be expressed within the same structure and for this reason expressed themselves as cancer. Unacknowledged or even unfelt emotions were therefore seen as key for the onset of diseases.

These assumptions were met with severe criticism even among those who did not look askance at psychosomatic medicine. This became very obvious when in the 1960s the Heidelberg medical faculty discussed whether Kütemeyer's publications would meet the standard of a German habilitation or whether he should instead be awarded an honorary professorship—a position that offered prestige but no salary. Some of the requested internal and external reviews of the work were devastating. They mainly criticized the methodological shortcomings, setting aside the causal relations that were drawn between the different spheres of body, psyche, and social environment. They also criticized his disregard for the scientific notion of evidence displayed in the fact that he rested his entire argument upon a single case study.[63] Even the most prominent disciple of Weizsäcker, Alexander Mitscherlich, who was to become one of the most influential psychoanalysts and psychosomatic physicians in post-1945 West Germany, fiercely demanded that the Heidelberg faculty not award

---

[61] See Wilhelm Kütemeyer, *Die Krankheit Europas: Beiträge zu einer Morphologie* (Berlin, 1951).

[62] Wilhelm Kütemeyer, "Psychosocial Aspects of Cancer," paper presented at the Fourth International Conference on Psychosomatic Aspects of Neoplastic Disease, Turin, June 1965. Kütemeyer used the same example in his book *Die Krankheit in ihrer Menschlichkeit*, referring to the reports of Primo Levi and Jean Cayrol about their concentration camp experience. He argued here—as in *Körpergeschehen und Psychose* when talking about the "demonic character"—that one cannot deny "the similarity of these circumstances with the structure of totalitarian society, particularly regarding the reversal of good and bad, characteristic for melancholy." He also justified his references to the concentration camps by explaining that only the remembrance of this "school of suffering" would help in pursuing the all-encompassing medical and political reform that was needed. Kütemeyer, *Die Krankheit in ihrer Menschlichkeit* (Göttingen, 1963), 183.

[63] Reviews by Prof. Dr. C. Kuiper (Amsterdam), 8 March 1964, and Prof. Dr. Paul Martini (Bonn), 19 February 1964. More positive were the reviews of Prof. Dr. F. J. J. Buytendijk (Utrecht), 22 October 1963, and Prof. Dr. A. Jores (Hamburg), 29 October 1963, UAH, PA 10389, Dr. Wilhelm Kütemeyer (*18 April 1904).

any honor or position to Kütemeyer.[64] Mitscherlich's statement is particularly revealing, not only because he had received his own formative medical training in Heidelberg, but also because he had begun to investigate the emotional and somatic repercussions for both the individual and the social psychology of West Germany's
National Socialist past. Mitscherlich published three best sellers on the topic: *Auf
dem Weg zur vaterlosen Gesellschaft* (1963), *Krankheit als Konflikt* (1966), and, together with Margarete Mitscherlich, *Die Unfähigkeit zu trauern* (1967).[65] In his letter
to the dean of Heidelberg University's Department of Medicine, Mitscherlich criticized the fact that Kütemeyer had based his work on Weizsäcker's *Gestaltkreis*, trans-
ferring the latter's principle of the equivalence between perception and moving to
pathogenesis in general. More or less overtly, Mitscherlich distanced himself from
medical anthropology in general and claimed to belong to an international community of clinical researchers whose standards Kütemeyer ignored.[66]

The debate about Kütemeyer's scientific achievements—he was eventually awarded
an honorary professorship but nothing more—marks an important turning point in the
history of psychosomatics in West Germany. It was not only the moment when psychosomatic medicine gained the scientific respectability and public recognition that it did
not have before; it was also the moment when proponents of psychosomatic medicine
placed greater emphasis on defining boundaries and on reinventing themselves as part
of modern science based on evidence, standardization, and experimental practice.

THE "MORAL SUPERIORITY" OF PSYCHOSOMATIC MEDICINE
IN POST-1945 GERMANY

Given the perception of mainstream physicians' widespread involvement in National
Socialist euthanasia, human experiments, and extermination practices, Mitscherlich
helped give German psychosomatic medicine a kind of moral superiority over "traditional" scientific medicine. The West German Medical Association (*Ärztekammer*)
appointed him as head of an official observer commission to the Nuremberg doctors'
trials that were held in 1946 and 1947 in US military courts. Mitscherlich and the
physicians he selected for the commission regularly reported on the proceedings for
the West German media. The final report of the Medical Association was published
in three versions: first in 1947, then—after heated debate within the West German
medical community—in 1949 under the title *Wissenschaft ohne Menschlichkeit* ("Science without Humanity"), and then again in 1960.[67] The overall interpretation of the

[64] Alexander Mitscherlich to Friedrich-Wilhelm Brauss, Dean of Medicine, Heidelberg, 14 January
1964, UAH, PA 10389, Dr. Wilhelm Kütemeyer. On Mitscherlich's intellectual career, see Martin Dehli,
Leben als Konflikt: Zur Biographie Alexander Mitscherlichs (Göttingen, 2007). For his role as a critical
voice in West German medicine and society, see Tobias Freimüller, *Alexander Mitscherlich: Gesell-
schaftsdiagnosen und Psychoanalyse nach Hitler* (Göttingen, 2007).
[65] Alexander Mitscherlich, *Auf dem Weg zur vaterlosen Gesellschaft: Ideen zur Sozialpsychologie*
(Munich, 1963), and the English edition, *Society without the Father* (New York, 1969); Mitscherlich,
Krankheit als Konflikt—Studien zur psychosomatischen Medizin (Frankfurt am Main, 1966); Mitscherlich and Margarete Mitscherlich, *Die Unfähigkeit zu trauern: Grundlagen kollektiven Verhaltens*
(Munich, 1967), and the English edition, *The Inability to Mourn: Principles of Collective Behavior*
(New York, 1975).
[66] Mitscherlich to Brauss, 14 January 1964 (cit. n. 64).
[67] Alexander Mitscherlich and Fred Mielke, *Wissenschaft ohne Menschlichkeit: Medizinische und
eugenische Irrwege unter Diktatur, Bürokratie und Krieg* (Heidelberg, 1949). The first report of
1947 was translated into English as *Doctors of Infamy: The Story of the Nazi Medical Crimes* (New

Nazi doctors' crimes presented by Mitscherlich and his colleague Fred Mielke in the report was that "humanity and medical autonomy perish if science solely perceives of and treats human beings as objects."[68] Medical anthropology, which advocated the reintroduction of the feeling subject into the medical encounter, was clearly depicted as a morally superior alternative. This line of thought was backed up by Weizsäcker, who argued in 1947 that "a conception of medicine that views disease solely as a scientific-biological entity has to look for ethical standards outside the medical realm."[69] Even though the public response to these publications in the late 1940s was limited, they gained widespread attention after their republication in 1960, paving the way for a reevaluation of psychosomatic medicine. It was perceived to be a more humane form of medical practice immune to political aberrations because it had a critical potential and a moral self-assertiveness that "scientific-biological" medicine lacked.[70] Yet even though West German psychosomatic medicine was eager to present itself as an untainted alternative to the dominant "mechanistic" tradition, the younger generation of those interested in psychosomatic approaches was also determined to embrace the standards of what they perceived to be modern science. As with Mitscherlich himself, who visited the United States on a Rockefeller Fellowship in 1951, many younger physicians turned to American psychosomatic medicine, especially with regard to the cancer studies that were already under way at various clinical centers in the United States, in their quest to answer the two questions Mitscherlich and others would later ask when judging Kütemeyer's work: on what evidence could investigations into emotions as a cause of cancer be based, and what is the causal link that explains how emotions work on the body, not only with regard to the autonomic nervous system but also in terms of such "impervious" material structures as cancer?

THE TRANSATLANTIC CONNECTION: HOW AMERICAN PSYCHOSOMATIC CANCER MEDICINE CAME INTO PLAY

As in Germany, there had been some reluctance to study cancer in early American psychosomatic medicine. The first major contribution to the psychology of cancer patients had, however, been made as early as 1926 by Elida Evans, who delineated a

---

York, 1949). The 1949 German report was met with silence by German physicians and media. Consequently, Mitscherlich prepared a reissue that was published in 1960 and received much more media coverage: Mitscherlich and Mielke, *Medizin ohne Menschlichkeit: Dokumente des Nürnberger Ärzteprozesses* (Frankfurt am Main, 1960), published in English as *The Death Doctors* (London, 1962). On the Nuremberg doctors' trials, see Paul Weindling, *Nazi Medicine and the Nuremberg Trials: From Medical War Crimes to Informed Consent* (New York, 2005); Wolfgang U. Eckart, *Medizin in der NS-Diktatur: Ideologie, Praxis, Folgen* (Vienna, 2012), 403–7. On Mitscherlich's role in particular, see Dehli, *Leben als Konflikt* (cit. n. 64), 145–75.

[68] Alexander Mitscherlich and Fred Mielke, *Das Diktat der Menschenverachtung: Eine Dokumentation* (Heidelberg, 1947), cover page (trans. Bettina Hitzer).

[69] Viktor von Weizsäcker, "'Euthanasie' und Menschenversuche," *Psyche* 1 (1947): 5–39, on 38. By presenting himself as an impartial observer, Weizsäcker concealed his own controversial involvement in National Socialist medicine both as the author of the 1933 *Ärztliche Vernichtungslehre* ("medical doctrine of annihilation") and, from 1941 to 1945, as head of the Neurological Institute in Wrocław, where his colleague Hans-Joachim Scherer was examining the brains of murdered children.

[70] One of Wilhelm Kütemeyer's reviewers, F. J. J. Buytendijk from Utrecht University, argued in favor of Kütemeyer by suggesting that the "dangerous appreciation of so-called objective science . . . had contributed to the abominable conduct of so many professors during Nazism." See Prof. Dr. F. J. J. Buytendijk (Utrecht), 22 October 1963, UAH, PA 10389, Dr. Wilhelm Kütemeyer (trans. Bettina Hitzer).

personality profile of female cancer patients based on her analysis of hundreds of women with breast cancer. Evans, a Jungian psychoanalyst, argued that cancer patients were extroverted individuals who had lost an object, role, or person fundamental to their identity. Because of their personality, they did not have the internal resources to cope with the loss and thus developed cancer.[71] Yet her extensive study would remain the only one for at least a decade.[72] In 1935, a first step toward renewing and intensifying interest in psychological analyses of cancer was made by Helen Flanders Dunbar, who was somewhat controversial but nevertheless one of the most influential figures in American psychosomatic medicine of the 1930s and 1940s. She discussed cancer (among many other diseases) in her pioneering survey of psychosomatic literature, *Emotions and Bodily Changes*.[73]

In general, the American school of psychosomatic medicine was heavily influenced by Adolf Meyer's promotion of the merger of psychiatry with general medicine, which he began in the 1920s in the United States.[74] Developments in psychiatry explain the interest in psychogenic connections and a new observation of the relationships between mind and body, emotions and their physical expression, affective states, and somatic disorders.[75] In Germany, early psychosomatic medicine was more the domain of physicians trained in internal medicine or neurology, like Viktor von Weizsäcker or his contemporary Gustav von Bergmann, a clinical director in Berlin and Munich.

From the late 1930s onward, physicians belonging to the American psychosomatic movement proposed adding a psychic link to the etiologic chain or etiopathogenic causes of cancer. Fundamental to this decision was the assumption of "multicausality" and thus the necessity of looking at "the varying distribution of psychological and non-psychological factors from case to case."[76] Franz Alexander, considered one of the founders of American psychosomatic medicine, proposed a list of nine etiological

---

[71] Elida Evans, *A Psychological Study of Cancer* (New York, 1926). The book is discussed in Marco Balenci, "Il lavoro pionieristico di Elida Evans e l'approccio junghiano alla psicosomatica del cancro," *Giorn. Stor. Psicolog. Dinam.* 14 (1990): 195–217.

[72] The American psychosomatic movement ignored Evans for twenty-five years. See Patricia Jasen, "Breast Cancer and the Language of Risk, 1750–1950," *Soc. Hist. Med.* 15 (2002): 17–43, on 41; John I. Wheeler and Bettye McDonald Caldwell, "Psychological Evaluation of Women with Cancer of the Breast and of the Cervix," *Psychosomat. Med.* 17 (1955): 256–68.

[73] Helen Flanders Dunbar, *Emotions and Bodily Changes* (New York, 1935). On Flanders Dunbar, see Robert C. Powell, "Helen Flanders Dunbar (1902–1959) and a Holistic Approach to Psychosomatic Problems. I. The Rise and Fall of a Medical Philosophy," *Psychiat. Quart.* 49 (1977): 133–52.

[74] Adolf Meyer (1866–1950) was born in Switzerland and emigrated to the United States in 1892, settling in Chicago, after having worked at hospitals for mental patients elsewhere in Illinois (Kankakee State Hospital), and in Worcester, Massachusetts (State Lunatic Hospital). He contributed significantly to the improvement of the medical-scientific standards of these institutions and strove to introduce fellow physicians to the unified concept of "psychobiology." See Suzanne R. Karl and Jimmie C. Holland, "Looking at the Roots of Psychosomatic Medicine: Adolf Meyer," *Psychosomatics* 54 (2013): 111–4; Holland, "History of Psycho-Oncology: Overcoming Attitudinal and Conceptual Barriers," *Psychosomat. Med.* 64 (2002): 206–21, on 208.

[75] In the early years of the twentieth century, American psychiatrists struggled to become faculty members at medical schools and staff members at general hospitals, urging that psychiatry be taught more widely to medical students and that greater attention be paid to the patient. Adolf Meyer, e.g., tried to promote the cause of psychiatry nationwide from his position at Johns Hopkins by challenging the medically useless opposition of the mental and the physical. See Brown, "Rise" (cit. n. 7).

[76] Franz Alexander, *Psychosomatic Medicine: Its Principles and Applications* (New York, 1950), 52.

factors in disease. Emotions played a role in at least three of them but were closely related to other factors that shaped the constitution of the body and its organs.[77]

In varying proportions, all of these factors were assumed to play an etiological role in all diseases, including cancer, and their interaction offered a more complete causal picture than before. However, the psychosomatic point of view stressed the role of emotions over other factors: "In the last two decades," Alexander pointed out in 1939, "increasing attention has been paid to the causative role of emotional factors in disease and a growing psychological orientation manifests itself among physicians."[78]

The second important element in American psychosomatic thinking was that the psychological orientation was based on assumptions about the physiology of emotions, which provided an answer to the quest for a causal link between body and emotions. These physiological assumptions integrated the findings in the early twentieth century of the Harvard physiologist Walter B. Cannon, who emphasized the concept of homeostasis. Emotions were thought to be accompanied by physiological changes: fear led to heart palpitations, while anger increased cardiac activity, brought on higher arterial pressure, and induced changes in the metabolism of carbohydrates, and so on. Physiological phenomena resulted from complex muscular interactions caused by the influence of nervous impulses and the autonomic nervous system. Thus, facial muscles and the diaphragm were supposedly modified by laughter, the lachrymal glands by weeping, the heart by fear, the suprarenal glands and the vascular system by rage, and so on.[79]

According to theories of psychosomatic medicine, the physiological changes that usually accompanied emotions were transitory, but they did produce physical alterations. These could then lead to functional disturbances, particularly when strong emotions were felt for a substantial period of time. Pathological conditions appeared when a person tried to hold back emotions for too long.[80]

---

[77] Alexander's list enumerated (*a*) hereditary constitution, (*b*) birth injuries, (*c*) organic diseases of infancy that increase the vulnerability of certain organs, (*d*) nature of infant care (weaning habits, toilet training, sleeping arrangements, etc.), (*e*) accidental traumatic physical experiences of infancy and childhood, (*f*) accidental traumatic emotional experiences of infancy and childhood, (*g*) emotional climate of family and specific personality traits of parents and siblings, (*h*) later physical injuries, and (*i*) later emotional experiences in intimate personal and occupational relations. See ibid.; see also Roy R. Grinker and Fred P. Robbins, *Psychosomatic Case Book* (New York, 1954), 327.

[78] Franz Alexander, "Psychological Aspects of Medicine," *Psychosomat. Med.* 1 (1939): 7–18, on 7. The article was reprinted in M. Ralph Kaufman and Marcel Heiman, eds., *Evolution of Psychosomatic Concepts: Anorexia Nervosa, a Paradigm* (New York, 1964), 56–77. In 1950, Alexander stated that "theoretically every disease is psychosomatic, since emotional factors influence all body processes." Alexander, *Psychosomatic Medicine* (cit. n. 76), 52.

[79] Alexander, "Psychological Aspects" (cit. n. 78), 14. In this sense, the 1956 annual meeting of the American Psychosomatic Society focused on memorializing the contributions of Cannon to the study of the role of emotions in disease, quoting his famous article published twenty years earlier. See Walter B. Cannon, "The Role of Emotion in Disease," *Ann. Intern. Med.* 9 (1936): 1453–65. However, they drew a distinction between physiological and psychological approaches to emotions; see "Meeting for Members of the Society and Invited Guests Held in Honor of Professor Walter B. Cannon at the Harvard Medical School, Sunday, March 25, 1956," *Psychosomatic Medicine* 19 (1957): 179–81 (the issue memorializes Professor Walter B. Cannon and includes Dr. Binger's introductory remarks at the meeting).

[80] Alexander, "Psychological Aspects" (cit. n. 78), 15. At least initially, the American psychosomatic school noted that each emotion was related specifically to an organic disorder or disease. Alexander, *Psychosomatic Medicine* (cit. n. 76), 9, 14, and 16. However, from the very first issue of the journal *Psychosomatic Medicine*, the centrality of this point in the debate was obvious. See Pilar León-Sanz, "Resentment in Psychosomatic Pathology (1939–1960)," in *On Resentment: Past and Present*, ed.

In the 1940s and 1950s, the psychosomatic theories based on physiological effects were fortified through linkage to the emerging concept of stress. Known simply as the "stress concept," it had been formulated by Hans Selye, a physiologist at the University of Montreal, who framed the physiological reactions as being part of a "general adaptation syndrome," a response of the body to stressful events that was directed toward reestablishing a lost balance (adaptation) that could itself harm or exhaust the body if it lasted too long.[81] The concept of stress was not only broad enough to encompass all sorts of stress—which explains in part why the concept was so popular and so widely deployed in the years to come—but also quickly integrated into psychiatry, where the formerly physiological emotion "stress" was reinterpreted within a psychological framework.[82] This kind of physiological-psychological stress was also integrated into psychosomatic medicine: "Many emotions due to the complications of our social life cannot be freely expressed and relieved, through voluntary activities, but remain repressed and then are diverted into wrong channels."[83] In 1954, the former military psychiatrist Roy Grinker insisted that

> we have to search in the environmental family, school, work, social, and other areas to determine what has been significant in evoking an unhealthy response in a particular subject. This may be as simple and ordinary as a change in the family circle through birth of a child or death of a mother, or as complicated and as extensive as a general social upheaval. All of these environmental factors may strike a vulnerable spot in the patient's integrative capacity, stir up anxiety, and initiate a series of psychological regressions which may be more adaptive, although they are costly. With the regressive phenomena are associated internal events which are often accompanied by organ dysfunctions. Whether the vulnerability of the patient or the more stressful environment is the crucial or most recent factor in etiology, the response is a multiple series of interactions within the patient and his environment.[84]

During this period, the journal *Psychosomatic Medicine* published the results of numerous studies on people's reactions to environmental or social stress and their ability to adapt to environmental circumstances.

Bernardino Fantini, Dolores Martín Moruno, and Javier Moscoso (Newcastle upon Tyne, 2013), 131–64, on 139.

[81] In 1950, an article coauthored by Hans Selye and Claude Fortier, "Adaptive Reaction to Stress," *Psychosomat. Med.* 12 (1950): 149–57, was followed by numerous articles on this subject, some of which connected cancer and stress. See Harrington, *Cure Within* (cit. n. 41), 139–74; Lea Haller, "Stress, Cortison und Homöostase: Künstliche Nebennierenhormone und physiologisches Gleichgewicht, 1936–1960," *NTM* 18 (2010): 169–95; Patrick Kury, *Der überforderte Mensch: Eine Wissensgeschichte vom Stress zum Burnout* (Frankfurt am Main, 2012); David Cantor and Edmund Ramsden, eds., *Stress, Shock, and Adaptation in the Twentieth Century* (Rochester, N.Y., 2014); Lea Haller, Sabine Höhler, and Heiko Stoff, "Stress—Konjunkturen eines Konzepts," in "Stress!" ed. Lea Haller, Sabine Höhler, and Heiko Stoff, special issue, *Zeithist. Forsch.* 11 (2014): 359–81, http://www.zeithistorische-forschungen.de/3-2014 (accessed 31 August 2015).

[82] On the shift of adrenaline excitement from physiological to psychological emotion after 1945, see Otniel E. Dror, "What Is an Excitement?" in Biess and Gross, *Science and Emotions* (cit. n. 8), 121–38. See also Theodore M. Brown, "'Stress' in US Wartime Psychiatry: World War II and the Immediate Aftermath," in Cantor and Ramsden, *Stress* (cit. n. 81), 121–41; Tulley Long, "The Machinery and the Morale: Physiological and Psychological Approaches to Military Stress Research in the Early Cold War Era," in Cantor and Ramsden, *Stress*, 142–85; Mark Jackson, *The Age of Stress: Science and the Search for Stability* (Oxford, 2013).

[83] Alexander, "Psychological Aspects" (cit. n. 78), 18.

[84] Grinker and Robbins, *Psychosomatic Case Book* (cit. n. 77), 331.

Psychosomatic cancer studies published from the 1940s onward also introduced new methods into researching cancer psychosomatically and thus established new standards for what was defined as scientific evidence within psychosomatic thinking. While Elida Evans had based her entire book on case studies and anecdotal evidence arising from her clinical psychiatric practice, "new" studies now turned to using prospective personality tests, methodological tools developed by psychologists and psychiatrists in the 1930s. The tests most frequently employed were the then-established Rorschach test and the Minnesota Multiphasic Personality Inventory.[85] Both tests offered a form of standardization and "scientification" to psychosomatic cancer research that it had previously lacked.

One of the most cited studies on emotions and cancer in the United States and in West Germany was also based on the findings of these personality tests. The study was conducted by Milton Tarlau and Irwin Smalheiser at the New York City Cancer Institute on Welfare Island, a hospital for terminal cancer patients, and published in 1951.[86] Their examination subjects were twenty-two married women who had all previously been diagnosed with cancer. Half of them had breast cancer and the other half cervical cancer. In order to assess their personalities, Tarlau and Smalheiser first did a personal interview with each of them lasting from one to two hours. The interview was directed toward those factors supposed to influence the psychosexual development of the patients. These factors included early family life, age of the patient at the time of her parents' death, sex education and reaction to menstruation, as well as marital adjustment. Tarlau and Smalheiser then asked the women to interpret the inkblot drawings of the Rorschach Method of Personality Diagnosis. The Rorschach data were considered the most valuable, for while the patients were suspected of deliberately distorting interview material, their "true" functioning was assumed to be revealed clearly in their reaction to the unstructured inkblots. Finally, Tarlau and Smalheiser asked the women to draw various human figures.[87] These drawings were used to supplement and corroborate the diagnosis produced by the Rorschach images. In the end, Tarlau and Smalheiser correlated the interview material with the Rorschach data and interpreted it as giving a consistent picture for the two groups.

[85] The Rorschach test was developed by the Swiss psychologist Hermann Rorschach in the early 1920s. Its purpose was to examine an individual's personality characteristics and emotional functioning. The Minnesota Multiphasic Personality Inventory (MMPI) was developed in the late 1930s by psychologist Starke R. Hathaway and psychiatrist J. Charnley McKinley at the University of Minnesota. It is still one of the most frequently used testing instruments both in clinical settings and in psychological research. The MMPI continued to be recommended during the following decades when searching for predictors of psychosocial adaptation to cancer. See Roderick D. Buchanan, "The Development of the Minnesota Multiphasic Personality Inventory," *J. Hist. Behav. Sci.* 30 (1994): 148–61; Harry J. Sobel and J. William Worden, "The MMPI as a Predictor of Psychosocial Adaptation to Cancer," *J. Consult. Clin. Psychol.* 47 (1979): 716–24; Rebecca Schilling and Stephen T. Casper, "Of Psychometric Means: Starke R. Hathaway and the Popularization of the Minnesota Multiphasic Personality Inventory," *Sci. Context* 28 (2015): 77–98. Regarding the Rorschach test, see Peter Galison, "Image of Self," in *Things That Talk: Object Lessons from Art and Science*, ed. Lorraine Daston (New York, 2004), 257–94.

[86] Dr. Milton Tarlau (1910–91) was a psychiatrist and neurologist with practices in Manhattan and Eastern Pennsylvania. He held positions as a neurologist in Manhattan at Goldwater Memorial Hospital, Bellevue Hospital Center, and the Veterans Administration Hospital as well as at Easton Hospital in Easton, Pennsylvania. During the 1960s he became an expert encephalographer. See "Milton Tarlau, Neurologist," *New York Times*, 16 June 1991.

[87] Milton Tarlau and Irwin Smalheiser, "Personality Patterns in Patients with Malignant Tumors of the Breast and Cervix: An Exploratory Study," *Psychosomat. Med.* 13 (1951): 117–21, on 117.

The data revealed, they argued, a common general framework of mother dominance and sexual maladjustment for all female patients with cancer of the sexual organs. They concluded that, although the patients had experienced other problems prior to the illness, there was "some evidence here which suggests that the personality structure may play a role in the pathogenesis of cancer of primary or secondary sex organs in predisposed individuals."[88] Tarlau and Smalheiser thus assumed that the personality patterns they described were not the result of the disease but may have had some significance in the genesis or localization of the pathologic process.[89]

In 1954, James H. Stephenson and William J. Grace of Cornell University reported their finding that a higher proportion of severe maladjustment of a particular kind was found in a personality study of 100 women with cervical cancer, compared to a similar study of women with other forms of cancer.[90] Even though they were unable to pinpoint the mechanism responsible for this difference, the study was nonetheless held to have an indicative value. The parallels in the methodology used in these studies, and the scientific weight ascribed to the psychological testing as such, are obvious.[91] In general, the personality studies led to the establishment of psychosomatic profiles for diverse illnesses.[92]

THE SHORT-LIVED MOMENT OF A PSYCHOSOMATIC DISEASE: "TYPE C" IN 1960s AND 1970s GERMANY

The overall impression in clinical oncology during the early 1950s was that of a crisis because the available treatment options were insufficient. Surgery based on the theory that cancer had local origins had proved to be much less successful than hoped, even in its most radical form, which was thought to eradicate an invasive disease by

---

[88] Ibid., 121.

[89] Ibid., 117. Another group of investigators who interviewed women with breast cancer concurred that breast-cancer patients were extremely repressed sexually. Furthermore, they indicated that women with breast cancer seemed to be markedly inhibited about expressing aggression and tended to camouflage themselves with a facade of pleasantness. But these authors were cautious about ascribing any causal significance to the relationships they found. See Catherine L. Bacon, Richard Renneker, and Max Cutler, "A Psychosomatic Survey of Cancer of the Breast," *Psychosomat. Med.* 14 (1952): 453–60.

[90] James H. Stephenson and William J. Grace, "Life Stress and Cancer of the Cervix," *Psychosomat. Med.* 16 (1954): 287–94. See also Wheeler and Caldwell, "Psychological Evaluation" (cit. n. 72).

[91] Similarly, a study done in California, involving fifty cancer patients and presented at the 1952 meeting of the American Association for Cancer Research by Eugene M. Blumberg, Philip M. West, and Frank W. Ellis, undertook "a complete battery of psychological tests, including the Minnesota Multiphasic Personality Inventory (MMPI), the Rorschach, the Thematic Apperception Test, and the Wechsler-Bellevue Intelligence Test" with a view to observing the relationship between average life expectancy and patients' personality profiles. The resulting data suggested that "longstanding, intense emotional stress may exert a profoundly stimulating effect on the growth rate of an established cancer in humans." Blumberg, West, and Ellis, "A Possible Relationship between Psychological Factors and Human Cancer," *Psychosomat. Med.* 16 (1954): 277–86, on 286.

[92] Interestingly, these cancer personality patterns were usually ascribed to women, the preferred subjects of 1950s psychosomatic cancer studies. As Jasen notes, one can wonder whether this feminization of the cancer personality and the concomitant claim that this personality was a defective one reveals an underlying misogynic tendency in psychosomatic medicine and psychiatry at that time. This is particularly underscored by the fact that the studies were based on contested methodological settings. Jasen, "Malignant Histories" (cit. n. 6), 284–5. How assumptions about femininity, especially about the character and role of female emotions, influenced psychological and psychiatric concepts during that time can be followed up in Anne Harrington's essay, "Mother Love and Mental Illness: An Emotional History," in this volume.

extreme measures.[93] Radiation therapy using radium or mesothorium not only was very expensive but also often turned out to merely alleviate symptoms without curing the cancer in the long term. Thus, researchers were bent on rethinking cancer etiology and treatment, a tendency that had been prepared by the expansion of German cancer research during National Socialism, which had placed great emphasis on investigating the influence of hormones, vitamins, and chemical agents on carcinogenesis and its treatment.[94] Numerous oncologists argued that one should conceive of cancer as a multifactorial and systemic disease, a development that allowed for more theoretical openness, since emotions did not have to be the one and only cause, but merely one among others.[95]

Furthermore, the turn to American developments in psychosomatic approaches to cancer helped raise the reputation of this field within the West German medical community. It was not only the adoption of standardized personality tests in place of case studies; of crucial importance was the final integration of the stress concept into West German psychosomatic medicine and particularly into psychosomatic cancer medicine—precisely at the historical moment when stress became a kind of guiding concept in West German medical research and practice, as well as in the public discussion about the ills of modernity.[96] "Stress" facilitated the conceptualization of emotions as working on the body that was in line with contemporary mainstream medical thinking on cancer, based on the assumption that cancer was a form of "disregulation" on a cellular level.[97] In addition, the stress concept in itself was multicausal—a broad variety of agents could be regarded as stressful events. Its integration was therefore a step toward investigating emotions as correlated to the etiological process, and no longer necessarily as causal.[98]

Finally, the stress concept allowed for research into the carcinogenic effect of emotions within animal experiments. As long as psychosomatic theory held that complex emotions like melancholy or grief caused cancer, animal experiments were inconceivable. However, conceptualizing emotions as "stressors" made psychosomatic emotions both human and "animalistic" and therefore suitable for investigation using animal subjects. These subjects, mostly mice, were genetically predisposed to cancer or had been exposed to some known carcinogenic chemical agent, and they were then divided into two groups. A control group would then be "stressed" by electric shocks, forced to swim, or placed under other experimental stress conditions.[99] Thus, in a

---

[93] The surgeon Karl Heinrich Bauer, one of the leading oncologists in West Germany, admitted in an internal communiqué circulated in the mid-1960s that 80 percent of cancer patients could still not be cured. See Senatskommission für das Krebsforschungszentrum an der Universität Heidelberg (DFG), "Denkschrift betr. Anstalt für Geschwulstforschung und Geschwulstbehandlung an der Universität Heidelberg," sgd. K. H. Bauer (n.d.), Bundesarchiv BArch B 142/3434, sheets 186–96, on 190.

[94] Eckart, *Medizin* (cit. n. 67), 287.

[95] J. Kretz and O. Pötzl, "Die Psyche des Krebskranken," *Der Krebsarzt* 1 (1946): 19–29, on 19.

[96] Hans-Georg Hofer, "Labor, Klinik, Gesellschaft: Stress und die westdeutsche Universitätsmedizin (1950–1980)," *Zeithist. Forsch.*, online ed., vol. 11 (2014), http://www.zeithistorische-forschungen .de/3-2014 (accessed 31 August 2015).

[97] Norbert Paul, "Die molekulargenetische Interpretation des Krebs: Ein Paradigma, seine Entwicklung und einige Konsequenzen," in Eckart, *100 Jahre* (cit. n. 3), 95–100, on 96.

[98] This development is in line with what Nissim Mizrachi has described as a general tendency in American psychosomatic theory. Mizrachi, "From Causation to Correlation: The Story of Psychosomatic Medicine 1939–1979," *Cult. Med. Psychiat.* 25 (2001): 317–43.

[99] See, e.g., Marvin Reznikoff and David E. Martin, "The Influence of Stress on Mammary Cancer in Mice," *J. Psychosomat. Res.* 2 (1957): 56–60; M. B. Waller, P. Waller, and R. F. Strebel, "Effects of

lengthy review article in 1961, Hans-Joachim F. Baltrusch, a young West German medical psychologist and member of the First International Psychosomatic Cancer Study Group, could refer to the numerous animal experiments that had been done in the 1950s in light of Selye's stress concept, in order to justify the psychotherapeutic treatment of those "major sicknesses unto death"—like cancer—that had formerly been regarded as purely organic diseases.[100]

During the 1960s and 1970s, the assumption that certain stressful emotional events together with a specific personality structure—"type C"—could lead to cancer was investigated in several clinical and epidemiological studies.[101] Cancer as a "disease of the soul" ultimately even entered the popular magazines.[102]

SOLVING GEORG GRODDECK'S RIDDLE

Cancer posed a particular challenge for psychosomatic medicine—at least for what is considered to be "modern" psychosomatic medicine, which antedated the advent of cellular pathology and bacteriology. More than any other disease, cancer was thought to be organic in the most concrete sense. This was very much due to modern laboratory practices. Once researchers began to isolate and study "unfeeling" cancerous cells under the microscope, letting cancerous tissue grow in petri dishes regardless of whether it was of human or animal origin, most medical experts became convinced that emotions like grief or melancholy, of which only humans were thought to be capable, could not be involved in carcinogenesis.[103] With very few exceptions, the great majority of German physicians and researchers had no doubt that a tumor's materiality was impervious to the influence of emotions. This was also true for those physicians who were interested in psychosomatic theory and practice. Cancer conceived as a disease that started with a local and almost always irreversible growth did not fit into the framework of neurosis that most of them favored—a realization that was reinforced by the ostensible hopelessness of treating cancer psychotherapeutically. National Socialist health policy, which increased the emphasis on restoring productivity, had made all attempts to approach cancer in psychosomatic terms seem preposterous.

Stress on the Course of Development of Cancer in Mice," paper presented at the First International Conference of the International Psychosomatic Cancer Study Group, Amsterdam, August 1960.

[100] Hans-Joachim F. Baltrusch, "Leukämien und andere maligne Erkrankungen des haemotopoetischen, lymphatischen und retikulo-endothelialen Systems in psychosomatischer Sicht," pt. 1, *Z. Psychosomat. Med.* 7 (1961): 229–35, on 230, and "Leukämien und andere maligne Erkrankungen," pt. 2, *Z. Psychosomat. Med.* 8 (1962): 13–23. In 1981, Selye even wrote the preface to a medical textbook on cancer and psychosomatics that pointed to the significance of "psychosocial stress": Kurt Bammer, *Krebs und Psychosomatik* (Stuttgart, 1981). See also the chapter on cancer in the most important West German textbook of psychosomatic medicine that explained the "psycho-physiological links": Claus Bahne Bahnson, "Das Krebsproblem in psychosomatischer Hinsicht," in *Lehrbuch der Psychosomatischen Medizin*, ed. Thure von Uexküll (Munich, 1979), 685–98, on 692–5.

[101] The German Research Fund even decided to finance a long-term prospective study on psychosomatic interdependencies in the development of chronic illnesses, including cancer. This study was carried out by the medical sociologist Ronald Grossarth-Maticek between 1971 and 1978 in Heidelberg—a clear indication that psychosomatic medicine had achieved scientific recognition. The "type C-concept" has been developed mostly by Lydia Temoshok and Henry Dreher, *The Type C Connection: The Behavioral Links to Cancer and Your Health* (New York, 1992).

[102] "Krebs durch Seelenschmerz und soziale Qual?" *Der Spiegel* 45 (1977): 102–16.

[103] One such expert was Isaac Berenblum, who from 1938 to 1948 was leader of the cancer research group British Empire Cancer Campaign, from 1948 to 1950 was a researcher at the National Cancer Institute in Bethesda, and eventually became professor for cancer research at the Weizman Institute in Rehovot; see Baltrusch, "Leukämien," pt. 2 (cit. n. 100), 15.

As Viktor von Weizsäcker put it, the "granite" of the body withstood every effort of psychosomatic thinking to see the relation between emotions and disease.[104]

Cancer ultimately went psychosomatic in Germany during the 1950s, and three main factors contributed to this shift. First, psychoanalysis and psychosomatic medicine in general gained increasing recognition in West German medicine and society, both morally and scientifically. Second, this heightened recognition was gained by resorting to those "scientific" methods that American psychosomatic medicine had already applied before and which were perceived as part of modern international medical standards. Interviews, psychological tests, and statistical evaluations increasingly replaced the clinical observations and case studies favored by medical anthropologists like Weizsäcker and psychoanalysts like Groddeck.[105] The third reason was the eventual adoption of the stress paradigm following the introduction of the "general adaptation syndrome" by Selye, and the transformation of stress from a physiological emotion to a psychological one—a development that occurred during the 1950s. This shift ultimately provided an answer to the question of how emotions could act upon the material body, human and animal alike, and even permanently harm it, an answer that was convincing not only for psychologists but to a certain degree for general physicians as well. Since emotions as "stressors" could be converted into experimental emotions, cancer could finally be studied as a psychosomatic disease using modern scientific standards. In addition, emotions ceased to be conceived as the only and specific causes of cancer—they could be important within the stress concept even when they were shown to be only correlated to carcinogenesis and cellular dysregulation, rather than having a causal role.

Beyond these three factors, there might have been a fourth involved in the process of reinventing cancer as psychosomatic disease. If one considers the twentieth-century trends and trajectories of scientific research into emotion, one can argue that the immediate postwar years were marked by a marginalization of emotions as scientific objects in various disciplines, followed by a reevaluation of emotions in science from the 1960s onward.[106] At first sight, the story of psychosomatic cancer medicine that has been investigated here seems to slightly contradict this trend, since it reveals an earlier shift toward studying emotions, starting in the late 1930s in the United States and later in West Germany during the 1950s. However, if one considers the logic underlying the post-1945 marginalization of emotions, the ostensible contradiction makes perfect sense. Post-1945 science tended to ignore emotions and opted for models based on "rationality" because emotions were regarded as irrational, dangerous, and even pathological forces that had been responsible for the National Socialists' rise to power and ultimately the atrocities committed under their rule. The belief that emotions could induce cells to become malignant and make them run riot against the body's health fit well into this framework and was further supported by psychosomatic models that took the social sphere into account while also drawing parallels between the body

[104] Viktor von Weizsäcker, "Klinische Vorstellungen," *Psyche* 1 (1947): 258–93, on 291.

[105] Erwin W. Straus characterized the different approaches to psychosomatic medicine in the United States and Europe at the Fourth International Congress of Psychotherapy (1958) as a "meeting between an empirical science more or less aware of its metaphysical and epistemological background and one unaware of it." Straus, "The Fourth International Congress of Psychotherapy Barcelona, Spain, September 1 through 7, 1958," *Psychosomat. Med.* 21 (1959): 158–64.

[106] This is especially obvious in social and political science as well as in economic thought. See Biess and Gross, "Emotional Returns" (cit. n. 8), 1–6.

proper and the body politic. Moreover, the experimental methodology applied to the study of emotions also contributed to the positivization or objectivization of emotions. The broader scientific return to emotions in the 1960s further strengthened the position of psychosomatic cancer medicine in West Germany, which was in its very beginnings.

THE STORY CONTINUED: FROM PATHOGENIC TO SALUTARY EMOTIONS WITHIN MEDICAL PRACTICE

Psychosomatic cancer was a short-lived notion. Criticism emerged in the United States as early as the end of the 1950s. In 1959, George M. Perrin and Irene Pierce, both members of the American Psychosomatic Society, pointed out that

> the case reports [like the Tarlau/Smalheiser or Blumberg/West studies] were little more than suggestive, and they rarely allow the reader to distinguish among those psychological characteristics which might be causative factors, those which might be typical reactions to any serious disease, and those which might show purely accidental variations.[107]

Although in Germany this criticism was at first not interpreted as a fundamental critique of the idea of emotion-based carcinogenesis, it helped pave the way for a later shift that was influenced by simultaneous developments in the United States.

In the United States, the emphasis within psychosomatic medicine had shifted from looking into the etiological role of emotions to investigating the role of emotions within medical practice.[108] At the annual meeting of the American Psychosomatic Society in 1954, George Engel summarized the results of a questionnaire sent to physicians that year. He noted that "many emphatically said they were not interested in any discussion of psychogenesis or psychological triggering mechanisms, but felt that a consideration of psychological reactions to cancer might be worthwhile."[109] The respondents hoped that "detailed psychological knowledge of cancer patients inevitably will contribute to better care of these patients and their families."[110] For American psychosomatic physicians, understanding the influence of the emotional element in cancer was thus a problem not only of knowledge but of action, as they stressed the role of emotions in medical praxis. They frequently noted that medicine should lead to a change in the attitudes of professionals and patients toward the disease because their attitudes were believed to have a direct impact on therapeutic strategy.

Although emotions in the doctor-patient relationship had already been described outside the psychosomatic area, scientists investigated the curative power of emo-

---

[107] George M. Perrin and Irene R. Pierce, "Psychosomatic Aspects of Cancer: A Review," *Psychosomat. Med.* 21 (1959): 397–421, on 416. Perrin and Pierce regarded such studies as questionable, at the very least, since they were basically anecdotal, e.g., an early twentieth-century study of Native Americans. See Isaac Levin, "Cancer among the American Indians and Its Bearing upon the Ethnological Distribution of the Disease," *Z. Krebsforsch.* 9 (1910): 422–35.

[108] David Cantor, "Memorial's Stress? Arthur M. Sutherland and the Management of the Cancer Patient in the 1950s," in Cantor and Ramsden, *Stress* (cit. n. 81), 264–87. As Felicity Callard reflects, the development of psychopharmacology also contributed to the emergence of a new psychological approach to the study of emotions. See Felicity Callard's essay, "The Intimate Geographies of Panic Disorder: Parsing Anxiety through Psychopharmacological Dissection," in this volume.

[109] Quoted in Dorothy Levenson, *Mind, Body, and Medicine: A History of the American Psychosomatic Society* (Baltimore, 1994), 126.

[110] Ibid., 127.

tions in the encounter between physician and patient in a rapidly growing literature on this topic.[111] For these scientists, therapy was based on what Franz Alexander had, in an earlier context, called the principle of "corrective emotional experience."[112]

In the 1950s, several projects analyzed psychological responses to cancer on the part of hospitalized patients, providing the first opportunity for collaborative research with physicians. The first reports of psychological adaptation to cancer and its treatment were made by the psychiatric group at the Massachusetts General Hospital directed by Jacob E. Finesinger, who described guilt and shame as the most prominent psychological responses to the stigma of cancer.[113] Meanwhile, under the direction of Arthur Sutherland, also a psychiatrist, the psychiatric research group at the Memorial Sloan Kettering Cancer Center in New York stressed the importance of the values promoted by the cancer patient's cultural environment as well as the significance of diverse familiar structures in influencing the patient's emotions.[114] These and other similar findings resulted in a critique of the faulty training offered by medical schools on the emotional component of the medical encounter. New demands to investigate, systematize, and institutionalize the physician's training in managing both her or his own emotions and those of the cancer patient as a crucial element of the medical encounter were to determine the further development of psycho-oncology in the United States from the 1970s onward.[115]

Research and clinical practice in the United States thus shifted during the 1950s from investigating the pathogenic propensity of emotions to exploring the emotional repercussions of cancer and its treatment. By deploying emotions' healing capacities and the possibility of "managing" emotions, this shift was geared toward enhancing the options for curing cancer as well as toward bettering the patient's life. In West Germany, by contrast, scientists valued both understandings of emotions in the 1960s and continued to examine the capacity of emotions in generating as well as curing and enduring cancer.

[111] E.g., in 1950, Avery D. Weisman stated: "Despite the importance of the doctor-patient relationship, it remains poorly understood, except in the special circumstances of psychoanalysis and psychoanalytic psychotherapy." Weisman, "The Doctor-Patient Relationship: Its Role in Therapy," *American Practitioner and Digest of Treatment* 11 (1950): 1144–51, on 1144. See Bettina Hitzer, "Oncomotions: Experience and Debates in West Germany and the United States after 1945," in Biess and Gross, *Science and Emotions* (cit. n. 8), 157–78.

[112] Franz Alexander, "Individual Psychotherapy," *Psychosomat. Med.* 8 (1946): 110–15, on 112. Helen Flanders Dunbar said, "The physician's responsibility is to correct intraorganism dysfunction, but often this is possible only when he becomes the catalytic agent in restoring the patient's capacity for integration in society." Dunbar, *Emotions and Bodily Changes: A Survey of Literature on Psychosomatic Interrelationships, 1910–1953,* 4th ed. (New York, 1954), 685. Dunbar quoted Grinker and Spiegel regarding this question and pointed out that unless there was full connection and sympathetic understanding by the doctor, very little could be achieved. This was empirically proven in a 1957 study carried out by Margaret Thaler, Herbert Weiner, and F. Reiser Morton, "Exploration of the Doctor-Patient Relationship through Projective Techniques: Their Use in Psychosomatic Illness," *Psychosomat. Med.* 19 (1957): 228–39, on 230. See Pilar León-Sanz, "El carácter terapéutico de la relación médico-paciente," in *Emociones y estilos de vida: Radiografía de nuestro tiempo,* ed. Lourdes Flamarique and Madalena D'Oliveira (Madrid, 2013), 101–30.

[113] Ruth D. Abrams and Jacob E. Finesinger, "Guilt Reactions in Patients with Cancer," *Cancer* 6 (1953): 474–82.

[114] Arthur M. Sutherland, "The Psychological Impact of Postoperative Cancer," *Bull. N.Y. Acad. Med.* 33 (1957): 428–45. See also Cantor, "Memorial's Stress?" (cit. n. 108).

[115] Holland, "History" (cit. n. 74).

Mother Love and Mental Illness:
An Emotional History

by Anne Harrington*

ABSTRACT

Most scholarship on the medicalization of emotions has focused on projects that locate emotions, one way or another, within individual brains and minds. The story of mother love and mental illness, in contrast, is a medicalization story that frames the problem of pathological emotions as a relational issue. Bad mother love was seen as both a pathology (for the mother) and a pathogen (for her vulnerable child). Moreover, different forms of pathological mother love—smothering love, ambivalent love, love that masked an actual desire to dominate and control—were supposed to have different effects on children, ranging from lack of fitness for military service to homosexuality to juvenile delinquency to outright psychosis, especially schizophrenia. Understanding why mother love came to be associated with mental illness—and, equally, what led to this viewpoint's rapid decline into disrepute—requires us to go beyond simply invoking the trope of "mother blaming" and leaving things at that. This essay is a first effort at a richer narrative, one that blends perspectives from the history of emotions and the history of science and medicine.

> The mother-child relationship is so important for ensuing pathology that it has probably received more attention than any other aspect of child psychiatry.
> —Franz Alexander and Sheldon Selesnick, *The History of Psychiatry: An Evaluation of Psychiatric Thought and Practice from Prehistoric Times to the Present*, 1966[1]

> We failed to understand why parents of a child with leukemia were treated with sympathy and understanding, while parents of a child with schizophrenia were treated with scorn and condemnation.
> —Eve Oliphant, mother of an adult son with schizophrenia, addressing the American Psychiatric Association, 1977[2]

  * Department of the History of Science, Science Center 360, 1 Oxford Street, Harvard University, Cambridge, MA 02138; aharring@fas.harvard.edu.

  Early versions of these ideas were presented at the Susan and Donald Newhouse Center for the Humanities, Wellesley College, and at the Max Planck Institute for the History of Science, Berlin, Germany. I am grateful for helpful comments from colleagues at both these presentations.

[1] Alexander and Selesnick, *The History of Psychiatry: An Evaluation of Psychiatric Thought and Practice from Prehistoric Times to the Present* (New York, 1966).

[2] Quoted in Kate Cadigan and Laura Murray, "When Medicine Got It Wrong," 2009, available from Documentary Educational Resources, http://www.der.org/films/when-medicine-got-it-wrong.html (accessed 4 February 2016).

In May 2010, a documentary aired on various public television stations across the United States. It told a story about an effort by courageous families to challenge and eventually overthrow a pernicious doctrine that had taken hold of American psychiatry from the 1940s through the 1970s: a doctrine that said that schizophrenia, the most serious and tragic of mental disorders, is caused by bad parents, and especially by bad mothers. May was chosen as the month for multiple broadcasts of the documentary—in honor of Mother's Day. And the take-home message of the documentary was clearly summarized in its title: "When Medicine Got It Wrong."[3]

It may be easy for us to agree that "medicine got it wrong," but it does little to help our understanding. Yes, we wince at the kinds of things that people told mothers as recently as twenty-five years ago. We are appalled at therapists who insisted that some mothers, literally, drove their children psychotic. Nevertheless, if the effort to bring the history of emotions into more productive dialogue with the history of medicine is to yield real fruit, we need to move ourselves beyond our own initial emotional responses of outrage and to cease being content with historical narratives designed primarily to further stoke that outrage.[4] The question of why mothers' emotional lives became of such concern to so many psychiatrists—and with such profound consequences—cannot be answered by simply invoking the chestnut of "mother blaming" and leaving it at that. We need to start doing better.

GETTING ORIENTED: THE PLACE OF TOXIC MOTHER LOVE IN THE HISTORY OF EMOTIONS AND THE HISTORY OF MOTHERHOOD

Most scholarship on the medicalization of emotions has focused on projects that locate emotions, one way or another, within individual brains and minds.[5] The story of mother love and mental illness, in contrast, is a medicalization story that framed the problem of pathological emotions as a relational issue. Bad mother love was both a pathology (for the mother) and a pathogen (for her vulnerable child). At the same time, definitions of toxic love were unstable and, to an important degree, tracked cultural shifts in the expert consensus about ideal forms of mother love more generally. For example, the 1920s began to see a growing resistance within American and British society to a style of Victorian-era mothering that some have called "moral motherhood." This style of motherhood had been marked by, among other things, a "tight, indeed controlling bond" between the mother and her (usually) male child. The bond was supposed to be for his own good: it would act as a moral rudder that would guide him through his life.[6]

By the 1920s, however, a form of mother love marked by intensive emotional involvement was increasingly believed to set the child up for a life of dependency and

---

[3] Ibid.

[4] This approach is perhaps most widely associated with the work of Barbara Ehrenreich and Deirdre English, in *For Her Own Good: 150 Years of the Experts' Advice to Women* (New York, 1979). For a more recent example, see Diane Eyer, *Motherguilt: How Our Culture Blames Mothers for What's Wrong with Society* (New York, 1996).

[5] See Otniel E. Dror, "Techniques of the Brain and the Paradox of Emotions, 1880–1930." *Sci. Context* 14 (2001): 643–60; see also Dror, "Cold War 'Super-Pleasure': Insatiability, Self-Stimulation, and the Postwar Brain," in this volume; Felicity Callard, "The Intimate Geographies of Panic Disorder: Parsing Anxiety through Psychopharmacological Dissection," in this volume.

[6] In the words of Mary Ryan, a "tight, indeed controlling bond between mother and male child was at the very core of the cult of domesticity." Cited in Rebecca Jo Plant, *Mom: The Transformation of Motherhood in Modern America* (Chicago, 2010), 89. For more on the fall of Victorian-era moral motherhood, see Glenna Matthews, *"Just a Housewife": The Rise and Fall of Domesticity in America* (New York, 1987).

immaturity—or worse. Consensus had begun to change in part because, during this time, a new breed of Progressive-era experts had begun to lay out the conditions for what they saw as a more rational and scientifically grounded approach to child rearing—one marked by schedules, authorized child-care products, and continued medical advice. In that context, mothers who were judged to be too doting, too affectionate, and too emotionally invested came in for increasing censure.[7]

After World War II, we see another shift. As the postwar nations embraced a new kind of idealization of domesticity, the prewar ideal of the modern "scientific" mother began to give way to a vision of the ideal mother as someone who was naturally affectionate and available to her child, albeit this time without stifling his or her emotional growth. In other words, this "naturally" loving mother now acted less as a moral rudder for her son and more to help him internalize a sense of emotional security so that eventually he would be able to stand on his own two feet.[8] Much was at stake in this new postwar understanding of healthy mother love. As scholar Teri Chettiar has noted, "emotional health . . . forged in the monogamous nuclear family" was also understood to be "the basis for social mobility, self-governance, and responsible citizenship."[9]

Other scholars, like the ones I have cited, have persuasively analyzed the history of changing ideals of motherhood and mother love. What we sometimes lack in their accounts, however, is a more granular understanding of the specific ways in which each of these shifting ideals created its antithesis or shadow. Behind every idea of mother love was a specter of defective mother love—of mother love gone wrong. When the love was deemed sufficiently dysfunctional to risk the health and well-being of the offspring, the experts got involved. Understanding how this all worked, and with what consequences, is the goal of this essay.

BEYOND FREUDIAN FATHERS: THE TURN TO MOTHER LOVE

We might imagine that the story begins with misogynistic men, but it is not so. In fact, many of the deepest roots of clinical interest in mother love—adequate and pathological—lie with influential female psychoanalysts: people like Anna Freud, Helene Deutsch, and Karen Horney. Their agenda was not misogynist: on the contrary, it was marked by a positive interest in the psychology of women and what one might even consider a pro-mother sensibility.[10]

Classical Freudian theory had sometimes seemed to suggest that the only reason women might want to be mothers would be to compensate for the feelings of inferiority that resulted from not having a penis ("penis envy"). In the 1920s, a new generation of women analysts began to speak up and insist that this was nonsense.[11] The

---

[7] See also Rima Apple, "Constructing Mothers: Scientific Motherhood in the Nineteenth and Twentieth Centuries," *Social History of Medicine* 8 (1995): 161–78; Apple, *Perfect Motherhood: Science and Childrearing in America* (New Brunswick, N.J., 2006).

[8] Marga Vicedo, *The Nature and Nurture of Love: From Imprinting to Attachment in Cold War America* (Chicago, 2013).

[9] Teri Chettiar, "The Psychiatric Family: Citizenship, Private Life, and Emotional Health in Welfare-State Britain, 1945–1979" (PhD diss., Northwestern Univ., 2013).

[10] See also Janet Sayers, *Mothers of Psychoanalysis: Helene Deutsch, Karen Horney, Anna Freud, Melanie Klein* (1991; repr., New York, 1993).

[11] Karen Horney in particular challenged the classical Freudian idea that a woman's desire for a baby was best understood as a compensatory reaction to her disappointment at not having been born

experience of motherhood, they argued, was a fundamental and—for the most part—emphatically positive dimension of female psychology, one that was marked by powerful emotions distinct from sexual love. "Its chief characteristic," Helene Deutsch insisted, "is *tenderness*."[12] This was not to say that motherhood was always easy or without pain, as Deutsch also underscored: "There is hardly a woman in whom the normal psychic conflicts do not result in a pathological distortion, at some point, of the biologic process of motherhood."[13] But that was all the more reason, these women analysts insisted, for psychoanalysis as both science and clinical practice to take mother love seriously as a critical part of female psychology.

What about the children on the receiving end of this love? Classical Freudian psychoanalysis had also sometimes implied that the mother-child relationship was a rather uninteresting prelude to the great oedipal drama focused on the father, when all the important developmental issues were worked out. For most of his career, Freud had portrayed mothers rather instrumentally, as simply the means for satisfying the physiological needs of infants for food, warmth, and safety. Only at the very end of his career did Freud consider that the mother-infant relationship might be psychologically critical as an end in itself (he thought perhaps especially for girls). But even as he tried out this new (to him) idea, he confessed that he was unable to follow through on all of its implications: "Our insight into this early, pre-Oedipus phase in girls comes to us as a surprise, like the discovery, in another field, of the Minoan-Mycenaean civilization behind the civilization of Greece."[14]

It fell to others, therefore, to clarify ways in which the maternal emotional relationship mattered—and actually mattered most—for the development of both girls and boys. The initiative here was undertaken in particular by a new generation of analysts, who came of age in the 1930s and helped spearhead a series of major revisions of psychoanalytic theory in Europe and the United States, including ego psychology and object relations theory. While the rise of both ego psychology and object relations theory was a complicated affair, for our purposes the most important thing to know about it was that (in contrast to classical psychoanalysis) it gave great weight to the emotional effects of experiences that happened in the first years and even months of life. In so doing, it also turned attention away from imaginary castrating fathers and toward early maternal care. At the same time, it downplayed classical Freudian concerns with infantile biological drives and focused more on ways in which an infant is shaped by its early experience within the family, and especially interactions with the mother.[15] Donald Winnicott, one

---

a man with a penis. In an essay originally published in 1926, she also suggested that much of male psychology was actually motivated by what she called "womb envy"—envy of a woman's ability to bear life and give birth. See Horney, "The Flight from Womanhood: The Masculinity-Complex in Women, as Viewed by Men and by Women," in *Feminine Psychology*, ed. Harold Kelman (London, 1967), 54–70.

[12] Helene Deutsch, *The Psychology of Women: A Psychoanalytic Interpretation*, vol. 2, *Motherhood* (New York, 1945), 20; emphasis in the original. Cf. Karen Horney, "Maternal Conflicts," originally published in 1933 and reprinted in Kelman, *Feminine Psychology* (cit. n. 11), 175–81.

[13] Deutsch, *Psychology of Women* (cit. n. 12), v.

[14] Cited in Mari Jo Buhle, *Feminism and Its Discontents: A Century of Struggle with Psychoanalysis* (Cambridge, Mass., 2009), 135.

[15] See also Martin Halliwell, *Therapeutic Revolutions: Medicine, Psychiatry, and American Culture, 1945–1970* (New Brunswick, N.J., 2013); Eli Zaretsky, *Secrets of the Soul: A Social and Cultural History of Psychoanalysis* (New York, 2004).

of the British architects of object relations theory, went so far as to say in 1940, "There is no such thing as an infant . . . whenever one finds an infant one finds maternal care, and without maternal care there would be no infant."[16]

LIFE WITHOUT A MOTHER'S LOVE

Significantly, many of the clinicians who concluded that mother love was critically important to an infant's health and development did so against a backdrop of extensive direct work with young children in hospitals, clinics, or nurseries. And it happened that, during World War II, many of them also witnessed the bad things that can happen to young children when their mothers were not around to love them.

Some of them witnessed what happened in England, for example, when many babies and young children were sent away from the major English cities (to avoid the bombings, or because both parents were working). Living in special institutions where they all received adequate physical care, these children nevertheless regressed developmentally and showed other signs of dysfunction. Without their families, and especially without the love of their mothers, their emotional and cognitive development was threatened. Freud's daughter Anna, having moved with her father to England to escape the Nazis, was key in identifying the changing signs of what would become known as "maternal deprivation." Working with her companion Dorothy Burlingham at a war nursery set up in Hampstead, in north London, she observed how, by six months, the infant's need for a mother's affection is "as urgent for his psychological satisfaction as the need to be fed and taken care of is for his bodily comfort." And by the second year of life, the child responds to separation from the mother with all the signs of deep grief and distress.[17]

Meanwhile, in the United States, the Viennese Hungarian émigré Rene Spitz had been studying children who were separated from their mothers, not because of war, but because they were orphaned or required hospitalization for a prolonged period. He found that such children not only regressed cognitively and emotionally but also in some cases wasted away physically and died. Spitz saw this as a special kind of psychogenic "marasmus" or wasting away. When caused by hospitalization, he proposed calling the syndrome in question "hospitalism." His most influential work on this front compared the outcomes of infants being raised in two institutions: a prison nursery in which they could interact with their own mothers and a home for foundlings in which they were cared for by "overworked nursing personnel." Here is what happened:

> While the children in the "Nursery" [where they could interact with their own mothers] developed into normal healthy toddlers, a two-year observation of "Foundling Home" [where they were cared for by the nursing staff] showed that the emotionally starved children never learned to speak, to walk, to feed themselves. With one or two exceptions in a total of 91 children, those who survived were human wrecks, who behaved either in the manner of agitated or apathetic idiots.

[16] Donald Woods Winnicott, "The Theory of the Parent-Infant Relationship," *Int. J. Psychoanal.* 41 (1960): 585–95, on 587.
[17] See Dorothy Burlingham and Anna Freud, *Young Children in War-Time in a Residential War Nursery* (London, 1942), 181. See also Freud and Burlingham, *Infants without Families: The Case for and against Residential Nurseries* (New York, 1944).

The most impressive evidence probably is a comparison of the mortality rates of the two institutions. "Nursery" in this respect has an outstanding record, far better than the average in the country. In a five years' observation period, during which we observed a total of 239 children, each for one year or more, "Nursery" did not lose a single child to death. In "Foundling Home," on the other hand, 37 percent of the children died during a two years' observation period.[18]

After World War II, Ronald Hargreaves of the World Health Organization (WHO) commissioned the child psychiatrist John Bowlby, then head of the Department for Children and Parents at the Tavistock Clinic in London, to write a report on the mental health of displaced, orphaned, and refugee children in postwar Europe and make recommendations. Informed by his own earlier work on "affectionless youth," and drawing on the insights of Anna Freud, Rene Spitz, and others (such as Bill Goldfarb in the United States), Bowlby's 1951 *Maternal Care and Mental Health* would be touted as a landmark synthesis of prevailing knowledge on this subject.[19] For our purposes, its importance lies in the way in which, more than any other document of the time, it sealed the case that a specific kind of maternal love—nurturing, attentive, and available to the child on demand—was the emotional bedrock necessary to ensure a child's mental health and normal development.[20] In Bowlby's words,

> What is believed to be essential for mental health is that the infant and young child should experience a warm, intimate and continuous relationship with his mother (or permanent mother-substitute) in which both find satisfaction and enjoyment. . . . Prolonged deprivation of a young child of maternal care may have grave and far reaching effects on his character . . . similar in form . . . to the deprivation of vitamins in infancy.[21]

MOTHER LOVE APPRAISED: THE ROLE OF CHILD GUIDANCE CLINICS

"Warm, intimate, and continuous." It was not enough to love one's child; one must love in the right way. But how many mothers did? Even before World War II, mental health workers in child guidance clinics in the United States and Britain had been clear that many mothers fell short. Memories of these clinics have largely faded today, but

---

[18] Rene A. Spitz, "The Role of Ecological Factors in Emotional Development in Infancy," *Child Develop.* 20 (1949): 145–55, on 149. For the original study, see R. A. Spitz, "Hospitalism—An Inquiry into the Genesis of Psychiatric Conditions in Early Childhood," *Psychoanal. Stud. Child* 1 (1945): 53–74.

[19] In short order, the book was translated into fourteen languages. The English paperback edition alone sold more than 400,000 copies. A second, expanded edition of the book, titled *Child Care and the Growth of Love*, was subsequently published by Penguin Books in 1965.

[20] This early report also laid the groundwork for Bowlby's later development (with Mary Ainsworth) of what is still today known as "attachment theory," which insisted that an infant is primed to develop a primary attachment to an attentive, loving mother and that this is evolutionarily necessary for its physical and emotional survival and health. This later work was also shaped by new developments from ethology on the instinctive way in which baby animals like ducks "imprint" themselves on a caregiver, generally the mother. See Vicedo, *Nature and Nurture* (cit. n. 8); Mary Salter Ainsworth, "Object Relations, Dependency, and Attachment: A Theoretical Review of the Infant-Mother Relationship," *Child Develop.* 40 (1969): 969–1025.

[21] John Bowlby, *Maternal Care and Mental Health* (Geneva, 1951), 13. For more on the presumed parallels between emotional and nutritional deficiency, see Eduardo Duniec and Mical Raz, "Vitamins for the Soul: John Bowlby's Thesis of Maternal Deprivation, Biomedical Metaphors and the Deficiency Model of Disease," *Hist. Psychiat.* 22 (2011): 93–107.

in the first decades of the twentieth century, they were important new outposts of psychiatry's outreach efforts. Their origins lay, not so much directly in concerns about ensuring mothers were providing the right kind of love, but rather in more practical, Progressive-era efforts to apply a public-health, early-intervention sensibility (often called "mental hygiene") to the problem of mental disorder. Critical of older mainstream approaches that had seen most or all cases of mental illness as rooted in tainted biology, child guidance clinics were instead committed to the meliorist view that, by intervening early in the lives of troubled (and troublesome) children, it might be possible both to reduce juvenile delinquency and to forestall future instances of adult mental pathology.[22] As the American psychiatrist Frankwood Williams put it in 1923:

> Individuals are not born odd, or queer, or peculiar. Timid, sensitive, blustering, rebellious children are not born, they are made—and made by quite human agencies. . . . Children born well who later contract tuberculosis of the spine or infantile paralysis come eventually to plaster casts or braces from which after a time they are relieved, many improved and helped; but children born well who later contract certain habits of emotional reaction come eventually to courts and reformatories from which after a time they are relieved, few of them improved or helped. . . . [Some] are to fail entirely. . . . [And] within fifteen years [are] gathered to their mattresses on the floors of hospitals for the insane.[23]

Through the 1920s, as scholars like Kathleen Jones have shown, the focus of these clinics was on instructing poor and uneducated immigrant families in good hygiene, nutrition, and scientifically validated, modern methods of child rearing. By the 1930s, though, these clinics had begun increasingly to serve middle-class families who brought in children with problems ranging from bed-wetting to truancy to stuttering. At the same time, they also began to engage with the new psychoanalytic ideas about the importance of family life and of nurturing mother love to healthy child development. If the child was troubled, they came to believe, the family must be troubled. And, if the family was troubled, somewhere one would find a mother who was failing—for reasons that might in fact be understandable—to love and emotionally support her children in ways that they needed for good mental health.[24]

Indeed, through the 1930s, there was a tendency within these clinics to portray many of these mothers in a relatively sympathetic light. Women who produced troubled children tended to be deeply unhappy with their own lives and marriages. Some, it was felt, also suffered from what was called "primary affect hunger," which originated from deprivation experienced in their own childhoods. While some of these deprived mothers responded by emotionally rejecting their children, most were seen as compensating

---

[22] See also Kathleen Jones, *Taming the Troublesome Child: American Families, Child Guidance, and the Limits of Psychiatric Authority* (Cambridge, Mass., 1999); Alice Boardman Smuts, *Science in the Service of Children, 1893–1935* (New Haven, Conn., 2006).

[23] Frankwood Williams, cited in Ernest Rutherford Groves and Mary Blanchard Phyllis, *Introduction to Mental Hygiene* (New York, 1930), 185.

[24] Kathleen Jones, "'Mother Made Me Do It': Mother-Blaming and the Women of Child Guidance," in *"Bad" Mothers: The Politics of Blame in Twentieth-Century America*, ed. Molly Ladd-Taylor and Lauri Umansky (New York, 1998), 99–124, on 101. Jones explains in this chapter that most child guidance experts were less influenced by Freud's writings directly than by the new psychoanalytic theories associated with ego psychology and object relations theory. One particularly influential text for this community was *The Psychoanalytic Study of the Family*, by the British psychoanalyst J. M. Flugel, first published in 1921 by the Hogarth Press. For more on the larger history of the child guidance movement, see Jones, *Taming the Troublesome Child* (cit. n. 22).

for the emotional poverty of their own lives by becoming "overprotective." The over-protective mother was in fact an invention of the 1930s child guidance clinic.[25]

To help the children, it was concluded, the child guidance centers had to find a way to help the mothers, too. Thus, in many instances, the mother would be offered an opportunity to undergo counseling with a (female) social worker, while a psychiatrist would focus on treating the child. As Kathleen Jones has shown in her examination of clinical records from the Boston-based Judge Baker Guidance Center, in some cases the mothers would agree to such sessions, and in other cases they would resist. The stakes were high, but they remained a reasonably private matter, something that was to be negotiated between the social workers, the psychiatrists, the mothers, and the children.[26]

SISSIES, HOMOSEXUALS, RACISTS, AND DELINQUENTS: THE HEIGHTENED POSTWAR STAKES

After World War II, however, the stakes were raised, and the whole problem of deficient mothers and troubled children became a much more public—and much more political—conversation. To understand why, it is important to acknowledge the degree to which, in the immediate postwar years, a new generation of American and European social scientists had successfully defined virtually all the big social problems of the day not as problems of public policy or institutional failure, but as problems of individual psychology—of emotional inadequacy. If the United States failed to win the Cold War, if the country succumbed to Communism, if the black man failed to advance himself socially, if boys were seduced into homosexual lifestyles, if girls failed to protect their sexual virtue, and if cities were terrified by lawless youth gangs, the fault lay in the brittle or warped personality structures of all the individuals involved.

None of the individuals in question had been born bad or defective, the experts continued. Their dysfunctional personalities had been shaped by their family circumstances, by the neighborhoods in which they had grown up, and above all by the ways in which they had been loved and cared for by their mothers. As we have seen, belief in the environmental roots of deviance and mental disorder had functioned in the 1930s child guidance clinics as a progressive and rather hopeful doctrine (since what had been made could presumably be unmade). In the postwar era, however, it took an increasingly punitive and accusatory edge. To a first approximation, the failure of the mothers to love their children in the right way came to be seen as the witting or unwitting pernicious force behind virtually every form of deviant citizenship within society.

Consider, by way of example, the discussion of this time about a problem of great national concern: psychiatric breakdown during the recent world war. During the war, 1,825,000 young men had been preemptively rejected from service on "neuropsychiatric" grounds. Nevertheless, close to one million others broke down mentally during the war itself, sometimes even before they saw combat, or under circumstances that did not seem excessively adverse. What was wrong with America's young men? Why were they so weak, immature, and unstable?

In 1945, Dr. Edward Strecker, a former president of the American Psychiatric Association and a psychiatrist who served as a special consultant to the Secretary of War and the Surgeon General of the Army and Navy, answered this question with a descrip-

---

[25] David Mordecai Levy, *Maternal Overprotection* (New York, 1943), 37, 23.
[26] Jones, *Taming the Troublesome Child* (cit. n. 22).

tion of a certain sort of inadequate mother that he called a "mom" (picking up a term employed to great effect by the freelance author Philip Wylie in his 1942 trenchant critique of bad faith motherhood, *A Generation of Vipers*).[27] Although Strecker believed that "moms" came in a number of types (the "self-sacrificing," the "ailing," the "Pollyanna," the "protective," the "pretty addlepate," and the "pseudointellectual"), all of them shared one critical attribute: they were all high-octane versions of the overprotective mother, all narcissistically invested in keeping their sons emotionally dependent on them, for their own gratification. Put another way, a mom was a woman who had "failed in the elementary mother function of weaning her offspring emotionally as well as physically." The consequences were grave, to say the least. Strecker believed that these moms were more responsible than anything else for the "epidemic" of psychoneurosis that had been revealed in the recent world war. Their selfish immaturity was now threatening the capacity of the country to defend itself against new enemies. For this reason, he concluded, these "moms" must be recognized as "our gravest menace, a threat to our survival" as a democratic civilization.[28]

Strecker's attack on moms was no marginal affair. He first made his case in a 1945 lecture to medical colleagues, which was reported in the *New York Times*. A print version of the argument appeared several months later in the psychiatric journal *Mental Hygiene*, under the title "Psychiatry Speaks to Democracy." A book expanding the argument, *Their Mothers' Sons*, was then published in 1946 to extensive press coverage. The argument was also widely discussed or reported in the popular press: pieces appeared in the *Ladies' Home Journal* ("Are American Moms a Menace?"), the *Saturday Evening Post* ("What's Wrong with American Mothers?"), *Time* ("Mama's Boys"), and the *Washington Post* ("Momism").[29] Many were persuaded that the issue at hand was no joke: "Mom is bad for the son and, therefore, bad for the country," agreed one military-minded reviewer of Strecker's book in 1947. It was good that Strecker was giving this matter the public airing it deserved.[30]

Strecker's analysis of these narcissistic mothers and their weak, immature sons attracted so much attention in part because it tapped into a more general debate at the time about an alleged "crisis of masculinity."[31] His critique, however, was, only one of many. As the larger vision of ideal mother love shifted generally away from the

---

[27] For Strecker's wartime role, see "Dr. Strecker to Aid Army: Head of Psychiatric Association Made Consultant to Stimson," *New York Times*, 13 June 1943, 6. For Wylie's work, see Wylie, *Generation of Vipers* (New York, 1942). For a thoughtful analysis of the reception of *Generation of Vipers*, see Plant, *Mom* (cit. n. 6).

[28] Edward Adam Strecker, *Their Mothers' Sons: The Psychiatrist Examines an American Problem* (New York, 1946), 219.

[29] "'Moms' Denounced a Peril to Nation: Psychiatrist Blames Them as He Says 350,000 Men Tried to Evade Military Service Scores: 'Most Powerful Lobby' Blames 'Mom' for Rejections," *New York Times*, 28 April 1945, 11; Edward A. Strecker, "Psychiatry Speaks to Democracy," *Mental Hygiene* 20 (1945): 591–605; Abram Scheinfel, "Are American Moms a Menace?" *Ladies Home Journal*, November 1945, 36; Strecker, *Their Mothers' Sons* (cit. n. 28); Strecker, "What's Wrong with American Mothers," *Saturday Evening Post*, 26 October 1946; "Mama's Boys," *Time*, 25 November 1946; "Momism," *Washington Post*, 11 February 1951, B4.

[30] Jon Zimmerman, "Military Library," review of *Their Mothers' Sons*, by Edward A. Strecker, *Military Affairs: J. Amer. Mil. Inst.* 9–11 (1945–7): 191.

[31] See also Elaine Tyler May, *Homeward Bound: American Families in the Cold War Era* (New York, 2008); Kyle A. Cuordileone, "Politics in an Age of Anxiety: Cold War Political Culture and the Crisis in American Masculinity, 1949–1960," *J. Amer. Hist.* 87 (2000): 515–45; on the closely connected issue of maternal attitudes toward sons and homosexuality, see Roel van den Oever, *Mama's Boy: Momism and Homophobia in Postwar American Culture* (New York, 2012).

rational and scientific model of the interwar period and toward a more instinctual, nurturing model, other critics drew attention to other kinds of deviations from the ideal:
permissive mothers, rejecting mothers, seductive mothers, domineering mothers, ambivalent mothers, and more. These mothers were variously deemed responsible for creating deviant citizens ranging from racists to juvenile delinquents to homosexuals.
Mothers of "sissies" and homosexuals, for example, were said to "smother" their children with a sometimes seductive love that had roots in the fact that they were emotionally unsatisfied with their own husbands. Mothers of juvenile delinquents, in contrast,
were supposed to be insufficiently attentive or consistent in the way they expressed
their love, creating insecurity in their offspring that led them to act out or seek affirmation in gangs.[32]

While most of the mothers who came under scrutiny in this way were implicitly
white and middle-class, some analysts also trained their attention on the deficiencies
of black mothers (especially in single-parent households) and the alleged effects they
were having on the psyches of their children, especially their sons.[33] At the same time,
some anthropologists looked outside the American context and examined the allegedly
rigid and undemocratic child-rearing traditions of Japanese and German mothers for
insight into the root sources of the national characters of America's recent enemies.[34]
There were many kinds of emotionally deficient mothers that people worried about in
the years after World War II.

MOTHER LOVE AND SCHIZOPHRENIA: A CONVERSATION APART

During this time, however, one type of mother stood out from all of the others. This
mother literally drove her child crazy—made him psychotic. Not only did the conversation about this mother represent the most extreme edge of the argument that mothers who love wrong can harm their children's emotional development. It was also
the only conversation that aimed—provocatively and defiantly—to make mothers responsible for a specific class of disorders that, even at the time, had a long history of
being understood very differently: as a brain disorder, a disorder of defective biology. I
am speaking of schizophrenia.

Schizophrenia was (and still is) generally diagnosed in people who have lost the
ability to connect ideas together coherently, who hear voices (that often insult them
or tell them to do terrible things), and who cling to what society judges to be irrational, often paranoid beliefs. Overwhelming anxiety, loss of the ability to care for oneself and achieve one's goals, and withdrawal from normal human interactions are also
common symptoms.

Since the late nineteenth century, there had been a general consensus by non-Freudian
clinicians (generally working in mental hospitals with chronic, severely disordered pa-

---

[32] Ladd-Taylor and Umansky, *"Bad" Mothers* (cit. n. 24); Ruth Feldstein, *Motherhood in Black and
White: Race and Sex in American Liberalism, 1930–1965* (Ithaca, N.Y., 2000).

[33] See also Michael Daryl Scott, *Contempt and Pity: Social Policy and the Image of the Damaged
Black Psyche, 1880–1996* (Chapel Hill, N.C., 1997); for a prominent example of an argument in this
vein, see Abram Kardiner and Lionel Ovesey, *The Mark of Oppression: A Psychosocial Study of the
American Negro* (New York, 1951).

[34] For a good analysis of this anthropological and "national character" work, see Peter Mandler, *Return from the Natives: How Margaret Mead Won the Second World War and Lost the Cold War* (New
Haven, Conn., 2013).

tients) that schizophrenia (and its predecessor, dementia praecox) was a degenerative and incurable brain disorder. Over the decades, some had sought its origins in structural abnormalities of the brain that could be made visible through microscopic research. Others had seen the disorder as a form of autointoxication, caused by toxins produced in other parts of the body. Still others, in the early days of germ theory, had wondered whether it might be caused by infection.[35]

All this gave the argument relating schizophrenia to defective maternal love an edge of urgency and defiance that we do not see elsewhere. Here is why. When certain clinicians and social scientists made claims in the postwar years about the maternal roots of juvenile delinquency, racism, military unpreparedness, and even sexual deviance, they might have been met with occasional skepticism on various fronts, but they would not have encountered an entrenched alternative set of explanations for those same disorders. With schizophrenia, they did. For this reason, the conversation about schizophrenia was not only animated by the same wartime and immediate postwar concern with mothers that animated all the other conversations. It was also animated by a fierce desire to challenge and overcome a resistant and continuing bastion of biological thinking in psychiatry.

What was wrong with thinking biologically about schizophrenia? The answer we hear was usually twofold. First, it was proposed that the biological approach had failed intellectually: decades of efforts to find a relevant brain lesion or source of infection, or to nail down the hereditary nature of the disorder, had never produced anything other than contradictory or equivocal evidence (as even those who believed in this approach were sometimes willing to admit). Second, and perhaps more important, it was proposed that the biological approach had failed morally and clinically. When seen through a biological lens, schizophrenic patients, it was said, became nothing more than their broken brains, "a cerebral machine thrown out of gear" in the words of psychoanalyst Carl Gustav Jung. It was assumed that nothing could be done for them, and that no particular sympathy needed to be extended to them. Given this, it was a great moral advance, as Jung also insisted, when a new generation of clinicians with psychoanalytic sympathies began to see severely psychotic patients as fellow human beings, and to realize that they actually suffered from human problems. In Jung's words:

> Hitherto we thought that the insane patient revealed nothing to us by his symptoms save the senseless products of his disordered cerebral cells; but that was academic wisdom reeking of the study. When we penetrate into the human secrets of our patients, we recognize mental disease to be an unusual reaction to emotional problems which are in no way foreign to ourselves.[36]

If this was so, it might also follow that some schizophrenic patients might benefit from psychoanalytic psychotherapy—that it was not incurable after all. Tellingly, perhaps, Freud himself had cautioned against drawing that conclusion; he did not believe that schizophrenia was a brain disorder, but he did suggest that psychotic patients were too disconnected from reality to be able to establish a viable therapeutic

---

[35] Richard Noll, "Historical Review: Autointoxication and Focal Infection Theories of Dementia Praecox," *World J. Biol. Psychiat.* 5 (2004): 66–72.

[36] Carl Gustav Jung, "The Content of the Psychoses" (1905), in *Collected Papers on Analytical Psychology*, ed. and trans. Constance E. Long (New York, 1916), 312–51, on 322.

relationship with a clinician.[37] Others, though, were not prepared to give up so quickly. By the 1940s, a small number of people in Europe and the United States—Marguerite Sechehaye, Lilly Hajdu-Gimes and Sandor Ferenzci, Gertrude Schwing, Frieda Fromm-Reichmann, John Rosen—began to adapt psychoanalytic psychotherapy into forms that might be suitable as a treatment for schizophrenia.

By this time too, the stakes had been raised by the fact that biological psychiatry was starting to become significantly less fatalistic about therapeutic options for schizophrenia than it had once been. Over the course of the late 1930s and 1940s, a series of new somatic treatments for schizophrenia and other serious mental disorders had been introduced in short order: insulin coma therapy, electroconvulsive therapy (ECT), and later lobotomies (drugs were not available until the mid-1950s). No one really knew why or how well these new treatments worked, but many in the hospital system were convinced that they helped. The psychotherapists for their part countered that the somatic treatments were brutal and dehumanizing. If they worked at all, it was only by dulling the symptoms and not by actually addressing the cause of the disorder. What did address the cause and had the potential to result in a lasting cure, they believed, were talking treatments with a sympathetic therapist who was prepared to assume that his or her patient was struggling with human, rather than physiological, problems. As Frieda Fromm-Reichmann said in the early 1940s, during a clinical case conference where staff were considering using insulin, barbiturates, or Benzedrine on an unruly patient: "Do you want to knock him out completely or give him enough to relax and then be able to talk to you as he comes out of it? . . . It seems to me you should give [the medicine] but not deprive him of his doctor."[38]

Which takes us back to mothers and mother love. People like Fromm-Reichmann believed that patients with schizophrenia needed "their doctor"—someone they could talk to and be in a relationship with—because their problems were relational as well. Something had gone wrong—terribly wrong—for them in their early childhood. The mothers who should have loved them and kept them safe failed to do so. The wounds caused by these mothers, however, might be healed by the therapist. It is striking in this context how many of the therapists who thought this way were women, offering themselves (implicitly and on occasion explicitly) as the "good mother" that their patients never had.

Indeed, in the 1940s, the Swiss therapist Marguerite Sechehaye went so far as to encourage her young adolescent patient to call her "Mama," physically cradled her as if she were a baby, and talked to her as if she were a tiny child. At one point in the therapy, Sechehaye even went so far as to symbolically breastfeed the girl, when it emerged that her desperate pleas for "apples" were really a cry for a loving maternal breast. In Sechehaye's words:

> I grope about in the beginning, not knowing how to calm this need for apples. I bring Renee beautiful apples, the most beautiful I can find, and pounds of them, telling the landlady to give Renee as many as she desires. But Renee always refuses them. . . . Renee runs away and arrives at my house all alone, at nine in the evening, in terrible agony. I

[37] Sigmund Freud, "On Narcissism" (1914), in *The Standard Edition of the Complete Psychological Works of Sigmund Freud, Vol. XIV (1914–1916): On the History of the Psycho-Analytic Movement, Papers on Metapsychology and Other Works*, ed. James Strachey (New York, 2001), 67–102.
[38] Cited in Ann-Louise S. Silver, *Psychoanalysis and Psychosis* (Madison, Conn., 1989), 28.

persist in trying to understand the symbolism of the apples. To the remark that I gave her as many apples as she wanted, Renee cries: "Yes, but those are store apples, apples for big people, but I want apples from Mummy, like that," pointing to my breasts. "Those apples there Mummy gives them only when one is hungry." I understand at last what is to be done! Since the apples represent maternal milk, I must give them to her like a mother feeding her baby: I must give her the symbol myself, directly and without intermediary.[39]

Insight having dawned, Sechehaye sliced an apple and invited her patient to lay her head on her breast and "drink the good milk." The girl did so, while slowly eating the apple, finally at peace.

In a somewhat different register, Fromm-Reichmann, a U.S.-based refugee from Hitler's Germany, became renowned, even at the time, for her willingness to do whatever it took to make an emotional connection to her severely disturbed patients. She would sit in their urine to show them that she was not better than they were. She would accept a gift of feces from them to show them that she was not rejecting them. In the admiring words of one of her colleagues:

> Sooner or later the schizophrenic patient experienced that he was no longer alone, that here was [a] human being who understood, who did not turn away in estrangement or disgust. This rare moment of discovery—unpredictable and unforeseen, like a gift of grace—sometimes became a turning point in the patient's life. The gates of human fellowship were opened—and thereby the slow way to recovery was opened also.[40]

MOTHERS OF SCHIZOPHRENICS: THE WORST MOTHERS OF ALL?

What kind of mother caused the terrible suffering that was schizophrenia? Did this woman love her children at all, or did she love them in a way that was so toxic that it eventually shattered their personalities? By the late 1940s, we begin to see the emergence of some explicit efforts to answer this question. Fromm-Reichmann in some ways set the terms of the conversation to come in 1940, when she rather casually coined the memorable term "schizophrenogenic mother" in the course of a discussion about research on the role of mothers in family groups. At that time, she was influenced by sociological work undertaken by fellow refugees on the "authoritarian family" and so suggested that in the United States (in contrast to Europe) children fear their "domineering mother" rather than their father.[41] Trude Tietze, another woman refugee immigrant from Vienna, also suggested that these mothers were dominating figures, in a widely cited 1948 article called simply "A Study of Mothers of Schizophrenic Patients." Dominating women, yes, but also possessed of other personality

---

[39] Marguerite Sechehaye, *Symbolic Realization: A New Method of Psychotherapy Applied to a Case of Schizophrenia* (New York, 1951), 51. See also Sechehaye, *Autobiography of a Schizophrenic Girl* (New York, 1951), in which the patient also describes this incident with the apples from her own perspective. I cite both texts from Annie G. Rogers, "Marguerite Sechehaye and Renee: A Feminist Reading of Two Accounts of a Treatment," *Int. J. Qualitative Stud. Educ.* 5 (1992): 245–51.

[40] Edith Weigart, on Frieda Fromm-Reichmann, cited in Edward Dolnick, *Madness on the Couch: Blaming the Victim in the Heyday of Psychoanalysis* (New York, 1998), 90.

[41] Frieda Fromm-Reichmann, "Notes on the Mother Role in the Family Group," *Bull. Menninger Clin.* 4 (1940): 132–48 (including discussion, 145–8).

traits: "rejection of the child, dominance, overanxiousness, obsessiveness, perfection-ism, and oversolicitousness."[42] Right from the start, it was a somewhat confusing profile.

And it became more confusing. By 1948, influenced by the tendency of new American colleagues like Clara Thompson and Harry Stack Sullivan to find the roots of schizophrenia in disturbed interpersonal relations, Fromm-Reichmann herself began to focus less on the presumed authoritarianism of the mothers of her patients and more on their presumed rejection and lack of authentic love:

> The schizophrenic is painfully distrustful and resentful of other people, due to the severe early warp and rejection he encountered in important people of his infancy and child-hood, as a rule, mainly in a schizophrenogenic mother. During his early fight for emo-tional survival, he begins to develop the great interpersonal sensitivity which remains his for the rest of his life.[43]

Thus the conditions were set up for a situation in which, throughout the 1950s, the schizophrenogenic mother was variously described as rejecting, rigid, domineering, and anxious, and sometimes all of the above. One 1957 study, run out of a veteran's hospital, used the so-called F-scale test ("F" for "fascism") to test the personality traits of a group of mothers of (male) schizophrenic patients and found evidence that they scored higher than the control mothers.[44] John Rosen felt that the mothers of his pa-tients were "perverse" individuals who fundamentally lacked the emotional ability to feel true love, with devastating consequences for their children.[45] Harry Stack Sullivan, one of Fromm-Reichmann's teachers, focused on the overwhelming anxiety he believed was experienced by these mothers, and the ways in which it created a home environ-ment that deprived the child of a "favorable opportunity . . . for building a successful self-esteem system. . . . Early in life, the idea was in some way conveyed inescapably to him that he was relatively infrahuman."[46] And so it went.

Virtually every clinician managed to find something wrong with these mothers, but by the 1960s, the lack of consensus on the nature of the dysfunction began to under-mine the larger effort to relate dysfunctional maternal love to schizophrenia. A new approach was needed, and some clinicians found one by broadening their gaze. In-stead of focusing on the mother's relationship to her child, they began to look at the whole family. The new idea was that schizophrenia might be a reaction to pathological behaviors—and especially pathological ways of communicating—within a family that was now to be conceived not simply as a set of sick individuals but as a dysfunctional system (in ways that were influenced by so-called systems theory).[47]

---

[42] Trude Tietze, "A Study of Mothers of Schizophrenic Patients," *Psychiatry* 12 (1949): 55–65.

[43] Frieda Fromm-Reichmann, "Notes on the Development of Treatment of Schizophrenics by Psy-choanalytic Psychotherapy," *Psychiatry* 11 (1948): 263–73.

[44] Jack Dworin and Oakley Wyant, "Authoritarian Patterns in the Mothers of Schizophrenics," *J. Clin. Psychol.* 13 (1957): 332–38.

[45] John N. Rosen, "The Perverse Mother," reprinted in *Direct Analysis: Selected Papers* (New York, 1953), 97–105.

[46] Harry Stack Sullivan, *Clinical Studies in Psychiatry* (1956; repr., New York, 1973), 364.

[47] For a review of the field of family systems theory in relation specifically to schizophrenia, see John G. Howells and Waguih R. Guirguis, *The Family and Schizophrenia* (New York, 1985). For an important overview of the larger history here, see Deborah Weinstein, *The Pathological Family: Postwar America and the Rise of Family Therapy* (Ithaca, N.Y., 2013).

A 1956 paper by sometime anthropologist Gregory Bateson and his colleagues, "Toward a Theory of Schizophrenia," was particularly catalytic here.[48] In this paper, the authors proposed that people who become schizophrenic might do so because they had grown up in families where people (especially parents) communicate in inauthentic and logically inconsistent ways—generally in order to cover up the ugly truths of their actual feelings for one another. A mother might say, "Come hug me, sweetie," but then flinch or grimace when the child touches her (a "meta-level" communication that contradicts the verbal communication). Upon seeing the child retreat, the mother might then further challenge the child's (accurate) reading of her feelings by saying, "What's wrong? Don't you love your mother?" Bateson's group named this kind of mutually contradictory communication loop "the double bind." They compared "double binds" to the insoluble riddles that a Zen Buddhist adept is expected to solve but then pointed out that the Zen adept had ways of transcending his dilemma and achieving enlightenment. The schizophrenic patient did not—as a child, there was no way for him to challenge his family system and survive. His only recourse in the end would be a flight into madness—a break with reality.[49]

The new systems approach to schizophrenia was supposedly committed to a neutral moral attitude toward the individual members of the families being studied. The idea was that it was the system—and especially communication within the system—that was broken, and everyone suffered from that fact. In practice, however, attitudes toward the parents—especially the mothers—in these families remained startlingly harsh. The mothers were described as domineering and the fathers as weak and henpecked. Some therapists implied that these people were so unpleasant as to almost overwhelm any capacity for therapeutic empathy and caring. Speaking at a symposium in 1962, one clinician put the matter this way:

> The family of the schizophrenic patient does violence to our implicit attachment to 18th century liberal rationalism. . . . This is probably one reason why existentialism, phenomenology, and other forms of subjective philosophy have become so intimately related to thinking about schizophrenia. These families can confound our rational theories, dispel optimistic planfulness, and plunge us into . . . therapeutic despair.[50]

Jay Haley, an influential founder of so-called family therapy, concurred:

> The greatest challenge to any family therapist is the psychotic family. Whatever difficulties there are in neurotic families, they are exaggerated to the point of parody in the family of the schizophrenic. Various approaches have been used and schools established to deal with these families, from valiant therapists singly assaulting the schizophrenic family citadel to expeditions in twos and threes. An attempt was even made to surround the whole family with a hospital.[51]

One of the most challenging things about these families, many therapists said, was their insistence that they were normal. They would say that they were just like any other

---

[48] In addition to Bateson's paper, there are several other more or less independent narratives that some people tell about the origins of family systems theory in relation to schizophrenia. See also C. Christian Beels, "Notes for a Cultural History of Family Therapy," *Family Process* 41 (2002): 67–82.

[49] See Gregory Bateson, Don D. Jackson, Jay Haley, and John Weakland, "Toward a Theory of Schizophrenia," *Behav. Sci.* 1 (1956): 251–64.

[50] Jerry Osterweil, "Discussion," *Family Process* 1 (1962): 141–5.

[51] Jay Haley, cited in Paul H. Glasser and Lois N. Glasser, *Families in Crisis* (New York, 1970), 188.

family and that they could make no sense of the sudden descent of one of their children into madness. The therapists, of course, knew better. One popular article from 1962 shared the following anecdote from the clinical archives:

> "As I told you," Mrs. Jones declared, "we just don't have any problems." "Except," the doctor noted, "for Judy, who's sitting here worrying about Communists from Mars." "But I explained that she's never acted like that." At that point, eight-year-old Betsy interrupted, "She was too like this before, mother. Remember last Christmas, when daddy didn't come home from his business trip, and you drank too much again and got sick."[52]

From the beginning, many family therapists saw themselves as a rebellious element within psychiatry and adopted a self-consciously avant-garde attitude. Of these, probably no one took that attitude further than the Scottish psychiatrist Ronald D. Laing. Over the course of the 1960s, Laing moved in his writings from an initial existentialist analysis of schizophrenia as a disorder with roots in profound "ontological insecurity" ("I have never known a schizophrenic who could say he was loved"),[53] to a hard-hitting analysis of the family's emotional dynamics as the source of the patient's insecurity,[54] and ending with a radical political analysis of the schizophrenic as a kind of thwarted mystic struggling to tell the truth about the ugly state of the world. By this time, too, Laing had concluded that the truly dangerous family was not the so-called psychotic family but the so-called normal family who successfully brainwashed its sons and daughters into supporting the mad political and social policies of the time:

> The ["normal"] family's function is to repress Eros; to induce a false consciousness of security; to deny death by avoiding life; to cut off transcendence; to believe in God, not to experience the Void; to create, in short, one-dimensional man; to promote respect, conformity, obedience; to con children out of play; to induce a fear of failure; to promote respect for work; to promote respect for "respectability."[55]

MOTHER LOVE AND THE END OF THE SCHIZOPHRENOGENIC MOTHER

Then, in the late 1970s, all of this finally began to unravel. The conventional explanation is that, just when things were getting completely ridiculous, psychiatry rediscovered the brain and biology. A slew of new research, along with effective new biological treatments (drugs), finally put the mother-blaming psychotherapists—psychoanalytic and family oriented—out of business.

In support of this claim, scientists often cite Seymour Kety's adoption studies in the early 1970s, which showed, among other things, that children born to a schizophrenic mother and reared in an adoptive family become schizophrenic at the same rate as siblings reared by their schizophrenic biological mother. This finding seemed to demonstrate a hereditary basis for schizophrenia.[56] One of Kety's colleagues and admirers was unequivocal: "As a result of these studies we no longer hear shrill voices proclaiming that schizophrenia arises from toxic interpersonal family environments."[57]

---

[52] Milton Silverman and Margaret Silverman, "Psychiatry inside the Family Circle," *Saturday Evening Post* 235 (28 July 1962): 46–51.

[53] R. D. Laing, *The Divided Self: A Study of Sanity and Madness* (London, 1960), 39.

[54] R. D. Laing, *Sanity, Madness, and the Family* (New York, 1964).

[55] Ronald D. Laing, *The Politics of Experience* (New York, 1967), 65.

[56] Seymour S. Kety, David Rosenthal, Paul H. Wender, and Fini Schulsinger, "Studies Based on a Total Sample of Adopted Individuals and Their Relatives," *Schizophrenia Bull.* 2 (1976): 413–28.

[57] Philip Holzman, "Seymour S. Kety and the Genetics of Schizophrenia," *Neuropsychopharmacology* 25 (2001): 299–304.

Other scientists point to the influence of work in the 1960s on neurotransmitters, especially animal research showing that the newly discovered drug Thorazine (which helped the symptoms of schizophrenia) depleted available dopamine levels in laboratory animals.[58] This work led some to suggest that schizophrenia was a disorder caused by an excess of dopamine in the brain.[59] And still other scientists point to new brain imaging work at this time that seemed to identify such things as enlarged brain ventricles as a reliable marker distinguishing the schizophrenic from the normal brain.[60]

These research developments and new pharmaceutical options emphatically mattered. On their own, though, they were not decisive. As we have seen, both drugs and biological research had coexisted with Freudian and other psychosocial perspectives on schizophrenia for more than a generation (since the 1950s), and, to a first approximation, no minds had been changed that did not want to be changed.

What else might have been important then? Part of the answer, but only part, was feminism's discovery of the politics of "mother blaming." As Betty Friedan noted in her 1963 *Feminine Mystique*, "Under the Freudian microscope . . . it was suddenly discovered that the mother could be blamed for almost anything. In every case history of a troubled child . . . could be found a mother."[61] It was high time, she said, to cultivate a profound skepticism toward this Freudian habit. By 1970, a group of radical women psychotherapists who called themselves the San Francisco Redstockings had begun distributing literature to sympathetic colleagues at the American Psychiatric Association, which included the following pointed suggestion: "Mother is not public enemy number one. Start looking for the real enemy."[62]

Feminists may have helped cultivate growing skepticism toward the assumption that mother was always to blame, but they did little if anything to cultivate the alternative view that broken brains were the problem instead. On the contrary, all of them broadly accepted that the roots of mental illness, including schizophrenia, would indeed be found in political and even familial dysfunctions—just not ones that made mothers responsible for everything. In her 1972 *Women and Madness*, for example, Phyllis Chesler conceded that mothers of schizophrenic patients probably were on some level being accurately described by clinicians. However, she argued that they should not be blamed for the damage they caused because, struggling to survive in an oppressive patriarchal society, they were as psychologically damaged as their children. "Perhaps the mothers are as hospitalized within their marriages as their daughters are within the asylums."[63] Similarly, psychologist Pauline Bart, in her famous 1971 feminist essay "Sexism and Social Science," condemned the mother-blaming tendencies of psychoanalytic psychiatry but then went on to welcome the more recent moves in her time toward system thinking—because such moves would start to hold fathers and other relatives

[58] Arvid Carlsson and Margit Lindqvist, "Effect of Chlorpromazine or Haloperidol on Formation of 3-Methoxytyramine and Normetanephrine in Mouse Brain," *Acta Pharmacol. Toxicol.* 20 (1963): 140–44.

[59] J. M. Van Rossum, "The Significance of Dopamine-Receptor Blockade for the Mechanism of Action of Neuroleptic Drugs," *Arch. Int. Pharmacodynam. Thérapie* 160 (1966): 492; Ian Creese, David R. Burt, and Solomon H. Snyder, "Dopamine Receptor Binding Predicts Clinical and Pharmacological Potencies of Antischizophrenic Drugs," *Science* 192 (1976): 481–3.

[60] E. C. Johnstone, T. J. Crow, C. D. Frith, J. Husband, and L. Kreel, "Cerebral Ventricular Size and Cognitive Impairment in Chronic Schizophrenia," *Lancet* 2 (30 October 1976): 924–6.

[61] Betty Friedan, *The Feminine Mystique* (1963; repr., New York, 2010), 276.

[62] Cited in Ellen Herman, *The Romance of American Psychology: Political Culture in the Age of Experts* (Berkeley, Calif., 1995), 288.

[63] Phyllis Chesler, *Women and Madness* (New York, 1972), 95.

also accountable for the problems of children: "Only recently," she noted approvingly, "have psychiatrists been talking about schizophrenogenic families."[64]

If both science and feminism fall short in helping us understand the fall of "mother blaming" and the rise of "brain blaming," then what additional elements can bridge the gap? My answer takes us back to mothers—but in this case, not the rendering of mothers offered by clinicians and critics, but real mothers speaking in their own increasingly politicized voices. It was these women who, more than anyone else, successfully challenged the portrayal of themselves as bad mothers by invoking the alternative understanding of schizophrenia as a disease of bad biochemistry instead.

Who were these mothers, and what catalyzed their emergence as a voice with which to be reckoned? The answer to the latter question is deinstitutionalization: the radical social experiment of the 1960s and 1970s that had involved releasing enormous numbers of patients from state mental hospitals, including chronically psychotic patients.

Complex economic, political, intellectual, and therapeutic forces lay behind this experiment. Since the late 1940s, a view had grown that mental hospitals were so badly broken that they were more likely to make mentally ill people worse rather than better. Spending years even in an acceptable institution, it was said, tended to encourage dependency, cognitive sluggishness, and lack of initiative.[65] What if it were possible to treat patients in the community, where they could have a meaningful, near-normal life, close to loved ones? Wouldn't that be infinitely better?[66]

Most agreed that it would be, but, before the late 1950s, the idea that severely psychotic patients could be treated in a community setting would have seemed foolhardy. The arrival of new drugs in the mid-1950s, however, changed that thinking. The new view was that, even if the drugs did not cure schizophrenia, they might stabilize many patients sufficiently to allow them to continue to receive care on an outpatient basis.[67]

That was the hope; the reality turned out to be more complicated. Once released, many patients actually went off their medications, did not refill their prescriptions, and either lacked access to a community center (many failed to be built) or failed to show up for appointments at the one to which they had been assigned. Instead, many cycled in and out of emergency rooms, became homeless, or ended up in prison. The lucky ones, though, had parents who tried to step into the breech. And these parents—especially the mothers—found themselves in a crisis of their own as they tried to navigate a mental health care system that no longer seemed to have resources for their children.[68]

The irony here, of course, is that these family caregivers were the same people who had been censured for decades as unloving, failed parents. Now, unexpectedly, they were supposed to be the frontline caregivers of their mentally ill adult children. Some of them began meeting to discuss their sense of frustration and anger. Calling themselves Parents of Adult Schizophrenics, they also talked about the failures of dein-

---

[64] Pauline B. Bart, "Sexism and Social Science: From the Gilded Cage to the Iron Cage, or, the Perils of Pauline," *J. Marriage Family* 33 (1971): 734–45.

[65] Russell Barton, *Institutional Neurosis* (Bristol, 1959).

[66] Albert J. Glass, "Military Psychiatry and Changing Systems of Mental Health Care," *J. Psychiat. Res.* 8 (1971): 499–512.

[67] For a fascinating early conversation about this novel possibility, transcribed and published in the journal *Mental Hospitals* in 1956, see Addison M. Duval and Douglas Goldman, "The New Drugs (Chlorpromazine & Reserpine): Administrative Aspects," *Psychiat. Serv.* 51 (2000): 327–31.

[68] William Doll, "Family Coping with the Mentally Ill: An Unanticipated Problem of Deinstitutionalization," *Psychiat. Serv.* 27 (1976): 183–5; Edward H. Thompson Jr. and William Doll, "The Burden of Families Coping with the Mentally Ill: An Invisible Crisis," *Family Relations* 31 (1982): 379–88.

stitutionalization, their fears for their children, their own feelings of blame, and their inability to get help from the medical system.[69]

And at some point—and this is the important part—they discovered that there existed—and had always existed—an entirely different perspective on what was wrong with their children: that they were sick not because of a failure of love but because of a defect in brain functioning. They made contact with various biological psychiatrists, who gave talks at their meetings and told them that schizophrenia was a disease like any other, and that it was no one's fault, least of all theirs.[70] They were galvanized.

They were also growing in numbers. By 1978, the original Parents of Adult Schizophrenics organization had grown to 200 family groups. More organizations started forming across the country. By September 1979, 100 of these individual groups came together to form the National Alliance for the Mentally Ill (NAMI). By 1982, NAMI had nearly 250 affiliates. It sent representatives to testify before government officials about the need for more funding, better research, and better care for their children. It collaborated with several other health consumer groups to create the first (and, to this day, still the largest) private charity—NARSAD—devoted exclusively to supporting biological research into mental illnesses. It developed a powerful public relations engine that used newspaper, radio, television, and billboard advertisements to spread (in their eyes) correct biological understandings of schizophrenia and (increasingly) other serious mental disorders. Chemistry, not character. Bad brains, not bad mothers. This became the bedrock of NAMI's whole approach to health advocacy, and it was stunningly successful in promoting it.[71] In 1977, one of its founders, Eve Oliphant, stood up before a large crowd of clinicians at the World Congress on Psychiatry and put it to them straight: "We failed to understand why parents of a child with leukemia were treated with sympathy and understanding, while parents of a child with schizophrenia were treated with scorn and condemnation."[72] Put that way, the Freudians and family therapists were hard put to respond.[73]

[69] Cadigan and Murray, "When Medicine Got It Wrong" (cit. n. 2).

[70] "Biochemical Imbalance Cited," *San Mateo Times*, 21 May 1976, 2. See also E. Fuller Torrey, *Surviving Schizophrenia: A Manual for Families, Patients, and Providers*, 5th ed. (1982; repr., New York, 2006).

[71] Agnes B. Hatfield, "The Family Consumer Movement: A New Force in Service Delivery," *New Directions for Mental Health Services* 1984 (1984): 71–79, on 71.

[72] Quoted in the documentary by Cadigan and Murray, "When Medicine Got It Wrong" (cit. n. 2). See also Duval and Goldman, "The New Drugs" (cit. n. 67).

[73] Britain did not experience the same kind of family consumer movement that the United States did, but it did pursue deinstitutionalization policies of its own (catalyzed by the 1962 Mental Health Act). In 1985, an investigative journalist named Marjorie Wallace caused a national furor when *The Times* published her searing three-part series, "The Forgotten Illness," on families with patients suffering from schizophrenia. "Should having a schizophrenic relative be a life sentence for the entire family?" she asked. She went on: "The family cannot win. If they seek help for their disturbed relative they are often told by doctors that they are being over-protective and that unsatisfactory family relationships are to blame. 'When did you last sleep with his father?' is [a] standard question to mothers. But if they close the door on their son or daughter, they not only risk his or her suicide but are also accused of neglect." The articles catalyzed the establishment of a new mental health consumer group in England, SANE—short for Schizophrenia: A National Emergency. Its goal was "to help raise awareness, to help sufferers and their families, and to carry out research for an eventual cure" of schizophrenia (today, its brief is broader). Wallace recruited Charles, the Prince of Wales, as SANE's royal patron, and within six months, she had become its chief executive. It remains a major force in the United Kingdom today. For the article series, see Marjorie Wallace, "The Forgotten Illness," *The Times*, 16–18 December and 23 December 1985; for discussion of its impact, see Linda Christian and Molly Derrett, "The Forgotten Illness," *The Times*, 30 January 1986.

Then, in 1981, social workers Carol Anderson and Gerard Hogarty published impressive outcome findings for a new approach to family therapy in schizophrenia that they called psychoeducation. In sharp contrast to the dominant family therapy approaches of the time, their therapy concentrated on building emotional alliances with family members. It turned out that, when families were approached sympathetically and taught skills for coping better with the profound emotional and interpersonal challenges of being the frontline caregivers of a schizophrenic child, there was a significant reduction in the relapse rates of the children themselves. There might or might not be any scientific evidence to support the idea that families could cause schizophrenia (and, in fact, Anderson and Hogarty were skeptics of this idea). What was clear, though, was that blaming as a practical approach—especially in the 1980s, a time when so many more families were directly involved in the care of such patients—did not work.[74]

The 1980s saw more and more signs of a shift in the dominant cultural understanding. In 1982, the first handbook, written by biological psychiatrist E. Fuller Torrey, designed specifically to help patients' families "survive" schizophrenia was published.[75] In 1984, a multipart American television documentary on the brain was widely viewed. The program on schizophrenia, called simply "Madness," included scenes of interactions with severely incapacitated schizophrenic patients and discussions of the likely biological basis of their profound disabilities. At the same time, it showcased presumptively normal, loving families struggling to come to terms with the shock of their children's terrible illness. In the words of one of the psychiatrists interviewed in this documentary, "They are like people who have been struck by lightning." Clinicians apologized on the air on behalf of their profession for having once supposed that families like these could have caused this disorder. There was not a "shred of evidence," they said, in support of this idea.[76]

Finally, in 1986, an article appeared in a small American regional newspaper, the *Oelwein Register*. "Every Family's Story" told the tale of an unnamed generic family, one that any reader might identify with, who found themselves dealing with a bewildering tragedy. Their teenage son, suddenly and without any obvious cause, was losing touch with reality, believing he was being persecuted by witches but that he had paranormal powers to resist them:

> They could not believe that this was happening to Mike . . . to them. Their kid wasn't really—the word did not come easily—really . . . crazy. Craziness was something that happened to other people, to awful people, to families that were abnormal, to families that treated their kids badly. It couldn't happen to them. Except that it was happening to them, in a nightmare they couldn't wake up from. Maybe they had made a mistake in thinking he was mentally ill. After all, who among them knew anything at all about such things. They reconsidered the evidence. Was Mike really crazy? There was no getting around it. He was.[77]

[74] Carol M. Anderson, Gerard Hogarty, and Douglas J. Reiss, "The Psychoeducational Family Treatment of Schizophrenia," *New Directions for Mental Health Services* 1981 (1981): 79–94. See also Judith A. Cook, "Who 'Mothers' the Chronically Mentally Ill?" *Family Relations* 37 (1988): 42–9.

[75] Torrey, *Surviving Schizophrenia* (cit. n. 70).

[76] "Madness," *The Brain*, episode 7, WNET (New York, 1984).

[77] Maryellen Walsh, "A Family Meets Schizophrenia," *Oelwein (Iowa) Register*, 7 October 1986, 12.

The mothers of schizophrenic children had always insisted that they and their families were normal and that they could not understand why their son or daughter had suddenly descended into madness. By the mid-1980s, it seemed that psychiatry and society were finally inclined to believe them.

CONCLUSION

The story of the mother who literally drove her offspring crazy was the radical, ragged edge of a larger history that entailed first a medicalization and then a politicization of the emotional relationship between mothers and their offspring. The larger story began with psychiatry validating mother love as critical to the healthy development of children; it moved to a focus on ways in which unhealthy forms of mother love might thwart the development of troubled youngsters; and then it moved again, to increasingly alarmist claims about ways in which unhealthy forms of mother love put democracy itself at risk.

In the case of the specific story of mother love and mental illness, however, something novel happened: a new politicization of mother love; but this time, one transformed into a movement that would force a reorientation of the field back to its biological roots and catalyze a crisis of conscience over its former casual cruelty toward suffering families. On its own terms, this final politicization of mother love was an astonishingly successful strategy. By the 1990s, psychoanalytically oriented clinicians who once were seen by their peers as serving the cause of love (for their patients) had been transformed by new critics into misogynists who, wittingly or not, had sown hatred (of parents). Apologies were made. Fervent desires for a new beginning were affirmed.[78]

To be sure, none of this meant that blame and guilt vanished from the world of schizophrenia entirely. The biological framing of schizophrenia brought its own kind of guilt, as parents now wrestled with the idea that a genetic vulnerability toward schizophrenia ran through their family line.[79] Nevertheless, the fact remains that in our own time even the most moderate psychosocial approaches to schizophrenia have become, in the eyes of many, not only scientifically wrong but positively unethical. Most people take for granted that the gold-standard solution to both the suffering and the stigma of schizophrenia is more and better research into its neurobiological basis, and improved drugs. They fail to realize, however, that the genetics and brain science that provided the original intellectual justification for a turn back to a strict biological approach to schizophrenia was always, at best, provisional and incomplete. There was always more to be said and done. As anthropologist Tanya Luhrmann has written in a recent (2012) article:

> In the early days of the biomedical revolution, when schizophrenia epitomized the pure brain disorder, the illness was said to appear at the same rate around the globe, as if true

[78] See, e.g., Harriet P. Lefley and Dale L. Johnson, *Families as Allies in Treatment of the Mentally Ill: New Directions for Mental Health Professionals* (Washington, D.C., 1990); Johnson, "Quality Services for the Mentally Ill: Why Psychology and the National Alliance for the Mentally Ill Need Each Other," in *New Directions in the Psychological Treatment of Serious Mental Illness*, ed. Diane T. Marsh (Westport, Conn., 1994), 31–49.

[79] Felicity Callard, Diana Rose, Emma-Louise Hanif, Jody Quigley, Kathryn Greenwood, and Til Wykes, "Holding Blame at Bay: 'Gene Talk' in Family Members' Accounts of Schizophrenia Aetiology," *BioSocieties* 7 (2012): 273–93.

brain disease respected no social boundaries and was found in all nations, classes, and races in equal measure. This piece of dogma was repeated with remarkable confidence from textbook to textbook, driven by the fervent anti-psychoanalytic insistence that the mother was not to blame. No one should ever have believed it. As the epidemiologist John McGrath dryly remarked, "While the notion that schizophrenia respects human rights is vaguely ennobling, it is also frankly bizarre." In recent years, epidemiologists have been able to demonstrate that while schizophrenia is rare everywhere, it is much more common in some settings than in others, and in some societies the disorder seems more severe and unyielding. Moreover, when you look at the differences, it is hard not to draw the conclusion that there is something deeply social at work behind them.[80]

We should not have to choose between biological and social understandings. It should be axiomatic that all human beings, even those with biological vulnerabilities that put them at risk for schizophrenia, are embedded in a social and interpersonal world that gets under their skin and affects them. By and large, though, people fail to see this, and they fail to see this in part because of legacies from the history recounted here. On the one side, the history of mother love and mental illness offers a rich opportunity for new kinds of intellectual cross-fertilization between historians of science and historians of the emotions. On the other side, it is also a history that has been difficult to tell well because we do not stand outside of it emotionally ourselves; it still stirs strong passions. Historians of emotions may understand that this is one of the occupational hazards of their trade; it is a good lesson for historians of science and medicine as well.

[80] Tanya Marie Luhrmann, "Beyond the Brain," *Wilson Quart.* 36 (2012): 28–34, on 31.

EMOTIONS INTO PRACTICE

Affected Doctors:

Dead Bodies and Affective and Professional Cultures in Early Modern European Anatomy

*by Rafael Mandressi**

ABSTRACT

From the end of the thirteenth century, when the practice of human anatomical dissections emerged in Europe, the dead body became part of the cultural economy of knowledge. This had epistemic, technical, and social consequences, in which the affective dimension played a crucial role. The type of manipulations the corpse underwent brought into play affective phenomena of unusual intensity. To a great extent, anatomy owed its repertoire of gestures, spaces, and instruments to the need to control these affects, and this repertoire contributed to the discourse that shaped the professional identity of anatomists. Rather than being simply knowledge trapped in a web of preexisting sensibilities, anatomy was, in early modern Europe, a locus where affective cultures were produced and negotiated among several professional and social groups.

Opening a dead body and exploring its interior in order to obtain knowledge is neither a historically neutral nor a trivial act. The anatomical dissection of human corpses was never formally prohibited, nor was it hampered by inviolable religious or cultural obstacles.[1] However, since its emergence in Europe toward the end of the thirteenth century, the practice had special connotations that endowed it with a remarkable status, full of fear and fascination. To slit open inert tissue and root around inside is, indeed, a grave action. And a daring one. The engravings that adorn the title pages of anatomy treatises capture this gravity: stretched on the dissecting table, the corpse offers its crude bareness to the anatomist's scalpel and the voracious gaze of the spectators. From the 1450s on, and even more so in the following century, dissections became more common in European universities, but the anatomical act never became an innocuous event: a glance at the iconographical production itself is enough to show that the circumstances were always extraordinary, that one is attending an exploration from which one does not return unscathed. The cadaverous matter hides *mirabilia* to which we cannot be indifferent—just like death.

* Centre National de la Recherche Scientifique, Centre Alexandre-Koyré, 27, rue Damesme, 75013 Paris, France; rafael.mandressi@cnrs.fr.
[1] Rafael Mandressi, *Le Regard de l'anatomiste: dissections et invention du corps en Occident* (Paris, 2003), 19–60.

From the dawn of the early modern period, anatomical research, based on the methodical and systematic dismemberment of human corpses, established a system of knowledge that was intimately associated with a set of strong affective consequences that had to be contained and controlled. Horror, shame, repulsion: implicit or explicit regulations were aimed at combating them. My intention in this essay is to analyze these regulations, taking them not as an anecdotic or accessorial feature in the study of training in the knowledge of anatomy, but as one of its constitutive aspects, one of its intrinsic characteristics. In other words, the techniques for the control of the inherent affects of the manipulation of dead bodies fully comprehend the ways in which anatomical learning is produced and thus influence how the latter is shaped. We shall also see how the discursive and operative strategies adopted by the anatomists became one of the elements that forged their professional identity. On this point, it can be said that the anatomy of the early modern period created an affective culture among its promoters, and its codification was decisive in the consolidation of a professional culture. This affective culture, understood as a set of interrelated techniques, gestures, and values associated with the management of affectivity, shows a remarkable stability over time. As we shall see through our sources, its basic elements do not change significantly throughout the early modern period: already present at the beginning of the sixteenth century, they can be found in anatomy texts at the end of the eighteenth century. This continuity is perhaps one of the most noticeable features that emerges from the study of the affective dimension of anatomical practice and discourse, along with its geographical scope. Indeed, the study of early modern European anatomical literature also makes clear that the affective culture of the discipline does not vary if we consider the Italian, French, or British contexts. In this regard, the aspects under study represent a *longue durée* phenomenon, covering, in spatial terms, Western Europe as a whole.

The affects brought about by anatomical practice are not unique or even exclusive to it, but both the actions and the objects that provoke them, alongside the circumstances in which they appear and their means of regulation, do involve, as a whole, a specificity. In this case, this specificity, together with a specialized knowledge, an institutional sense of belonging, and an exercise of an activity, defines a group of individuals who see themselves as professionals and are socially recognized as such. From the early sixteenth century, anatomy began to hold a central position in the theoretical and technical knowledge of physicians and surgeons, and by the early seventeenth century a new professional group—anatomists—had emerged. Anatomy was also a key component of their training, as dissections were a pedagogical tool: apart from delving into the structure of the human body, the opening of corpses was carried out with a view to teaching students. The process of teaching and learning, then, included passing on the values, the rules, the demands, and the resources of an affective culture implied in the manipulation of dead bodies.

The business of anatomy was not limited to a specialized community. In the early modern period in Europe, anatomy, as a social and cultural phenomenon, was at its peak, involving a much larger and more diverse public than university physicians, surgeons, and students. This public facet, within the analysis of the affective phenomena linked with the practice of dissection, poses other important issues. I will mention two. The first issue refers to the social status of the corpse as an object of knowledge that is determined in a cultural framework where the affective phenomena have a crucial role to play. The second issue, closely linked to the first, pertains to the ways in which the dead body was given visibility within a public space—before hundreds of spectators in the anatomy amphitheaters or by means of its representation in images—which show the rules governing the management of affectivity in relation to the exhibition of the anatomized corpse.

In what follows, I will analyze the rhetorical constructions of the scientific uses of the corpse as an affective challenge to be overcome, the specific answers anatomists developed when confronted with this challenge, and how these answers played in the context of the social consumption of dissections. My aim is to show, concisely, the core ideas of what could be called the affective economy of anatomy—that is, the actual organization of the constitutive aspects of its affective culture—in the early modern period, so as to explain how it shapes the identity of the discipline in its three fundamental aspects, that is, epistemic, professional, and social.

"CADAVERA IMMUNDA SUNT"

The wait for the delivery of the bodies, the rotting that lurked and followed the rhythm of dissection, the growing stench, and the practical contiguities between the spaces of death and the spaces of science all accompanied the activity of anatomists for five centuries. Confronted with the flesh and blood of others, prying through touch into the very materiality of the corpse, anyone who devoted themselves to digging into its livid reality realized the price to be paid for this raw research material: access to knowledge demanded that one lean into the abyss of human remains—a stinking abyss, as the Italian anatomist and philosopher Alessandro Benedetti (d. 1512) stressed in his *Historia corporis humani sive Anatomice* in 1502. In order to study the bones, Benedetti comments, some boil half a corpse at a time in a large pot, after removing the muscles and flesh; others, conversely, scrutinize the bodies by visiting tombs, "not without nausea" [*non sine nausea*], because corpses "are disgusting" [*immunda sunt*].[2] Around the same period, Leonardo da Vinci (1452–1519) cautioned that the anatomist's task is awful and often hard to face. He writes in his *Quaderni*, "though possessed of an interest in the subject you may perhaps be deterred by natural repugnance, or, if this does not restrain you, then perhaps by the fear of passing the night-hours in the company of these corpses, quartered and flayed and horrible to behold."[3] Another contemporary of Benedetti and Leonardo, Michelangelo Buonarroti (1475–1564), having devoted himself to anatomy for a long time,[4] "stopped cutting bodies up" because, according to his disciple and biographer Ascanio Condivi (1524–74), "having manipulated them at length had upset his stomach so much that he could no longer eat or drink properly."[5]

There is no shortage of statements of this kind in the anatomical literature or in other writings that refer to anatomy throughout the early modern period. In the second half of the eighteenth century, Louis XVI's *Premier médecin*, Joseph Lieutaud (1703–80), declared, in his *Anatomie historique et pratique*, that anatomy, "so nec-

---

[2] Alessandro Benedetti, *Historia corporis humani sive anatomice*, ed. Giovanna Ferrari (Florence, 1998), 326–8. My translation.
[3] Leonardo da Vinci, *Leonardo da Vinci's Notebooks*, ed. Edward MacCurdy (New York, 1923), 84.
[4] "Infinite volte fece notomia," indicates Giorgio Vasari (1511–74) in his *Vite de' piu eccellenti pittori, scultori e architettori*, 2d ed. (Florence, 1568), 3:774.
[5] Ascanio Condivi, *Vita di Michelangelo Buonarroti raccolta per Ascanio Condivi da la Ripatransone* (Rome, 1553), 41.

essary for physicians," should "hold high rank among human learning." However, he adds, "it is often neglected, and . . . few devote themselves to it; this is not surprising, if we remember that most people lack the time and the corpses, not forgetting how terrible and repellent such work is, nor the many other obstacles that one cannot always be sure to overcome."[6] Ten years later, his colleague and compatriot, Félix Vicq d'Azyr (1748–94), states more emphatically:

> Anatomy is perhaps, of all the sciences . . . the one whose study presents the greatest difficulties: research into it not only lacks that pleasure which attracts, moreover it is accompanied by circumstances that repel; broken and bloody members, filthy unhealthy emanations, the horrid apparatus of death, are the objects presented to those who cultivate it. . . . It is only by descending into tombs and defying the laws of men, in order to discover those of nature, that the anatomist has painfully and perilously laid the foundations of his useful knowledge.[7]

The artist who accompanies the anatomist in this disagreeable task also deserves recognition. Hence, Vicq d'Azyr praised his assistant, Briceau, who was deterred "neither [by] the dangers of a prolonged stay in the midst of putrid exhalations, nor the repugnance of a spectacle to which he was not accustomed; he corrected and began his drawings anew with extreme docility as often as necessary in order to imitate nature."[8]

Many other similar comments could be added to the above paragraphs. The descriptions use many adjectives, and their emphasis on distaste, disgust, and fear is a leitmotif whose expression can be found in writings from an earlier period than that of Leonardo or Benedetti. In his treatise *De nobilitate legum et medicinæ*, written in 1399, the Florentine humanist Coluccio Salutati (1331–1406) concluded that showing human entrails "per anathomiam" is a "horrid, disgusting thing" that cannot be viewed without feeling horror and shedding tears, even though it is most worthy of being known [*quæ licet dignissima scitu sit*].[9] Nevertheless, it would be hasty to conclude that the recurrence of these opinions and valuations shows unanimous feelings and attitudes that are similar among those who practice anatomy and those who comment on any of its features without practicing it—as far as we know, Coluccio Salutati never carried out a dissection. It is advisable, then, to examine more deeply the functions these statements may have, depending on who makes them; in this way, we may be able to establish the role played, in each case, by the affects to which they refer.

In his description of the methods used to remove the soft tissues attached to the bones, in the same text where he mentions the "nausea" caused by corpses, Benedetti tells of seeing a doctor chew a piece of human flesh after boiling a corpse. When asked about the taste, he said that it tasted "of beef" ("Bubulæ"). Benedetti describes the scene without adding any comments or referring to either his own or others' par-

---

[6] Joseph Lieutaud, *Anatomie historique et pratique . . . Nouvelle Edition, Augmentée de diverses Remarques historiques et critiques, et de nouvelles planches; par M. Portal* (Paris, 1776), 1:vii–viii.
[7] Félix Vicq d'Azyr, *Traité d'anatomie et de physiologie avec des planches coloriées représentant au naturel les divers organes de l'Homme et des Animaux* (Paris, 1786), 1:1.
[8] Ibid., 12.
[9] Coluccio Salutati, *Tractatus insignis, et elegans . . . de Nobilitate Legum, et medicinæ in quo terminatur illa quæstio versatilis in studijs: utrum dignior sit scientia legalis, vel medicinalis* (Venice, 1542), chap. 12, fol. Eiiijr.

ticular reactions to such behavior. However, in the following sentence, he says that the corpses "are disgusting." They are disgusting when examined in tombs—Benedetti mentions this precise situation—but clearly less so when boiled, that is, processed following a certain technique that may involve steeping them in running water or dissolving them in quicklime.[10] The anecdote also shows the capacity "the doctor" had to vanquish his repugnance, even priding himself on it. In a few short lines, the Italian anatomist's text does several things at once: it introduces an affective description of the fresh corpses. In other words, it suggests that the transformation of the corpse by the proper means leads to the neutralization of those affective properties, and shows, in one episode, how a physician acquires the capacity to control his affective responses through the exercise of his profession. Even in its extreme synthesis, Benedetti's brief passage in his *Historia* reveals more varied and complex rhetorical uses than the mere inventory of the unpleasant feelings that the dead body gives rise to in the person who dislocates it, dismembers it, "destroys" it. According to the seventeenth-century physician Jean Riolan Jr. (1580–1657), when anatomists dissect, they follow a certain order "in the destruction of the body."[11]

The vocabulary is crude, even violent, and Riolan's formula, which associates dissection with destruction, is far from extreme. This persistent harshness of language becomes an important ingredient in the production of heroic narratives of anatomy in the eighteenth century. The more repellent the task, the greater the effort, and this is the merit of overcoming repugnance in the name of the progress of a science that requires its practitioners to contend with dead matter. As Vicq d'Azyr writes, the mission to find "useful knowledge" compels the anatomist to coexist with the "horrid apparatus of death," to support "filthy unhealthy emanations," to expose himself to "putrid exhalations," and to be in contact with "broken and bloody members." Anatomy, in the end, is "painful and perilous," but if it is seen as a mission, if the thirst for knowledge is sufficient, then the dread and repulsion can be endured.

Frequent attendance at anatomical operations led to a familiarity that helped anatomists deal with feelings of dread and repulsion. It is much more comfortable, says Lieutaud, in his *Essais anatomiques*, to consult others who have written about anatomy than to practice dissection: "it is less onerous to read a book than to dig in a corpse; the former is not at all repulsive, and only with difficulty can one become familiar with the latter."[12] This affective learning is what the young Goethe (1749–1832) proposed to begin in Strasbourg around 1770. While he was studying law there, Goethe decided to attend the courses of Johann Friedrich Lobstein (1736–84), anatomist, surgeon, and rector of the university, with the aim of "freeing myself from any apprehension of repulsive sights. And I actually succeeded so well, that nothing of this kind ever made me lose my self-possession." Anatomy, as Goethe later wrote in his autobiography, "was of twofold value to me, as it taught me to tolerate the most repulsive sights, while satisfying my thirst for knowledge."[13] The familiarity with the corpse,

[10] Benedetti, *Historia corporis humani* (cit. n. 2), 326–8.

[11] "Similiter nos in distructione corporis per administrationem anatomicam"; Jean Riolan Jr., *Encheiridium anatomicum et pathologicum: In quo ex naturali constitutione partium, recessus à naturali statu demonstratur, ad usum Theatri Anatomici adornatum* (Paris, 1648; Leiden, 1649), 417.

[12] Joseph Lieutaud, *Essais anatomiques, contenant l'histoire exacte de toutes les parties qui composent le corps de l'homme, avec la manière de disséquer* (Paris, 1742), v.

[13] Johann Wolfgang Goethe, *Poetry and Truth: From My Own Life*, trans. Minna Steele Smith (London, 1908), 1:334.

which was what Goethe attempted and, as he stated, achieved, is one of the attributes of any good anatomist. Bernard de Fontenelle (1657–1757), Perpetual Secretary of the Royal Academy of Sciences in Paris, in his eulogy of Jean Méry (1645–1722), emphasized that Méry, who became the *Premier chirurgien* of the Hôtel-Dieu in 1700, dedicated all his time to anatomy, to such an extent that "he only dealt with the dead." During his surgical studies in Paris in the 1660s, Méry, "not satisfied with his diurnal exercises, . . . cunningly stole a dead body when he could, took it to his bedroom, and spent the night dissecting it in great secrecy."[14] In a eulogy to another member of the academy, Alexis Littré (1658–1725), Fontenelle says that he advanced in his studies and became a great anatomist "thanks to the link he made with a surgeon in La Salpêtrière, who has all the corpses in the hospital at his disposal. Littré shut himself away with this surgeon during the winter of 1684, which fortunately was very long and very cold, and together they dissected over 200 corpses."[15]

In the early modern period then, a system of topoi—a set of commonplaces—about the affects ran both through the anatomical-medical literature sensu stricto and through numerous other texts dealing with this discipline that based its research procedures on the "destruction" of corpses. We shall return to this second category later. In the anatomical treatises, compendiums, and commentaries written by physicians and destined to circulate in the public sphere, accounts of the pathos dissectors experienced fulfill, as we have seen, diverse functions that change in intensity and emphasis over time. In spite of these transformations, there is a sense of coherence throughout the anatomy texts produced from the late fifteenth century at least until the close of the eighteenth century. During these three centuries, leading anatomical texts incorporated affect into the defining characteristics of medical and natural philosophical knowledge of this period.[16] This discourse created a public image of anatomy that underscored the affective difficulties and obstacles, methods to overcome these obstacles, and the overall thanklessness of the commerce with dead bodies. In fact, it is a public expression, addressed to a readership made up of physicians, medical students, philosophers, and, more broadly speaking, the cultured curious who were interested in knowing the "secrets" and the "marvels" of the human body. Apart from this printed material, there are, however, other texts not destined for public circulation that present a different perspective.

WRITING THE EXPERIENCE

Between 15 and 28 January 1540, Andreas Vesalius (1514–64), *chirurgiæ explicator* in Padua since December 1537, was invited by the students of the University of Bo-

---

[14] Bernard de Fontenelle, *Eloge des académiciens: avec l'histoire de l'Académie royale des sciences en M. DC. XCIX; avec un discours préliminaire sur l'utilité des mathématiques* (The Hague, 1740), 2:166, 163–4.

[15] Ibid., 237.

[16] According to Andrea Carlino, anatomy is, with reference to medicine, "surplus" knowledge that produces "conoscenze in sovvrapiù." See Carlino, *La fabbrica del corpo: Libri e dissezione nel Rinascimento* (Turin, 1994), 10. On the relation between medicine, anatomy, and natural philosophy, see also Jerome J. Bylebyl, "The Medical Meaning of Physica," *Osiris* 6 (1990): 16–41; Mark D. Jordan, "The Construction of a Philosophical Medicine: Exegesis and Argument in Salernitan Teaching on the Soul," *Osiris* 6 (1990): 42–61; Ian MacLean, *Logic, Signs and Nature in the Renaissance: The Case of Learned Medicine* (Cambridge, 2001); Rafael Mandressi, "Dire la nature: La médecine et les frontières du surnaturel (XVIᵉ–XVIIᵉ siècles)," *Corpus* 54 (2008): 141–82; Mandressi, "Médecine et discours sur l'homme dans la première modernité," *Rev. Syn.* 134 (2013): 511–36.

logna to carry out a series of anatomical demonstrations. We know of these demonstrations thanks to the class notes taken by one of these students, the Silesian, Baldasar Heseler (d. 1567). Here we see that neither the anatomist nor the students were in complete agreement with the kind of behavior described in the great majority of printed texts, nor did they express exactly the same sentiments. On the morning of 14 January, Heseler wrote, "our anatomical subject had been hanged. He was a very strong, muscular and fat man, perhaps 34 years of age."[17] The dissections were to begin the following day, in the hall where the rector of physicians [*Rector medicorum*] was elected. In spite of the recommendations of the beadle Pellegrinus, over 150 students headed there, said Heseler, "in great disorder, as the mad Italians do."[18] The corpse was lying on the dissection table, "cut up and prepared beforehand, already shaved, washed and cleaned."[19] The first demonstration was devoted to the skin, to flaying the body and dissecting the abdominal muscles; in the second demonstration, on the afternoon of the same day, Vesalius showed the spectators the abdominal organs and later anatomized the muscles of the left arm. "Certainly this was very beautiful to see," noted Heseler.[20] On 16 January, the "subject" was already stinking, and Vesalius "said, because the intestines of this body already stink, therefore, I cannot demonstrate them to you now, but they will be cleaned and inflated, and tomorrow you will see them better." Meanwhile, he had to be content with a dog.[21] In spite of the stench and the advancing decomposition of the intestines, on 18 January, in the evening, in his seventh demonstration, Vesalius included the stomach and the liver.[22] On 19 January, the corpse was "too dried and wrinkled" to be used to show the veins, arteries, and nerves. To do so another body was needed, which would not take long because, according to the anatomist: "tomorrow we shall have another body—I believe they will hang another man."[23] However, they had to wait three days: "our next subject was hanged this morning (there are two of them)," Heseler wrote on 22 January.[24] For the anatomy of the brain, which was demonstrated after supper, Vesalius used the head of the first subject; it was all that was left of him. Once the scalp had been removed, Vesalius extracted and showed the students the top half of the cranium, previously sectioned and separated from the brain matter. "Then the Italians cried: O what a beautiful cap," Heseler recorded. This operation uncovered the *dura mater*. The anatomist removed it together with the *pia mater* and completed the demonstration, "inserting the fingers of one hand through the suture or partition of the brain."[25] The two new subjects should have been available on 23 January. However, when the students arrived to attend the sixteenth demonstration, there was only one body. Another anatomist, Jacobus Erigius, had appropriated one of them so that he could dissect it himself. The rector had to intervene and ordered the immediate return

[17] Baldasar Heseler, *Andreas Vesalius' First Public Anatomy at Bologna, 1540: An Eyewitness Report by Baldasar Heseler medicinæ scolaris, Together with His Notes on Matthaeus Curtius' Lectures on Anatomia Mundini*, ed. Ruben Eriksson (Uppsala, 1959), 71.
[18] Ibid., 85.
[19] Ibid., 87.
[20] Ibid., 95, 97.
[21] Ibid., 115, 117.
[22] Ibid., 147.
[23] Ibid., 177.
[24] Ibid., 209.
[25] Ibid., 219, 221.

of the stolen corpse. Unfortunately, Erigius had already begun the dissection with an assistant. "A butcher," exclaimed Vesalius on realizing the damage caused by the incisions of his dishonest colleague. "And he exposed several other incongruities and mistakes of those dissectors. And all laughed mightily at this 'anatomicus'!"[26]

Heseler and his companions did not seem especially fearful or disgusted by the three "subjects" that the Bolognese authorities offered to the university for the purposes of instruction. Neither their dismemberment, nor the putrefaction, nor the smells caused any repugnance or squeamishness worthy of mention by Heseler in his notes. Nor did Vesalius's manipulations, when he used candles to burn the skin of the first body, razors to remove the skin and fat from all three, saws to cut bones, or his own fingers to dig into the cerebral matter or take hold of the intestine with his left hand, unwind it until he found the rectum, and then grasp the latter with his right hand and remove the whole lot. In contrast, Heseler did point out his own admiration for the anatomized body, which "was very beautiful to see," and the appreciation of the delighted Italian students on seeing the lid of the cranium sawed off by the anatomist. He also tells of the racket that repeatedly surrounded the dissection, so much so that, on 22 January, Vesalius "was very confused, upset and bewildered owing to the noise and disorder that the students then made."[27] The scenes found in this Silesian student's notes do not show an atmosphere of quiet and seclusion; on the contrary, when Vesalius ridicules his colleague Erigius because of a half-dissected corpse, the reigning mood is hilarity.

Naturally, it would be unwarranted to conclude that the reality of the anatomical work completely contradicts the descriptions that the anatomists themselves give in their printed treatises. Rather than belying these descriptions, the manuscript sources help to enrich them, in particular when dealing with the range of possible attitudes in a situation of close physical and sensorial proximity to dead objects. We must bear in mind the wavering between reactions of aversion, counteracted or not by willpower, and of euphoria. And we must not forget that whether the text is a printed book, notebook, or other type of document, the act of writing implies translating the facts and experiences into verbal forms in such a way that what we use as sources will never offer these facts or, even less so, these experiences, but rather a discursive formalization the contents of which are the result of a set of options that fit in with the intentions, resources, and circumstances of the writer. From this perspective, the nature of the texts will, sometimes substantially, vary the elements narrated and the style of narration. If there is a discrepancy between one and the other, rather than using one to belie the other, it is important to be aware of the conditions of enunciation that direct the selection and the expression of affective phenomena.

One example of a personal diary is that of the Swiss physician Félix Platter (1536–1614) during his years as a medical student in Montpellier in the mid-sixteenth century. Just because it is a private piece of writing, it must not automatically be considered of particular value, superior to that of a university treatise; it is more interesting to identify how the elements connected with the topic of the corpse and the affects we have found in printed literature are shown. Among other points, a diary allows more free evocation of illicit episodes, such as clandestine dissections implying grave rob-

[26] Ibid., 221, 223.
[27] Ibid., 221.

bing, usually from cemeteries.[28] Instead of treating these episodes as anecdotes—which medical historiography has done too often—they may be analyzed as evidence of what has earlier been called affective learning. Platter writes: "Not only did I never miss any of the dissections of men or animals held at the college, but I also participated in all the autopsies of corpses that were done in secret, and I went so far as to wield the scalpel myself, despite the revulsion I had earlier felt. I even placed myself in danger, with other French students, to obtain subjects for myself." The group slipped into the cemeteries of the cloisters, stole a corpse, and then took it to the house of a certain Gallotus, bachelor in medicine, who lent his house for the dissection of the booty. "Watchmen," adds Platter, "informed us of the burials and at night led us to the grave."[29] The dissections in the homes of students or masters were not necessarily clandestine, however. In December 1650, three criminals were executed in Paris. According to Guy Patin (1601–72), the body of one of these criminals was requested by Renier, a professor on the university faculty, in order to carry out "surgical operations in his home." Patin, at that time dean of the faculty of medicine in Paris, signed the request that allowed Renier to acquire the corpse, in which he discovered "something extraordinary, the liver was on the left side and the spleen on the right. Everyone went to see this peculiarity, including [Jean] Riolan who is delighted to have seen it."[30]

Riolan was "delighted" to have seen a corpse with the organs reversed, at a duly authorized home dissection. Platter, having exposed himself "to more than one danger," wielded the scalpel in Gallotus's home, or in the Augustinian convent, in order to get over the disgust he had felt at the beginning of his studies. Perhaps someday, by virtue of training, he would feel "delighted" as did old Doctor Riolan. These two cases, with a hundred years between them, allow us to again raise the issue, in a different context and situation, of affective learning and its role in the acquisition of important characteristics of the professional culture of physicians and anatomists. In this sense, these privately written stories are convergent, leaving aside the different discursive registers—few oratorical precautions, an absence of the "heroic" tone and of claims of the dissectors' abnegation—with what can be read in the printed material in public circulation. Must we therefore think that Baldasar Heseler's notes are an anomaly in this sense? Other passages in Platter's diary lead us to answer in the negative because they show the same unconcerned excitement found in the Bologna student.

TECHNIQUES FOR DISSECTION AND TECHNIQUES FOR CONTROL

On 10 November 1555, in Montpellier, Platter attended an anatomy session "where an old woman who died of apoplexy was dissected. On opening the bones of the cra-

---

[28] This does not mean that some great treatises do not also refer to nocturnal outings to cemeteries. Andreas Vesalius tells how his eagerness to perfect his anatomical knowledge led him to spend a whole night, in 1536, outside the gates of Louvain: he had gone to the cemetery late at night, in the company of his friend and fellow student Reiner Gemma Frisius, to seize corpses that he had taken down from the gallows. See Vesalius, *De humani corporis fabrica* (Basel, 1543), 161–2.

[29] *Félix et Thomas Platter à Montpellier, 1552–1559, 1595–1599: notes de voyage de deux étudiants bâlois, publiées d'après les manuscrits originaux appartenant à la bibliothèque de l'Université de Bâle, avec deux portraits* (Paris, 1995), 90–1. Félix Platter relates several of these expeditions, which were carried out in 1554 and 1555, at length.

[30] Guy Patin, *Lettres choisies de feu Mʳ. Patin . . . , Augmentées de plus de trois cens Lettres cette derniere Edition* (The Hague, 1707), 1:152. Jean Riolan the younger (1580–1657) was one of the leading anatomists in Paris in the first half of the seventeenth century.

nium and the brain casing, the brains spilled out like a starchy porridge and overran the face."[31] A year earlier, in the clandestine dissection of two bodies that he had dug up with his companions in the cemetery of the convent of Saint-Denis in Montpellier, the autopsy "revealed serious damage: the lungs were rotten and gave off a frightful stench, despite the vinegar we sprinkled them with."[32] The reference to the "frightful-ness" of the smell is, in these two tales—normal dissection in one case and clandes-tine in the other—the only sensorial point that is negative. In both episodes, however, illness and decomposition do increase the repulsive characteristics that a corpse may have. This is why the cause of death is important when choosing the body. In his an-atomical demonstrations in 1540, Vesalius was forced to show the Bologna students the larynx of an ox, as that of the hanged human subject had been destroyed by the noose. Thus, individuals will be chosen who have "drowned in water, rather than [those who have been] strangled or decapitated or killed by a wound or by some dis-ease," states Jacques Dubois (1478–1555), one of Vesalius's masters in Paris in the 1530s. In individuals who have been asphyxiated, he adds, all the parts are complete, while in "those who were strangled, the whole neck is damaged, and the head and thorax are so full of blood, that you will not easily explore those parts." As for the decapitated, obviously "the continuity of several parts has been lost, and moreover, the great shedding of blood makes the veins and arteries flaccid, and so they cannot be seen properly." Something similar occurs in "those who have died of a great wound," while "those who die of disease, often tend to have several parts that are rot-ten or so altered in comparison to their natural state, that you will neither wish to look at them nor be able to do so due to the stench."[33]

Platter and his companions sprinkled vinegar on rotting body parts to counteract the fetid emanations. There were other methods, such as those praised by the Parisian doctor and printer Charles Estienne (d. 1564) in 1545: "Incense or other good smells must not be omitted to avoid the corrupt vapors affecting those who are present."[34] The perfumes were used to alleviate nausea but were also part of the "small cares that may preserve from grave infirmities," as described by Joseph Lieutaud: a hand-kerchief soaked in an odorant liquid helped anatomists avoid inhaling pernicious fumes.[35] It is not, then, merely a question of disgust, but also of danger, an aspect that is further emphasized in the eighteenth century. Dead matter can be lethal, so clean-liness is essential, although it is difficult to maintain during a dissection. Even though the corpse is washed, shaved, and cleaned, it does not take long to again exude; the fluids and excrements continue to escape from the cavities and canals. Some soft parts suddenly spill out—remember the brain of the old woman who died of apoplexy that Platter compared to "porridge." Half-dissected corpses must never be left, warns Lieutaud, "to continue later, without having washed and dried them well: without this precaution there are no guarantees against infection. The guts should also be removed after the first opening of the lower belly, leaving only a small portion of the rectum, which must be emptied and carefully washed, before tying up the cut end. Eau de vie

---

[31] *Félix et Thomas Platter* (cit. n. 29), 123.

[32] Ibid., 94.

[33] Jacques Dubois, *In Hippocratis et Galeni physiologiæ partem anatomicam Isagoge* (Paris, 1555), fol. 57v.

[34] Charles Estienne, *La dissection des parties du corps humain, divisée en trois livres* (Paris, 1546), 372.

[35] Lieutaud, *Anatomie historique* (cit. n. 6), 2:2.

should not be skimped on; it is a good idea to sprinkle the uncovered parts with it, from time to time."[36]

Together with the tools for cutting and others that help to dismember a body—"razors, scalpels, hooks, fine probes, saws, drills, mallets, needle, thread, buckets, and sponges," according to the list given by Jean Fernel (1497–1558)[37]—substances like vinegar, alcohol, or perfumes and, of course, incense, are also part of the arsenal of the dissector. The sponges mentioned by Fernel, which appear on every list of advisable or prescribed tools in anatomy books, are used to absorb the liquids and clean the corpse; the buckets are there to wring out the sponges, and the baskets, which are always present also, are for throwing away the parts that have already been examined, cut out, and extracted. The dissection of a body, as already laid down in 1316 by the Bologna professor Mondino de' Liuzzi in his *Anothomia*, begins with the abdomen, the "lower belly," so as to extract its contents immediately, as these parts are the first to rot [*quia primo illa membra fetida sunt*].[38] The "anatomical administration" is a struggle against decomposition, filth, risk, and repugnance; it outlines the sequence of operations and influences how the technical movements are decided. Knowledge of dissection means carrying out a series of procedures that simultaneously aim to optimize the use of the corpse as the object of exploration and to control the sensorial and affective effects that such an object can cause. In other words, the anatomical system, in its concrete, material, technical facet, the structure of which shapes the production of knowledge,[39] is closely linked to the conditions in which affectivity is managed around the accomplishment of the scientific task. The intellectual aspects—that is, the order of thought that anatomy established from the dawn of early modernity—are inseparable from the technical aspects. The intellectual aspects of anatomy are also inseparable from the repertoire developed to regulate the affects that derived from the invasive methods and intensive manipulation of the dead body.

Within the set of technical aspects of dissection, we can also include the distribution of tasks for its organization. The changes that came about on this point in the first half of the sixteenth century are directly related to a redefinition of the role of sensory perception and contact with corpses. Until the 1540s, there were three individuals involved in carrying out anatomies: the professor captained its progress and read and commented on the texts of the authorities; his second-in-command was a *demonstrator* or *ostensor*, in charge of letting the spectators see what the professor explained; the preparation of the corpse was almost always left in the hands of a *prosector*, in general a surgeon or barber. The iconography shows numerous images of this kind of anatomy lesson, for example, the dissection scene illustrating the first edition of the *Anothomia* by Mondino in Italian, part of a compendium of medical texts, the *Fasciculo di medicina*, printed in Venice in 1494.[40] A similar picture adorns a French

---

[36] Ibid.

[37] Jean Fernel, *De naturali parte medicinæ libri septem, ad Henricum Francisci Galliæ Regis filium* (Paris, 1542), 42. Fernel, professor at the Faculty of Medicine in Paris, was also physician to King Henri II; he soon became, through his works, one of the most influential medical authorities in Western Europe in the second half of the sixteenth century.

[38] Ernest Wickersheimer, ed., *Anatomies de Mondino dei Luzzi et de Guido de Vigevano* (Geneva, 1977), 8.

[39] Rafael Mandressi, "Zergliederungstechniken und Darstellungsweisen: Instrumente, Verfahren und Denkformen im Theatrum anatomicum der Frühen Neuzeit," in *Theatrum Scientiarum V: Theatrum Anatomicum*, ed. Helmar Schramm, Ludger Schwarte, and Jan Lazardzig (New York, 2011), 54–74.

[40] *Fasciculo de medicina* (Venice, 1494), in *Medicina Medievale*, ed. Luigi Firpo (Turin, 1972).

version of the same treatise, published in 1532:[41] the master dictates the lesson from his chair, with an open book in his hand, while the operator prepares to make the first incision on a corpse stretched on a table. In 1535, the frontispiece of a Venetian edition of the *Isagogæ breves* by Jacopo Berengario da Carpi (d. 1530),[42] professor of surgery in Bologna from 1502 to 1526, also presents an anatomy lesson that includes the book of the *magister*, the pointer of the *demonstrator*, and the knife of the *prosector*.

What the images suggest is confirmed by some texts, the most well known of which is undoubtedly the preface to *De humani corporis fabrica*, in which Vesalius laments the "detestable custom of entrusting the dissection of the human body to some individuals and the description of the body parts to others" and denounces those who, from the height of their professorial chairs, "speak of things they have never seen up close."[43] The Flemish anatomist campaigns for the unification of these functions, which until then had been segmented, into a sole protagonist responsible for cutting, showing, and explaining. The progressive adoption of this system meant the consolidation of the "sensorial program," which the texts presented and demanded more and more, and which consists of raising direct perception, particularly the visual and tactile, to the level of an exclusive methodological principle in order to obtain reliable knowledge: one must dissect with one's own hands and see with one's own eyes.[44] The hierarchical organization of the tasks fundamentally based on touch, which was in force for two and a half centuries, thus underwent a significant change. The spectators clearly did not disappear, nor did the anatomist begin to work alone, without the participation of other actors, but the demand for proximity marked new parameters and new constraints in the affective economy of anatomical activity. This change in professional skills led to a change in the repertoire of techniques to control affect, even as anatomists were in the process of acquiring these skills.

A PASSION OF THE MASSES

As is well known, the business of anatomy did not remain confined, in the early modern period, to members of a learned community. In fact, it developed an extraordinarily significant public facet. Apart from the private dissections, often performed outside the institutional framework regulated by university statutes, "public anatomies" were carried out annually in specially prepared spaces, sometimes with hundreds of onlookers. In other respects, they were organized as described above. Much has been written on the subject of these public anatomies,[45] sometimes excessively

---

[41] *Cy est l'anathomie de maistre Mundin boullonoys* (Paris, 1532), fol. 2v.

[42] Jacopo Berengario da Carpi, *Isagogæ breves perlucidæ ac uberrimæ in anatomiam humani corporis* (Venice, 1535).

[43] Vesalius, *De humani corporis fabrica* (cit. n. 28), sign. *3.

[44] For the "sensorial program" in early modern anatomy, its effects, and its limits, see Rafael Mandressi, "De l'œil et du texte: preuve, expérience et témoignage dans les 'sciences du corps' à l'époque moderne," in "Figures de la preuve," ed. Rafael Mandressi, special issue, *Communications* 84 (2009): 103–18; see also Mandressi, *Le Regard de l'anatomiste* (cit. n. 1), 83–95.

[45] See, in particular, Giovanna Ferrari, "Public Anatomy Lessons and the Carnival: The Anatomy Theater of Bologna," *Past & Present* 117 (1987): 50–106; Cynthia Klestinec, "A History of Anatomy Theaters in Sixteenth-Century Padua," *J. Hist. Med. Allied Sci.* 59 (2004): 375–412; Klestinec, "Civility, Comportment, and the Anatomy Theater: Girolamo Fabrici and His Medical Students in Renaissance Padua," *Renaiss. Quart.* 60 (2007): 434–63; Alvar Martínez Vidal and José Pardo Tomás,

referring to the "theatricality" of the dissections, or even to a "sacred ritual."[46] Apart from the epistemic bases that gave rise to such demonstrations,[47] it would be more appropriate to emphasize the function fulfilled by the protocol followed during these ceremonies of exhibition of anatomical knowledge and, thus, of the anatomized body. The opening of a corpse, in full view of a numerous and varied audience, offered an opportunity to attend an exceptional event, while it simultaneously showed the gestures and acts that directed the experience, made sense of it, and regulated it.

Public dissections were prepared and performed following precise rules, with the objective of demonstration, in venues designed to satisfy the proper requirements of what "was offered to the public" [*vulgo proponenda*], as stated by Estienne.[48] In his *Historia corporis humani*, Benedetti was the first to describe this type of venue, the anatomy amphitheater. As Benedetti states, these spaces should be circular [*circumcaveatis sedilibus*] and large enough to admit numerous spectators. The spaces in the stands were to be assigned in accordance with the dignity of the persons attending [*pro dignitate*], had to have a "*præfectus*" responsible for controlling and maintaining order, together with some guards to stop awkward individuals, and two trusted *quæstores*, who would provide whatever was needed. The corpse was to be placed in the center of the theater, on a raised bench, in a well-lit place that was comfortable for the dissector.[49] After Benedetti, descriptions of anatomy theaters became commonplace in the treatises. They sometimes corresponded to structures that were actually constructed, but they could be purely normative descriptions, that is, indications of how they should be built. This is the case with Estienne, or Guido Guidi (d. 1569) from Florence, the first professor of medicine of the Royal College in Paris, who, in chapter eight of Book I of his *De anatome corporis humani*, elaborates on a long series of provisions and recommendations for the conditioning of the "building for seeing performances" [*ædificium videndis spectaculis*].[50]

During the second half of the sixteenth century, the anatomy theaters were in sections that could be dismantled and were set up especially for public dissections that were held once a year for a week or two in winter. The first permanent amphitheater was probably the one built in Salamanca in 1554, close to the cemetery of the church

---

"Anatomical Theatres and the Teaching of Anatomy in Early Modern Spain," *Med. Hist.* 49 (2005): 251–80.

[46] The use of dramatic theater as a conceptual tool in order to describe public dissections implies converting a particular historical form, theatrical representation, into an analytical category, i.e., reducing the specificity of all other types of performances; see Rafael Mandressi, "Le corps des savants: sciences, histoire, performance," in "Performances: le corps exposé," special issue, *Communications* 92 (2013): 51–65. The allusion to the "theatricality" of public dissections can be found mainly in a classic medical historiography that is neither possible nor useful to quote here. As for the term "sacred ritual," Andrew Cunningham uses it to establish a comparison between anatomy and fox hunting in England from the mid-nineteenth century. Such an isomorphism, rather than supporting the idea that public dissections constituted a sacred ritual, speaks of the difficulty in considering them or interpreting them in themselves rather than by referring to other practices. See Cunningham, "The End of the Sacred Ritual of Anatomy," *Can. Bull. Med. Hist.* 18 (2001): 187–204 (191–4 for the comparison with fox hunting).

[47] Mandressi, "Zergliederungstechniken und Darstellungsweisen" (cit. n. 39).

[48] Estienne, *La dissection* (cit. n. 34), 373.

[49] Benedetti, *Historia corporis humani* (cit. n. 2), 84.

[50] Guido Guidi's book, written around 1560, was not published for half a century; Guidi, *De anatome corporis humani libri VII. Nunc primum in lucem editi. Atque LXXVIII Tabulis in aes incisis illustrati et exornati* (Venice, 1611), 12–3.

of Saint Nicolas.[51] In Barcelona, an "Aula de les Anatomies" has existed since 1573;[52] in 1584, Girolamo Fabrici d'Acquapendente (1533–1619) inaugurated the permanent anatomy amphitheater of Padua, where he was the professor of anatomy and surgery for almost half a century (between 1565 and 1613); in 1586, one was built in Saragossa, la "Casa de Anathomia," beside the hospital Nuestra Señora de Gracia; in 1593, the university of Leyden was granted its own, at the request of the anatomist Pieter Paaw (1564–1617). In the seventeenth and eighteenth centuries, new permanent anatomy amphitheaters were built in other European cities: Paris (1616, 1620, 1673), Delft (1614) and Amsterdam (1619) in the Netherlands,[53] Copenhagen (1643), Bologna (1649), Uppsala (1662), Berlin (1720), Halle (1727), and Bern (1735), to mention but a few. The proliferation of these venues in Europe, together with their permanency, demonstrates the success of anatomy at both a scientific and cultural level, and that of knowledge conceived as an action, as a demonstrative practice. Anatomical knowledge was tested by being offered to the public as a performance, displaying, apart from the dead body, the skills, sequences, and gestures through which the professionals of dissection scrutinized and revealed the structures of the body. It was a matter of showing. The celebration of anatomy in the amphitheaters was addressed to certain people, whose presence defined the very nature of the event.

These individuals made up an audience that, as we can see from the capacity of the amphitheaters, could number several hundred. These numbers are confirmed by witnesses such as the French surgeon Pierre Dionis (1643–1718), who recalls that the number of spectators who attended his demonstrations in the Jardin du Roi between 1672 and 1680 "always came to four or five hundred people," which was proof that his demonstrations "were liked, and were of use to the Public."[54] Apart from being multitudinous, the audience was heterogeneous. On 14 November 1552, in Montpellier, the dissection of a young man who had died of an abscess in the chest was carried out. Platter, who was present, wrote that "as well as the students, among the public were many individuals from the nobility and the bourgeoisie, and even young ladies, although the autopsy was done on a man. Even monks attended."[55] Far from being exceptional, the composition of this assembly described by Platter was the norm. Aristocrats, civil authorities, local dignitaries, churchmen, and the curious frequently attended these public events, driven by their desire to see a dissection, sometimes from up close. Indeed, Estienne notes that some parts of the body, such as the heart or the womb, once removed, would be "taken to the stands of the theater and shown to each one [of the spectators] as clear evidence."[56]

Confronted with the visible, even palpable reality of the corpse, not all spectators reacted in the same way. The French physician Guillaume Lamy (1644–83), who re-

---

[51] Teresa Santander, "La iglesia de San Nicolás y el antiguo teatro anatómico de la Universidad de Salamanca," *Rev. Españ. Teol.* 43 (1983): 253–73; Martínez Vidal and Pardo Tomás, "Anatomical Theatres" (cit. n. 45), 256.

[52] Alvar Martínez Vidal and José Pardo Tomás, "El primitivo teatro anatómico de Barcelona, "*Med. & Hist.* 65 (1996): 5–28.

[53] J. L. Rupp, "Matters of Life and Death: The Social and Cultural Conditions of the Use of Anatomical Theatres, with Special Reference to Seventeenth Century Holland," *Hist. Sci.* 28 (1990): 263–87.

[54] Pierre Dionis, *L'Anatomie de l'homme suivant la circulation du sang et les dernières découvertes, démontrée au Jardin-royal* (Paris, 1690), preface, n.p.

[55] *Félix et Thomas Platter* (cit. n. 29), 30.

[56] Estienne, *La dissection* (cit. n. 34), 373–4.

lates how he went to the Jardin du Roi to listen to an anatomy lecture given by his colleague Pierre Cressé, complained that he had only been able to hear Cressé's introduction, because of the noise made by "several ruffians from the slums, attracted by the pointless curiosity of seeing a body dissected."[57] In Bologna, in the seventeenth and eighteenth centuries, the *gran fontione* of public dissections was held during Carnival, as also occurred in other cities in Italy, such as Pisa, Ferrara, Rome, and Turin. These "useful shows" [*utilia spectacula*], as a Dutch chronicler called them, were attended by many masked spectators.[58] Public anatomy may then be festive, although its execution is carried out following a series of rigorous steps and rules that give the event a framework of seriousness and circumspection: the anatomist makes his entrance once the spectators have occupied their respective seats in the stands, the assistants light large candles, pieces of music are played before beginning the demonstration—this custom was adopted in Padua, for example, in 1597. There is really no contradiction between the Carnival masks and the unhurried solemnity of the dissection, or between the "ruffians from the slums" and the churchmen or aristocrats. From both a social perspective and an affective one, the anatomy demonstrations in early modern Europe are not places of antagonism, but of coexistence of attitudes, identities, and reactions. They are simultaneously places of science and celebration, of entertainment and retreat, of exhibition and edification. It would be delusional to think that the public dissections in the amphitheaters had produced a sort of affective acculturation of the nonprofessional spectators who attended them, if we understand this to mean an integral assimilation, by these audiences, of the professional *ethos* and *pathos*. However, it seems undeniable that, while the anatomy performance contributed, through its ceremonial qualities, to accentuating some of the most noticeable traits of its affective content, it also managed, due to the same ceremonial qualities, to guide affects and keep their eruption within certain limits.

CURIOSITY, PITY, AND SHAME

The dissections in amphitheaters were undoubtedly the most emblematic expression of the public face of anatomy, but certainly not the only one. Other social practices took shape around a discipline that, for centuries, held a central position in the system of European knowledge and contributed to representations and discourses far outside university circles. Already in the sixteenth century, the anatomized corpse as the focus of apprehension, suspicion, admiration, embarrassment, or rejection had gone beyond the gates of schools of medicine; in the epoch of Dionis, dissections had become an object of social consumption that stimulated an extraordinary desire, particularly among the aristocracy, learned or not. On 3 February 1667, a young woman from Tours was hanged in the Grève square in Paris, "because, they said, she had got rid of her child"; immediately, a carriage arrived and took the body to the Palace of the Louvre, "where some Grandee wanted a demonstration performed."[59] This curiosity of the "Grandees" can be seen in the "Anatomical History of an Extraordinary Womb," written by Dionis. The womb referred to in the title, which had "a particular and most surprising struc-

---

[57] Guillaume Lamy, *Discours anatomiques . . . Avec des Reflexions sur les Objections qu'on luy a faites contre la maniere de raisonner de la nature de l'Homme, et de l'usage des parties qui le composent, et cinq Lettres du mesme Autheur, sur le sujet de son Livre* (Rouen, 1675), 120.

[58] Ferrari, "Public Anatomy Lessons" (cit. n. 45), 52.

[59] Patin, *Lettres choisies* (cit. n. 30), 3:219.

ture," was that of a twenty-year-old woman who died in June 1681, in the sixth month of her pregnancy. "The sound of this accident spread throughout the Court," says Dionis, who received the order from the queen and the dauphine to open the body of the unfortunate young woman "to find the cause of such a sudden death." The autopsy, performed in the presence of Antoine Daquin (d. 1696) and Guy-Crescent Fagon (1638–1718), physicians to the king and queen, respectively, showed the existence of "two similar parts at the base of the womb, which differed in that the part on the right seemed to be stretched, and [to] have held a baby not long before; and the bottom of the other was almost of natural thickness." There was then "a supernumerary part located on the left side of the ordinary base of the womb." Once the dissection was finished, Dionis wrapped this strange womb in a cloth and had it sent to his chambers, in order to examine it more thoroughly. The following day, the queen wished to see it, and "was curious enough to examine it for a long time. M. Daquin and M. Fagon told her and the Dauphine and some other Ladies of high quality what they thought." Sometime later, the queen asked Dionis to again show her "that part." Her majesty, concluded the surgeon, "did not feel the same repugnance that most Ladies feel for anatomy demonstrations."[60]

Marie-Thérèse d'Autriche (1638–83), who felt more curiosity than repugnance, did not experience what the theologian Fénelon (1651–1715) thought to be a natural sentiment that should be woken by the "inner body of a man." To know God, that body should however be studied. It is true, writes Fénelon, "that the inner parts of man are not pleasant to see," but one should be warned that "they are not made to be seen." In addition, it was necessary "that they should not be uncovered without horror; and that in this way a man could not uncover them . . . without feeling violent repugnance. . . . That interior of man, at the same time so repellent and so admirable, is precisely as it should be to show the mud worked by a divine hand. In it can be seen at the same time the fragility of the creature, and the art of the creator."[61] Fénelon insists on this: the body is "kneaded with clay" (mud), the "vile" material with which "the Worker" has made a masterpiece. This is apparently a paradox that invites us to exalt the glory of God, and to withdraw when facing the evidence of the imperfect, transitory, corruptible nature of human existence. Anatomy reminds man that he is mortal. Between 1639 and 1646, the physician, poet, and professor of theology, Caspar van Baerle (1584–1648), composed three poems on the subject of anatomy, two of which had the anatomy theater of Amsterdam as their subject. The shorter, *In locum anatomicum recens Amstelodami exstructum*, celebrates the inauguration of the premises; the other, *In Domum Anatomicam, quæ Amstelodami visitur*, describes the nature of anatomy, more moral than scientific: it brings to light what is hidden in the interior of the body, home of the soul and tabernacle of the spirit, an insignificant receptacle for the Divine; "what we have been and no longer are," and when the light goes out, putrefied man is dismembered. The subject of the third poem is the dissecting table: *In mensam Anatomicam*. On it lies the "afflicted" human condition, its desolation, the naked limbs exposed to the view of miserable mortals: "Qui jacet afflictæ conditionis Homo."[62]

---

[60] Pierre Dionis, *Histoire anatomique d'une matrice extraordinaire* (Paris, 1683), 9–10, 14, 19–20.

[61] Fénelon (François de Salignac de La Mothe), *Traité de l'existence et des attributs de Dieu*, in *Œuvres de Fénelon* (Versailles, 1820), 1:57–8.

[62] Caspar van Baerle, *Poematum collectio, Elegiarum et miscellaneorum carminum* (Amsterdam, 1646), 207, 249.

Here the discourse becomes normative, affectivity is prescribed, and, even more, it is hierarchized: the disgust and horror are but a record for edification, for that moral meditation that should result from them and should become affliction, humility, perhaps anguish—memento mori. This is not incompatible with curiosity, the will to see and inspect the corpse or its parts. Nor is it incompatible with the merrymaking that can be seen at public dissections, nor, much less, with the solemnity that prevails in them. There is, in short, no condemnation for anatomical procedures; no doubt is cast on the treatment of dead bodies. Certain social and cultural groups perceive anatomy to be useful not only for the knowledge it offers but also for the sentiments it inspires. There have been objections to the practice of dissection elsewhere based on affects such as shame, which imposes a strict selection of the bodies that it is licit to dissect. In fact, one could not dissect just any body, only those that, because of their low social standing, could be submitted to mutilation and dismembering and then exhibited in broad daylight. In this sense, executed criminals supplied most of the raw material. But even with criminals there are restrictions. For dissections, explains Benedetti, one can only request the corpses of people "of common stock, unknown, from faraway regions" [*ignobiles, ignoti, ex longinquis regionibus*], so as not to injure local people or provoke the shame of relatives or friends.[63] In Bologna, the statutes of the school of medicine stated, in 1442, that only the corpses of foreigners, who resided over thirty *miliari* (ca. 45 kilometers) from the city, could be used for public anatomies. Similar regulations were adopted later in Padua.[64]

Criminals, foreigners, strangers, individuals on the margins of society, whose remains no-one would claim, whose corpses were unconnected enough from both those who were to dismember them and those who were to attend the demonstration. No one should recognize the deceased whose body was to be cut up, to preserve the honor of the relatives.[65] The identity of the dead thus defined a restricted universe of corpses to which the anatomists could have access. This demand for anonymity in the practice of dissection has its parallel in the texts, more precisely in the tens of thousands of printed images that publicized the likeness of the anatomized body in the sixteenth, seventeenth, and eighteenth centuries. With very few exceptions, the corpses shown in this vast iconography have no visible distinguishing marks. In some cases, we may observe the presence of elements that indicate a singularity: as the historian Martin Kemp has observed, "a particular, individual example" and not "the typical or exemplary model."[66] But individualization does not imply identification. In book V of Vesalius's *Fabrica*, for example, figure XXV shows the internal parts of the abdominal cavity of a woman, where certain deformities of the ovaries, perhaps of pathological origin, suggest that this is a particular body, except that this female anatomy is inserted into an ancient sculpture, with no head and so no face.[67] Here lies the key to identification: the bodies shown in the anatomy images are not portraits. When the face is shown, the features correspond to a "typical or exemplary model." In the over-

[63] Benedetti, *Historia corporis humani* (cit. n. 2), 84.

[64] See Heseler, *Andreas Vesalius' First Public Anatomy* (cit. n. 17), 304n35.

[65] Katharine Park, "The Life of the Corpse: Division and Dissection in Late Medieval Europe," *J. Hist. Med. Allied Sci.* 50 (1995): 129–30.

[66] Martin Kemp, "'The Mark of Truth': Looking and Learning in Some Anatomical Illustrations from the Renaissance and Eighteenth Century," in *Medicine and the Five Senses*, ed. William F. Bynum and Roy Porter (Cambridge, 1993), 85–121, on 88.

[67] Vesalius, *De humani corporis fabrica* (cit. n. 28), 378.

whelming majority of other cases, the facial features are covered by a cloth or by some other part of the body (often the skin or a layer of muscle) or are simply outside the frame of the picture. In other words, the corpse does not have a face. This is the rule, almost without exception, that controls the anatomical image in the early modern period. The face is represented only if it does not give away the identity. The body is moved toward the abstract, polarized between the visualization of the dismembering, on the one hand, and the disappearance of any clue that might reveal the presence of a subject, on the other.[68]

AFFECTIVE CULTURES

From the end of the Middles Ages, when the practice of anatomical dissection emerged in Europe, the dead body became part of the cultural economy of knowledge. It did not take long to become its focus, causing very different effects: epistemic, technical, and social. In each of these, the affective dimension plays a crucial role. The concrete, material relationship with the corpse, including the type of manipulations it undergoes, is one aspect that involves affective phenomena of an extraordinary intensity. Another is offered by the symbolic position occupied by the dead body, the qualities associated with it, and the properties attributed to it. Disgust, horror, and the means of avoiding them or adapting to them dominate the concrete, material relation with the corpse, whereas in the symbolic position occupied by the dead body, admiration alternates with fear, and curiosity alternates with shame. To a great extent, anatomy owed its repertoire of gestures, spaces, and instruments to the need to control these affects, and this repertoire contributed to the discourse that shaped the professional identity of anatomists. It is also important to underscore that anatomy, the science of the body, is, even more specifically, the science of the dead body. As such, it cannot help but display apparently paradoxical strategies: exaltation of its mission and the efforts that the object of its exploration implies, while at the same time it muffles the evocation of death and removes the link between the individual whose remains lie on the dissection table and those very remains.

Here we have an anthropological nexus that is essential to anatomical knowledge, crossed by internal and external affective cultures. The former nourish a professional culture, composed of regulations, training, and exhibition. The latter give structure to an area of acceptance or acquiescence, but also of opposition and resentment, that reached one of its flash points in the eighteenth century, for example, in the debates caused by the trafficking of corpses to satisfy the growing demand for dissections in Paris, or the violent riots that occurred in London when surgeons appropriated the corpses of hanged criminals.[69] Whatever the case, this is not simply knowledge trapped in a web of preexisting sensibilities, but the coproduction of affective cultures, negotiated among several social groups around a discipline and its object, its methods, and its main subject, the corpse, where, in early modern Europe, science and death came together fully and freely.

---

[68] See Rafael Mandressi, "L'identité du défunt: Représentations du visage des cadavres dans les livres d'anatomie (XVIe–XIXe siècle)," *Corps* 11 (2013): 45–55.

[69] On this episode, see Peter Linebaugh, "The Tyburn Riot against the Surgeons," in *Albion's Fatal Tree: Crime and Society in Eighteenth-Century England* (London, 1975), 65–117. On the Paris controversies on the uses of anatomy in the eighteenth century, see Mandressi, *Le Regard de l'anatomiste* (cit. n. 1), 180–91.

Pain as Practice in Paolo Mantegazza's Science of Emotions

by Dolores Martín Moruno*

ABSTRACT

Paolo Mantegazza's science of emotions represents the dominant style of thinking that was fostered by the late nineteenth-century Italian scientific community, a positivist school that believed that the dissemination of Darwin's evolutionary ideas would promote social progress in that country. Within this collective thought, Mantegazza was committed not only to studying the physiological experience of pain by means of vivisection but also to completing an anthropological study that examined the differences between the expressions of suffering in primitive and civilized cultures. Thus, the meaning of pain appears throughout Mantegazza's research as a result of applying an ensemble of scientific practices integral to observation, experimentation, and the scientific self, which enabled its main physiological and psychological manifestations to be reproduced in the laboratory. Among these practices, photography allowed Mantegazza to mobilize pain as an emotion whose performativity shaped national identities, such as those that embodied the recently created Italian state.

MAPPING EMOTIONS IN LATE NINETEENTH-CENTURY ITALY

In 1897, the Italian physician Paolo Mantegazza (1831–1910) wrote a novel in which he imagined a future world made up not of countries but of what he called "the United Planetary states," whose capital, Andropolis, was a city in which cultural, social, and economic differences had disappeared thanks to the advances of science and technology. In his novel, *The Year 3000*, Mantegazza also imagined that physicians had successfully mitigated all human suffering resulting from disease.

* Institute for Ethics, History, and the Humanities, Geneva Medical School, Case Postale 1211, Geneva 4, Switzerland; dolores.martinmoruno@unige.ch.

I thank the Wellcome Trust for a grant in 2010 to fund my project, "Mapping Emotions in the Nineteenth Century: Paolo Mantegazza's Alphabet of Expression" (091377 WT, Medicine, Society and History), as well as Colin Jones, who supported my research during that period. I am also grateful to Eva Botella, who kindly sent me a copy of the exemplar of the *Atlante* from the Widener Library. Javier Moscoso's advice on Mantegazza was priceless during the early stages of this project. Victoria Diehl was my best traveling companion in Florence, when we visited the National Museum of Anthropology and Ethnology. Marco Tripodi and Jonathan Seaburne-May also helped me with the laborious task of translating Mantegazza's work from Italian into English. I am also in debt to all my colleagues involved in the *Emotional Bodies* project for their productive discussions about the historical performativity of emotions. Last but not least, I would like to acknowledge the editors of this volume and the anonymous peer reviewers for their useful suggestions.

> Once dying a natural death was an extremely rare exception; you could count one lucky case out of perhaps a thousand, and everyone else died of illness. Today, though, the general rule is to die of old age, painlessly, because, as you will see . . . illness, which is detected at its first appearance, can almost always be cured and stifled at its inception.[1]

This utopian vision perfectly illustrates Mantegazza's enthusiastic support of "science as the religion of the future," which would accelerate the progress of society, leading inexorably to a human victory against pain and, therefore, to a state in which everyone would enjoy everlasting happiness.[2] Like other late nineteenth-century Italian physiologists, such as Carlo Matteucci (1817–83) or Angelo Mosso (1811–68), Mantegazza believed that science had a clear political aim: namely, to complete Italian unification by creating a new secularized state, which would be possible through the implementation of educational reform involving all the citizens of the country.[3]

In post-unification Italy, positivism became something close to what the historian Ludwik Fleck termed *Denkstil*, a style of thinking characterized by a particular community of persons "exchanging ideas" and "maintaining intellectual interaction" that was not independent from emotions.[4] As the editors of this volume explain in their introduction, this "collective thought" concerns a "specific emotional *haltung*—an ethos, mindset, an attitude—that the researcher of a certain group, discipline, or science in general had internalized" and reenacts "every time he or she is doing research."[5] This was the case with Mantegazza, whose science was particularly emblematic of the way in which Italians conducted emotion research during the last decades of the nineteenth century.

Within this distinctive style of thinking, Charles Darwin's notion of evolution became a meaningful principle, which was related not only to the theory of the natural selection of species but also to the idea that selection meant a kind of social progress, orienting human history toward its perfectibility, as Herbert Spencer suggested throughout his works.[6] Indeed, Mantegazza believed that the completion of his own scientific project on the natural history of the moral man was a way of improving the health of future generations on Earth, by gradually replacing negative emotions such as pain with positive ones such as pleasure, until the utopia described in his science fiction book, *The Year 3000*, had been reached.

By taking Mantegazza's science on emotions as a starting point, this essay seeks to explore the relationship between his research on the physiological mechanisms regulating the experience of pain and his illustrated atlas, which focused on shedding light on the main bodily expressions of pain.[7] The isomorphism established by Mantegazza

---

[1] Paolo Mantegazza, *The Year 3000*, trans. David Jacobson (1897; repr., Lincoln, Neb., 2010), 134.

[2] On Darwinian science as the religion of the future, see Mantegazza's intervention at the Italian Parliament on 20–21 April 1866, included in Giuliano Pancaldi, *Darwin in Italy: Science across Cultural Frontiers* (Bloomington, Ind., 1991), 155.

[3] See Suzanne Stewart-Steinberg, *The Pinocchio Effect: On Making Italians, 1860–1920* (Chicago, 2007), 3.

[4] Ludwik Fleck, *Genesis and Development of a Scientific Fact*, ed. Thaddeus Trenn and Robert Merton, trans. Fred Bradley and Thaddeus Trenn (1935; repr., Chicago, 1979), 39.

[5] Otniel E. Dror, Bettina Hitzer, Anja Laukötter, and Pilar León-Sanz, "An Introduction to *History of Science and the Emotions*," in this volume.

[6] Pancaldi, *Darwin in Italy* (cit. n. 2), xiii. On the notion of the survival of the fittest, see also Herbert Spencer, *The Principles of Biology* (London, 1864–7), 2:48.

[7] Paolo Mantegazza, *Fisiologia del dolore* (Florence, 1880); Mantegazza, *Atlante della espressione del dolore: fotografie prese dal vero e da molte opere d'arte, che illustrano gli studi sperimentali sull'espressione del dolore* (Florence, 1876).

between his works dealing with the experience and the expression of pain enables us to interpret this emotion as a kind of social practice whose meaning emerges from bodily dispositions, which are always culturally situated.[8] Halfway between physiology and anthropology, this aspect of Mantegazza's work is a particularly interesting lens through which to examine the connections between the histories of science and medicine within late nineteenth-century European culture, by means of adopting a Bourdieuian approach to emotion history.[9]

On the one hand, Mantegazza's science of pain has some significant implications that go well beyond his own research and are directly connected with fin de siècle European cultural life. As some relevant studies on the history of pain have concluded, the idea that suffering either was a necessary punishment inflicted by God or had a positive function in the regulation of health would be replaced from the mid-nineteenth century onward by a more contemporary perception, following the introduction of anesthesia for surgical purposes.[10] Mantegazza's belief that human suffering should be completely banished by artificial means was clearly a result of this shift in the understanding of pain, a revolutionary opinion in a country like Italy in which the Catholic Church still had a strong influence on political institutions, as well as on the population.

On the other hand, Mantegazza's work can be better clarified from a history of emotions perspective because he conceived pain in a wider sense, not only as a way of knowing, but also of feeling, which implied an affective response from the point of view of the experimenter, who regarded others' suffering through the lens of sympathy, pity, or compassion.[11] In other words, Mantegazza did not exclusively define pain as the main subject of his study, a physiological experience that he reproduced by experimental means in order to determine its influence on animal warmth, heart movement, breathing, digestion, and nutrition. He was also aware of the fact that studying pain had an influence on the attitude of the scientist while he conducted experimental research in the laboratory and, specifically, when he practiced vivisection. Following the criticism that he would receive from the British antivivisectionist movement, Mantegazza was mired in a controversy about the emotions that the physiologist was expected to feel when injuring animals, a thorny issue that would lead him to introduce new techniques of observation in order to neutralize his emotional attitude when conducting research on pain.

With this aim in mind, Mantegazza asked his colleague, the photographer Carlo Brogi (1850–95), to shoot a series of photographs to capture fleeting expressions of pain while

[8] Monique Scheer, "Are Emotions a Kind of Practice and Is That What Makes Them Have a History? A Bourdieuian Approach to Understanding Emotion," *Hist. & Theory* 51 (2012): 193–220, on 193.

[9] A Bourdieuian approach in emotion history has been taken by Fay Bound Alberti, ed., *Medicine, Emotion and Disease, 1700–1950* (New York, 2006), xvii–xviii; and Scheer, "Emotions" (cit. n. 8), 193–220.

[10] On the shift in the cultural perception of pain as a result of the historical debate about the use of anesthetics, see Roselyne Rey, *Histoire de la douleur* (Paris, 1993), 320–42; Javier Moscoso, *Pain: A Cultural History* (Houndmills, 2014), 79–105; Joanna Bourke, *The Story of Pain: From Prayer to Painkillers* (Oxford, 2014), 272–90. In relation to the religious uses of pain, see also Esther Cohen, *The Modulated Scream: Pain in Late Medieval Culture* (Chicago, 2010), 168–222.

[11] On the notion of experience as "a way of knowing and feeling," see Lorraine Daston and Elizabeth Lunbeck, eds., *Histories of Scientific Observation* (Chicago, 2011), 2; Susan Lanzoni, "Empathy in Translation: Movement and Image in the Psychological Laboratory," *Sci. Context* 25 (2012): 301–27. On the differences between sympathy, pity, and compassion, see Luc Boltanski, *Distant Suffering: Morality, Media and Politics* (Cambridge, 1999); Ute Frevert, *Emotions in History: Lost and Found* (Budapest, 2011), 149–202.

he conducted experiments on various subjects, including himself. By introducing the camera for scientific purposes, he was attempting to distance himself from his object of study, avoiding any form of affective reaction that naturally moved humans toward sympathy or compassion. As in the case of Charles Darwin (1809–82), photography became a useful tool that provided Mantegazza with new evidence about the expression of emotions and that also helped him maintain control of his own feelings.[12]

As will be shown in the following sections, Mantegazza was less concerned with establishing a theoretical definition of pain than in elucidating its phenomenological laws through the reproduction of the experimental conditions, thus re-creating the body in pain. Like his Italian peers, he was not interested in speculating about what emotions were—as their definition was subject to change according to age, gender, race, and class differences—but rather to study what they could do when shaping physical bodies. Within this late nineteenth-century Italian style of thinking, the meaning of pain appears as a result of applying a wide range of scientific practices integral to observation, experimentation, and the scientific self, enabling its main physiological and psychological manifestations to be controlled in the laboratory.[13]

EVERYBODY HURTS[14]

Born in October 1831 in the Lombard city of Monza, Mantegazza studied medicine at the universities of Pisa, Milan, and Pavia. However, he discovered his true vocation as an anthropologist during his numerous travels around the world, which included trips to Argentina, Lapland, and India. Moved by the passion for understanding the cultural diversity of human beings from the point of view of their emotional life, he published widely about what he considered to be the four poles of human existence: pleasure, pain, love, and hate. At the age of twenty-two, he wrote his first book on emotions, *The Physiology of Pleasure*, a book that he would complement later with a study of the history of stimulants, paying particular attention to the properties of the coca leaf, a finding that he made when he traveled to South America in 1856.[15]

Once he returned to Italy, he worked first as surgeon at the Milan hospital and later as professor of general pathology at the University of Padua. In 1870, he was appointed professor of anthropology at the Istituto di Studi Superiori in Florence, a chair that was created specifically for him. During these years, he became fascinated with Darwin's theory of sexual selection, which led him to reinterpret his entire project in evolutionary terms and, notably, to focus on love, an affection that he claimed to study just as any other living force in nature that "measures, weighs and governs itself."[16] Mantegazza also

[12] Paul White, "Darwin's Emotions: The Scientific Self and the Sentiment of Objectivity," *Isis* 100 (2009): 811–26.

[13] Paul White, "The Emotional Economy of Science," *Isis* 100 (2009): 792–7.

[14] The title of this section is inspired by the song written by Bill Perry, Peter Buck, Mike Mills, and Michael Stipe, "Everybody Hurts," on *Automatic for the People*, by R.E.M., Warner Bros., 1992.

[15] Paolo Mantegazza, *Fisiologia del Piacere* (Milan, 1854). His study on the coca leaf and on nervine foods can be found in Mantegazza, "Sulle virtù igieniche e medicinali della coca e sugli alimenti nervosi in generale," *Ann. Univers. Med.* 167 (1859): 449–519; and Mantegazza, *Quadri della natura umana: Feste ed ebbrezza*, 2 vols. (Milan, 1871).

[16] Paolo Mantegazza, *The Physiology of Love and Other Writings*, ed. Nicoletta Pireddu (Toronto, 2007), 94. Mantegazza corresponded with Charles Darwin from 1868 to 1875. In his letters, Mantegazza referred to sexual selection and particularly to the notion of variation as "a sublime monument to human intelligence." The citation can be found in Frederick Burkhardt, James A. Secord, Sheila Ann Dean,

wrote the *Physiology of Hate* (1889) and, most notably, his work on pain, which appeared as a central theme throughout numerous publications such as *The Physiology of Pain* (1880), *Physiognomy and Expression* (1881), and a visual study entitled *The Atlas of the Expression of Pain* (1876).[17]

The physiological study of what he considered the "cancerous plague that corrodes the happiness of living beings" was a crucial step within Mantegazza's science of emotions because his experience as a physician had familiarized him with the suffering of patients enduring disease. For Mantegazza, pain was not only the most important enemy to combat but also "the most difficult emotion to study" because it was almost impossible "to keep the necessary self-control to observe an animal or a man suffering" without being troubled by emotion.[18] Bearing this obstacle in mind, he then undertook an ambitious project aimed at exploring pain in order to fight against its abominable effects with "all the arms of science and all the resources of the arts."[19]

At the beginning of his *Physiology of Pain*, he reviewed a large amount of literature that had been written about what pain was, in order to conclude that the majority of philosophers, artists, moralists, and scientists had provided false, incomplete, and incongruous definitions. Although everybody hurts, nobody agrees about the nature of pain because we are always dealing with an incontestable "subjective experience, a fact of consciousness," which seems, furthermore, to be incommunicable in all its intensity when everything goes wrong.[20] Luckily for us, Mantegazza added, "pain is healed better than it is understood," showing that his interest in studying this emotion went far beyond being a mere intellectual concern.[21]

In the same vein, his colleague, the physiologist Angelo Mosso, commented in his work *Fear* (1891) that "human pain is of such importance that all scientific curiosity becomes a trifling and ridiculous thing, and our mind rebels and feels an invincible repugnance to every desire which has not the alleviation of the sufferer for its object, to every act which does not spring from a lively and intense compassion."[22] Together, Mantegazza and Mosso imagined that the moral purpose of their physiology of emotions was to set mankind free from physical suffering by placing their science at the service of a future society in which health conditions would be radically improved and each human would henceforth be ready to enjoy the benefits of pleasure.[23]

This utilitarian dimension of the Italian physiology of emotions explains why Mantegazza provided only a working definition of pain, which he expressed as "an alteration in the sensitivity that results in extreme disgust to those who experience it," as his main objective was to lessen its negative impact in man.[24] He therefore preferred to explore the diverse manifestations of human suffering that ranged from a baby's cry to the hys-

---

Samantha Evans, Shelley Innes, Alison M. Pearn, and Paul White, eds., *The Correspondence of Charles Darwin*, vol. 16, pt. 2 (Cambridge, 2008), 922.

[17] Paolo Mantegazza, *Fisiologia dell'odio* (Milan, 1889); Mantegazza, *Fisiologia del dolore* (cit. n. 7); Mantegazza, *Fisonomia e mimica* (Milan, 1881); Mantegazza, *Atlante* (cit. n. 7).

[18] Mantegazza, *Fisiologia del dolore* (cit. n. 7), 437. All translations are my own.

[19] Ibid., 122.

[20] Ibid., 12.

[21] Ibid., 384.

[22] Angelo Mosso, *Fear*, trans. E. Lough and Frederich Kiesow (1884; repr., New York, 1896), 201. See also Otniel Dror, "Visceral Pleasures and Pains," in *Knowledge and Pain*, ed. Esther Cohen, Leona Toker, Manuela Consonni, and Otniel E. Dror (New York, 2012), 147–68.

[23] Mantegazza, *The Year 3000* (cit. n. 1), 109. See also Mantegazza, *Fisiologia del dolore* (cit. n. 7), 18.

[24] Mantegazza, *Fisiologia del dolore* (cit. n. 7), 12.

teria complicated with paralysis suffered by women, or the sublimated pain provoked by unrequited love, which had inspired geniuses such as Goethe when writing his *Werther*.[25]

The multifaceted nature of pain finally led Mantegazza to abandon the idea of measuring it by scientific means. In contrast to the criminologist Cesare Lombroso (1835–1909) who had proposed to measure pain with his algometer, an instrument that was designed to test tactile sensibility by inducing electrical currents to the body, Mantegazza was convinced that pain's experience could not be completely quantified by this kind of technology.[26] Rather than conceiving this emotion as the mere result of "physical and chemical changes in a nerve or in nerve cells," Mantegazza regarded pain as a form of practice culturally constructed in relation to the codes, norms, and institutions that prevailed in late nineteenth-century society.[27]

According to Mantegazza's anthropological approach, gender, race, and class differences led to variation in the way individuals experienced pain. Cultural factors, such as being highly intelligent, using stimulants and narcotics, or belonging to a European ethnic group, were thought to explain the increased sensitivity to pain in certain groups of people.[28] Furthermore, the severity of suffering intensified throughout an individual's life. While infants most often endured painful experiences related to not having their basic needs satisfied, young people learned to feel the wounds caused by love and friendship, and adults were prone to blows to their self-esteem. Finally, because of the fragility of the human physical constitution and the increasing fear of death, old age was the period of life during which humans suffered the most.[29] In Mantegazza's view, pain was the result of lifelong learning, as this emotion only acquired meaning in light of "our past experiences" that showed us "that there are never two identical pains in one's life."[30]

Despite all the particularities that make suffering an irremediably personal experience, Mantegazza thought that it was possible to develop a science of pain by classifying this phenomenon according to its physiological origins. As he had already done in his *Physiology of Pleasure*, he would explore pain's experience following a tripartite schema that analyzed its occurrence in relation to senses such as sight, and feelings like self-esteem and the intellect, when, for example, the act of reflection became painful.[31] Following this progressive understanding, Mantegazza first identified its main physiological manifestations by inflicting painful stimuli on rabbits, chickens, rats, birds, guinea pigs, and frogs, as well as on himself. He confessed that his *Physiology of Pain* was "the fruit of long and patient observations and of many cruel experiments" showing that, although everybody hurts, certain pains mattered while others did not, such as the pain that could be experienced by animals when physiologists practiced vivisection.[32]

---

[25] Ibid.

[26] On Lombroso's use of the algometer, see also Cesare Lombroso, *Criminal Man*, trans. Mary Gibson and Nicole Hahn Rafter (1876; repr., Durham, N.C., 2006), 389; and Peter Becker and Richard F. Wetzell, *Criminals and Their Scientists: The History of Criminology in International Perspective* (New York, 2006), 323–4.

[27] Mantegazza, *Fisiologia del dolore* (cit. n. 7), 11.

[28] Ibid., 160.

[29] Henry Rutgers Marshall, "The Classification of Pleasure and Pain," *Mind* 14 (1889): 511–36, on 520.

[30] Mantegazza, *Fisiologia del dolore* (cit. n. 7), 23, 153.

[31] Ibid., 36.

[32] Paolo Mantegazza, *Physiognomy and Expression*, ed. Havelock Ellis (1881; repr., London, 1890), 121.

DOING PAIN IN THE LABORATORY

Historians of emotions have pinpointed the late nineteenth-century Victorian controversy over vivisection as a meaningful context that enables us to study the changing definition of pain, compassion, and sympathy in the modern laboratory, as well as in Western society.[33] Vivisection implied a reflection, on the one hand, about animals' ability to feel pain and their potential rights when they were involved in this kind of experimentation and, on the other hand, about the emotional reaction of the scientist while confronted with the reproduction of pain in the laboratory. Vivisection addressed, furthermore, a question about the epistemological legitimacy of establishing analogies in order to compare animal and human pain in scientific research, as physiologists were not sure that they fully understood animal behavior from an anthropomorphic perspective. As the historian of emotions Paul White has observed, all these issues related to the vivisection controversy prompted a redefinition of the "nature of feelings and their place in medical practice," as well as about its role in late nineteenth-century society.[34]

Within this passionate debate, Mantegazza offered some key arguments in defense of vivisection as one of the most fecund and elemental methods that had been used in physiology since its beginnings. Indeed, he suggested that vivisection could be better regarded as the founding practice of modern experimental physiology, as shown through the work of physicians such as François Magendie (1783–1855) or Claude Bernard (1813–78), to whom Mantegazza professed a particular admiration.[35] Vivisection became, in this way, an essential practice for the physiologist, which was the veritable signature of his profession not only because it permitted the study of pain, but also because it had forged his moral character when facing surgery on animals for experimental purposes.[36]

In tune with this argument, Mantegazza thought that physiologists should resign themselves to studying animal pain "in all the cruelty of its manifestations."[37] The "anatomist, the physician or the chemist are obliged . . . to do things . . . that are quite awfully repugnant," but to which they became "indifferent" purely out of habit.[38] In particular, Mantegazza opined that the physiologist needed to be prepared to address "a profound study of all agonies," and, therefore, he was strongly encouraged not to use anesthesia as it could modify the animals' responses. Although he considered anesthesia to be "one of the most beautiful and most useful discoveries of our century," Mantegazza did not recommend that it be systematically applied to animal experimentation, as the only ethical subjects on whom the physician should focus his "charitable action" were humans facing a painful surgery.[39] Thus, in Mantegazza's experimental research,

---

[33] Paul White, "Sympathy under the Knife: Experimentation and Emotion in Late Victorian Medicine," in Alberti, *Medicine, Emotion and Disease* (cit. n. 9), 100–25; Tiffany Watt-Smith, *On Flinching: Theatricality and Scientific Looking from Darwin to Shell Shock* (Oxford, 2014), 114–9; Rob Boddice, "Species of Compassion: Aesthetics, Anaesthetics, and Pain in the Physiological Laboratory," *Nineteen* 15 (2012), http://19.bbk.ac.uk/articles/10.16995/ntn.628/ (accessed 11 December 2013).

[34] White, "Sympathy" (cit. n. 33), 101.

[35] Mantegazza, *Fisiologia del dolore* (cit. n. 7), 48.

[36] To further explore how early modern anatomists were involved in a similar discussion about their identity in relation to dissection techniques, see Rafael Mandressi, "Affected Doctors: Dead Bodies and Affective and Professional Cultures in Early Modern European Anatomy," in this volume.

[37] Mantegazza, *Fisiologia del dolore* (cit. n. 7), 83.

[38] Ibid., 186.

[39] Ibid., 83.

animal pain became merely a medium to explore this emotion as experienced by humans.

Mantegazza's opinion differed from that of other late nineteenth-century physiologists, such as Moritz Schiff (1823–96). After conducting vivisection experiments at the Istituto di Studi Superiori in Florence, Schiff modified his position by publicly recommending the use of anesthesia to render animals completely impervious to pain.[40] In contrast, Mantegazza believed that the physiologist should restrain his feelings of disgust while practicing vivisection in order to engage in the noble pursuit of eliminating suffering from the future of mankind.[41] Mantegazza viewed the unpleasant task of the physiologist, confined in his laboratory while inflicting painful stimuli on animals, as compassionate dedication that would one day find a chloroform capable of removing "all the torment of physical pain, and silence all the agonies of the heart."[42]

The high moral aims of Mantegazza's physiology of pain contrasted with the cold enumeration of the experiments that he carried out in the laboratory of experimental pathology of the University of Padua, in order to measure the physiological changes experienced by animals subjected to physical suffering, and even death. Among his findings, he noted that pain was always related to muscular contractions, a decrease in body temperature, a decrease in heart rate, and difficulties related to vital bodily functions such as breathing, nutrition, and digestion. He also observed that "pain could not be felt in animals without provoking violent muscular contractions . . . and the action of suffering." This assertion showed that he believed that animals felt pain even though they expressed their suffering in a different way than humans.[43] For instance, frogs manifested their agonies by closing their eyes, salamanders by swelling, tarantulas by groaning, and pigs by attempting to bite the experimenter.[44]

Mantegazza arrived at these conclusions after planning and carrying out a series of violent experiments on animals, which he nevertheless claimed to have conducted with "much love and a lot of patience."[45] These experiments involved perforating the chest muscles of birds in order to cause inflammation, amputating the limbs or removing the brains of frogs, and subjecting rats and rabbits, including some in advanced stages of pregnancy, to starvation and then exhaustively studying their agonies while they were dying. His laboratory was therefore the site of some horrific scenes.

Mantegazza also designed a device of special interest for vivisection purposes, which the famous British antivivisectionist activist Frances Power Cobbe (1822–1904) referred to as "the tormentor" because it "surpasses all that Dante has imagined."[46] This machine had "in the centre a large cylindrical glass box or bottle, in which lies a rabbit. . . . A handle, terminating in iron pincers and claws, [descends through the cover and moves

[40] For more on Schiff's activities in Florence, see Patrizia Guarnieri, "Moritz Schiff (1823–96): Experimental Physiology and Noble Sentiment in Florence," in *Vivisection in Historical Perspective*, ed. Nicolaas A. Rupke (New York, 1987), 105–24, on 106.

[41] Marco Piccardi, "L'espressione del dolore nel fondo Mantegazza dell'Archivio fotografico del Museo Nazionale di Antropologia e Etnologia (Museo di Storia Naturale) di Firenze," *Arch. Antropol. Etn.* 128 (1998): 83–103, on 97.

[42] Mantegazza, *Fisiologia del dolore* (cit. n. 7), 83.

[43] Ibid., 38.

[44] Ibid., 282.

[45] Ibid., 101.

[46] Susan Hamilton, *Animal Welfare and Anti-vivisection 1870–1910: Provivisection writings* (New York, 2004), 38. An illustration of this device is reproduced in Mantegazza, *Fisiologia del dolore* (cit. n. 7), 99.

freely] so . . . that the presiding physiologist may grip at pleasure any part of the animal's body."[47] The vision of these atrocities had, however, an aesthetic dimension close to what some scholars have termed the "pornography of pain," that is, the spectacle of suffering that the physiologist enjoyed while practicing vivisection.[48] Indeed, Mantegazza used to say, without any remorse, that he "sought to put the beautiful on an anatomical table and make an autopsy of it."[49]

Mantegazza's account of the unmentionable tortures that he inflicted on animals would bring him considerable criticism from the recently created British Society for the Protection of Animals Liable to Vivisection, better known as the Victoria Street Society, an animal advocacy group that Cobbe founded in 1875. In the first issue of *Zoophilist*, the journal published under the auspices of the society, Mantegazza's *Physiology of Pain* was reviewed in the following terms:

> The picture of a vivisector's mind, the record of a vivisector's life. That picture and that record are the best possible evidence of what the supposed "ennobling pursuit of science" really means. It means an exceptionally able, cultivated, eloquent and far-travelled gentleman spending his days in putting helpless little animals into a glass box where he pinches them, tears them, and crushes them for hours . . . noting with keen interest and satisfaction every detail of agony.[50]

In this review, the main criticism of Mantegazza's work was focused on his perverted moral character, that of someone taking great pleasure in injuring animals, rather than on evaluating the suffering that he provoked in them. As Anne DeWitt has pointed out, the focus on the vivisector's personality was a great topos in the Victorian novel, which was exploited by antivivisectionist supporters in order to demonstrate how cultivated men like Mantegazza were transformed into "hard, cold and remorseless" scientists who were not able to feel compassion toward the living beings subjected to experimentation in their laboratories.[51]

In this sense, what seemed to be at the heart of the vivisection controversy was the very meaning of compassion, which was used to either discredit or justify the moral ethos of the physiologist. While antivivisectionists understood this affective reaction to others' pain as being closely related to the notion of natural sympathy, as theorized by late eighteenth-century philosophers such as Adam Smith, vivisectionists like Mantegazza were making an abstraction from its effects "in the immediacy of the laboratory" in order to apply them to the wider scope of humanity.[52]

As the historian of emotions Rob Boddice has observed, late nineteenth-century physiologists were consciously redefining the meaning of compassion within the framework of evolutionary ethics in order to justify vivisection as a legitimate prac-

---

[47] Hamilton, *Animal Welfare* (cit. n. 46), 38.

[48] On the pornography of pain, see Karen Halttunen, "Humanitarianism and the Pornography of Pain in Anglo-American Culture," *Amer. Hist. Rev.* 100 (1995): 303–34. On the connections between the vivisector and the sexual predator, see Nigel Rothfels, ed., *Representing Animals* (Bloomington, Ind., 2002), 29.

[49] Mantegazza's citation appears in Benedetto Croce, *Guide to Aesthetics*, trans. P. Romanell (1965; repr., Indianapolis, 1995), xi.

[50] Anne DeWitt, *Moral Authority, Men of Science, and the Victorian Novel* (Cambridge, 2013), 126–7; Mantegazza, *Fisiologia del dolore* (cit. n. 7), 42.

[51] DeWitt, *Moral Authority* (cit. n. 50), 42.

[52] Boddice, "Species of Compassion" (cit. n. 33), 1.

tice that was crucial to bringing progress to the whole of society.[53] Conversely, animal rights supporters stressed the moral dimension inscribed in human sight when it was confronted with the pain of other living beings, in order to claim that the moral aims of compassion were to immediately alleviate their suffering.

We can appreciate the echoes of this debate about the changing meaning of compassion in Mantegazza's *Physiology of Pain*, particularly when he wrote about the role played by this affective component in the laboratory. In spite of his efforts, Mantegazza would finally acknowledge the effects of this emotion when he practiced vivisection. However, he did so not primarily because of the potential rights of animals, but because of the impossibility of being an objective observer while seeing others suffering.

> At that moment, our moral activity is more inclined to lessen the pain of the others instead of studying it. . . . When we have undergone the misfortune of suffering, we feel ill-disposed to observe. Once I had hoped to collect a rich treasure of observations in the animal kingdom but I confess that, when I was obliged to provoke them by means of experiment, I felt too much repugnance to repeat these cruel studies and I could only collect a few results.[54]

Vivisection seemed to reveal the limits of objectivity in Mantegazza's science of pain. He was not the only Italian physiologist who admitted the difficulties involved in the study of this emotion. Angelo Mosso arrived at a similar conclusion when he explained that "even men who had been hardened and accustomed to the sight of blood and of human misfortunes are . . . moved at the terrible picture of pain breaking its rude will on a sensitive organism."[55] Both scientists recognized the role that pain played in experimental research, as a sort of automatic response in the observer, who could not avoid feeling pity for the animals subjected to experimentation, though the physiologist was expected to restrict his own personal feelings within the laboratory.

With the aim of neutralizing this affective reaction, Mantegazza advised working with species that display few similarities to us, as our pity toward animals varied according to our similarity to them. Although he recognized that each living being could potentially suffer harm, we—as humans—are not able to fully understand the way in which other beings express their feelings. Pain was, in this sense, an emotion deeply rooted in the idea that a hierarchy of sensitivity existed among living beings.[56] Drawing on these considerations, Mantegazza established an analogy that allowed him to compare animal and human pain within his experimental research by arguing that lower species were less susceptible to feeling pain than highly organized beings.

> When we overstep the great fraternity that guides us with the vertebrates, we can understand the pains of worms and molluscs, insects, crustaceans and infusoria by using a distant cause and effect analogy. The uncertain criterion of the analogy gets gradually more obscure, just as by the same proportion, our pity decreases. When we descend little by little from the man to the protist, we can neither hear nor share the pain of the minor species, so

---

[53] Ibid., 15.
[54] Mantegazza, *Fisiologia del dolore* (cit. n. 7), 280.
[55] Mosso, *Fear* (cit. n. 22), 201.
[56] On the hierarchies of pain, see Joanna Bourke, *What It Means to Be Human: Reflections from 1791 to the Present* (London, 2011), 71–92; Rob Boddice, *Pain and Emotion in Modern History* (Houndmills, 2014), 4–7. On the notion of pity and its differences from compassion, see Frevert, *Emotions in History* (cit. n. 11), 181.

much so, that we are completely indifferent to the sufferings of the plants, and only a delicate lady may be moved at the thought of the pain that a violet feels when its stem is broken or the pain that a plant experiences while suffering from a killing thirst because of her metaphysical and transcendental pity.[57]

Mantegazza was convinced that even plants were able to feel emotions, a subject to which he would devote an entire book entitled *The Legends of Flowers* (1890).[58] Nevertheless, he also believed that their harm would solely arouse women's compassion, because the female sex was biologically prone to sentimentality. In contrast, the male scientist was able to keep control of his sympathetic feelings, as he had been trained so as to guarantee the objectivity of his research in the laboratory. The implicit definition of the modern laboratory as a man's world was not only the result of Mantegazza's sexist bias. In the Victorian mind, the antivivisectionism movement was strongly associated with the campaign led by suffragette activists, such as Cobbe, as many of them understood that the defense of animal rights was a part of the feminist agenda.[59] Therefore, when Mantegazza discredited female compassion as being exaggerated by nature, he was consciously debunking the antivivisectionist cause as the result of "a sentimental malady proper of his times," which touched mainly women because they were able "to suffer other's pain as well their own."[60]

This kind of gender distinction between rational men of science and innately sensitive women was also commonly used by late nineteenth-century experimental researchers in order to defend the public validity of a new observational paradigm based on seeing emotions rather than feeling them.[61] As Otniel Dror has pointed out, observation in terms of seeing required the scientist to maintain control of his emotions in order to safeguard the objectivity of his research. Therefore, late nineteenth-century physiologists introduced a whole spectrum of technologies in order to distance themselves from the emotions that they were scrutinizing in the laboratory. Machines came to replace old techniques of observation, such as sympathy or innate sensitivity, two capacities that had been historically associated with women, and that became obsolete ways of studying emotions in the modern laboratory because of their imprecision.

This desire for objectivity explains why Mantegazza emphasized the advantages of using new technologies, such as photography, for scientific purposes. He argued that this medium was less harmful, but of no less interest to the scientist wishing to study human pain:

> I believe I have found a new source of observation that can be useful both for the analysis of the scientist and for the inspiration of the artist. . . . We can produce in the human being, without cruelty and too much suffering, many pains. . . . The images collected by this method are a faithful representation of nature and should not be confused in terms of scientific value with those obtained from models or dramatic artists who exaggerate . . . and

---

[57] Mantegazza, *Fisiologia del dolore* (cit. n. 7), 20.

[58] Paolo Mantegazza, *The Legends of Flowers; or, 'Tis Love That Makes the World Go Around*, trans. J. Alexander Kennedy (1871; repr., Edinburgh, 1909).

[59] Sally Mitchell, *Frances Power Cobbe: Victorian Feminist, Journalist, Reformer* (Charlottesville, Va., 2004), 278. See also White, "Sympathy," 107; Boddice, "Species of Compassion" (both cit. n. 33); Bourke, *What It Means to Be Human* (cit. n. 56), 95–6, 107.

[60] Mantegazza, *Fisiologia del dolore* (cit. n. 7), 436.

[61] On the redefinition of observation in the late nineteenth-century science of emotions by means of introducing new technologies, see Otniel Dror, "Seeing the Blush: Feeling Emotions," in *Histories of Scientific Observation*, ed. Lorraine Daston and Elizabeth Lunbeck (Chicago, 2011), 326–48.

distort nature, even when they deeply touch the hearts of the spectators who contemplate them, or inspire the painter or the sculptor who should reproduce them.[62]

Mantegazza believed that he could contribute to the study of pain by improving the use of the camera as a tool of scientific observation. Photography would not only provide him with new evidence on the expression of emotions; it would also become a practice aimed at neutralizing his own feelings when studying pain. In contrast to the photographic tradition representing the emotions, as performed by models or dramatic actors, Mantegazza was convinced that he could use the camera in order to reproduce a true picture of human suffering by recording the infliction of painful stimuli on various individuals, including himself.[63] While vivisection practices had enabled him to establish the main conditions for the physiological production of pain in the laboratory, his study on physiognomy made it possible to approach the irremediably personal experience of suffering through the Darwinian laws of expression.

THE ALPHABET OF PAIN

On 23 December 1872, Mantegazza wrote a letter to Darwin in which he praised the British naturalist as the modern reformer of physiognomy following the publication of his book, *The Expression of Emotions in Man and Animals*. As Phillip Prodger has suggested, this letter probably accompanied the photographs on the expression of pain that he sent to Darwin.[64] In the letter, Mantegazza addressed the British naturalist in the following flattering terms:

> Your latest work has overjoyed me; you have made a new science out of physiognomy: up to now it has been an alchemy with Lavater as its founder, you are its Lavoisier. . . . Some days ago now, I posted your two pamphlets on pain, which accord perfectly with what you say in your admirable work. . . . For a long time I have been busy with a work on the physiognomy of pain and I ask your permission to dedicate it to you.[65]

Although Mantegazza never dedicated a book on the physiognomy of pain to Darwin, he openly declared his admiration of Darwin's theory on the expression of emotions on many occasions.[66] In particular, he paid special attention to understanding the physiognomy of pain according to the Darwinian laws of expression: "the principle of

---

[62] Mantegazza, *Fisiologia del dolore* (cit. n. 7), 281.

[63] On photographs representing facial expressions mimed by actors, such as those made by Oscar Gustave Rejlander, see Anne Schmidt, "Showing Emotions, Reading Emotions," in *Emotional Lexicons: Continuity and Change in the Vocabulary of Feeling 1700–2000*, ed. Ute Frevert, Monique Scheer, Anne Schmidt, Pascal Eitler, Bettina Hitzer, Nina Verheyen, Benno Gammerl, Christian Bailey, and Margrit Pernau (Oxford, 2014), 62–90, on 83–4.

[64] Phillip Prodger, *Darwin's Camera: Art and Photography in the Theory of Evolution* (Oxford, 2009), 213. Prodger incorrectly pointed out that Brogi's original photographs can be found in the Allinari Collection (Florence); in fact, they are archived at the Museo di Storia Naturale di Firenze. Darwin's photographic collection at the Cambridge University Library preserves fifteen of these photographs, from CUL MS.DAR.53.1: C.11 to MS.DAR.53.1: C.25 inclusive. They are available at http://darwin-online.org.uk (accessed 21 June 2016).

[65] Frederick Burkhardt et al., eds., *The Correspondence of Charles Darwin*, vol. 20 (Cambridge, 2013), 646–7.

[66] A beautiful example of Mantegazza's devotion to Darwin is the eulogy he wrote for Darwin. See Paolo Mantegazza, *Commemorazione di Carlo Darwin celebrata nel R. Istituto di Studi Superiori in Firenze: Discorso del Professore Paolo Mantegazza, 1882, 21 maggio* (Florence, 1882), 97.

serviceable associated habits," in which he explained why some complex expressions are advantageous under certain conditions; "the principle of antithesis," which referred to the tendency of performing contrary expressions in opposite conditions; and "the principle of actions due to the constitution of the nervous system," which was independent of the will and, up to a certain point, also of custom.[67]

Even though Mantegazza seemed to profess an unlimited faith in Darwin's work, he did not hesitate to criticize Darwin's contribution because he believed that his laws of expression were formulated in a confused manner.[68] He proposed, alternatively, to deal with the problem of the laws of expression from two other synthetic principles: first, that there was "a useful expression of emotions" with defensive purposes explained through evolutionary biology and, second, that there are also some "facts of expression," that he considered to be complementary because their reproduction depends on imitation, mechanical or muscular reactions, the sympathy of the functions, and the obscure sympathies of the nerve centers.[69] For Mantegazza, "defence and sympathy . . . govern all expression" and "are always more automatic in the animal than in man," as much "in the child" as "in the adult."[70]

With regard to pain, Mantegazza briefly classified its main expressions according to the classification that he had already used throughout his experimental research, such as those of reaction, paralysis, and mingled expressions of suffering with various other feelings.[71] While the expressions of reaction were most commonly related to minor injuries, paralysis frequently indicated a most violent pain, which could lead to syncope, or death in the most extreme cases. However, paralysis was rare, and the majority of the expressions of pain implicated the "contractions of the facial muscles, agitation of the limbs and of the trunk, complaints, sighs, sobs, erection of the hair, threatenings to real beings either present or absent, or even to imaginary beings."[72] This combination of movements had a twofold object: "to release the nerve centres from the excessive tension which afflicts them, and to struggle with the pain," even when pain gestures were only destined to "excite in those who listen to us a compassion, which may be of aid to us."[73]

Pain gestures had obvious rhetorical effects, such as exciting compassion in the subject who was contemplating the suffering of another living being. Mantegazza could not avoid the influence that this dramatic component played in his research on the mimicry of pain. As in the case of the laboratory experiments that he carried out, compassion seemed to play a crucial role in the study of the physiognomy of pain, because the sole act of regarding the suffering in others' faces involved a kind of natural imitative capacity in the observer. A good example of the performativity of this emotion was the tendency for those who observed the pain of those they loved to mimic their pain:

> The painful mimicry of love is among the most contagious, and when we are dealing with one of these scenes, we imitate the expression of those who are suffering, without knowing or wanting to, and we reenact the movements of those whom we would like to

[67] Charles Darwin, *The Expression of Emotions in Man and Animals* (London, 1872), 28.
[68] Mantegazza, *Physiognomy* (cit. n. 32), 90.
[69] Ibid., 89.
[70] Ibid., 92.
[71] Ibid., 123.
[72] Ibid.
[73] Ibid., 114.

console by assuming a sympathetic expression, even when our pain cannot be of the same degree.[74]

Sighs, groans, cries, moans, and yawning were expressive elements used in the mimicry of pain, with the aim of discharging the nervous centers, communicating our sorrows, and arousing the sympathy of our fellow man, who by imitation would resent our suffering. However, as Mosso remarked on his study of the physiognomy of pain, "in acute states of suffering the human face inspires fear," resulting not from compassion, but rather from "the selfish thought that this palpitating flesh might be our flesh."[75] Thus, the intensity with which the expressive elements of pain gestures were articulated was directly related to the modulation of the other's response to suffering within a whole range of emotional possibilities, from sadness to terror.

In contrast to the expressions of reaction and paralysis, Mantegazza would define "mingled expressions" of pain as those provoked by one's own feelings. This type of mimicry of pain helps us to induce "by the gestures of a man . . . whether he is suffering from a tooth or a corn" thanks to the particular dramatization performed by the individual.[76] This mimicry of pain is especially useful when we are trying to interpret specific pains of the senses because they "take their form from the special nature of the offended organ" but not the expressions of moral suffering, such as those related to feelings like fear and thoughts like vanity.[77]

In this respect, Mantegazza was particularly proud of having discovered some fundamental laws with which to study the expression of moral pain, which showed the common nature of all forms of suffering as modifications of the nervous system. Mantegazza's *Atlante* completed what he considered to be his main contribution to psychology with a series of photographs shot with the famous Florentine photographer Carlo Brogi in order to show the analogies that could be established between the expression of physical and moral pain.

PHOTOGRAPHING PAIN

Carlo Brogi inherited Brogi Edizioni, the studio that his father, Giacomo Brogi, had opened in Florence in 1864. Both father and son were active members of the Italian Photographic Society, created by Mantegazza in 1889.[78] Brogi's studio specialized in re-creating Italian traditional scenes, monuments, and most notably portraits that captured industrialists, artisans, and men of letters. Brogi's predilection for portraits was explained by his belief that this genre could contribute to the popularization of this new medium among the Italian population. The fashion of photographic portraiture became, in this way, a powerful ideological strategy aimed at reinforcing "the proper figural representation of bourgeois cultural norms" in late nineteenth-century Italian society.[79]

[74] Mantegazza, *Fisiologia del dolore* (cit. n. 7), 348.
[75] Mosso, *Fear* (cit. n. 22), 202.
[76] Mantegazza, *Physiognomy* (cit. n. 32), 123.
[77] Ibid., 128.
[78] On the role of Florence as one of the most important Italian photographic centers in the late nineteenth century, see Elvira Puorto, *Fotografia tra arte e storia: il "Bullettino della Società fotografica italiana (1889–1914)"* (Naples, 1996), 24. Mantegazza's interest in the application of photography to anthropology was so great that he also created a photographical collection at the Museo di Storia Naturale di Firenze, now known as the Fondo Mantegazza, which preserves images from all over the world.
[79] Pasquale Verdicchio, *Looters, Photographers, and Thieves: Aspects of Italian Photographic Culture in the Nineteenth and Twentieth Centuries* (Plymouth, 2011), 87; Silvia Paoli, "Italy," in *Encyclopedia of Nineteenth-Century Photography*, ed. John Hannavy (New York, 2008), 752–58, on 755–6.

This idea also runs through *Practical Notes on Posing for the Photographic Portrait* (1895), which Brogi wrote in order to give some advice to both the photographer and the person photographed, about how to capture what he called the "beautiful profile of Italians."[80] Curiously, this book included a foreword by Mantegazza, who wrote insightfully about the political implications of photography, the mechanism representing democracy in the recently created Italian state.

> Photography has another valuable feature, that of being democratic. Science is certainly always the most faithful and true of democracies, that is, [it is] of the kind that tends to elevate those at the bottom and not to lower those who already are at the top. The railroad carries today, with the same speed, the prince and the proletarian, and electric light shines on our roads over the heads of the poor as well as the rich. For photography it is the same; today it allows everyone to preserve the physical aspect of the dearest person with very little money, which once was granted only to the great lords. So bless science, which broadens the horizon of the human eye, and grants to the heart of all, that which was once the privilege of few; bless photography, which is the youngest and most lovely daughter of science.[81]

In Mantegazza's words, photography became the symbol of the new democratic Italian parliamentary system because this new technique made it more affordable for the entire population to obtain a souvenir of their loved ones. When Mantegazza referred to the population [*il popolo*], he was, however, less concerned with the working class than with the bourgeoisie, whom he considered to be the most outstanding social agent to emerge after Italian unification.[82] Neither the upper nor the lower classes represented the national identity of the recently created Italian state. The prototypical Italian citizen, a figure that was being consciously constructed in post-unification Italy by politically committed scientists, such as the socialist Lombroso and the liberal Mantegazza, was just the type of man that Brogi asked to pose for his photographs, namely, the middle-class Italian who embodied stability and progress in late nineteenth-century society.[83]

Nevertheless, Brogi's book on portraits was not the only opportunity these two personalities had to collaborate and exchange their ideas on photography. About twenty years earlier, the *Atlante della espressione del dolore* was published. Brogi and Mantegazza had been working together for several months on this study, which was aimed at completing Mantegazza's physiological research with a visual study of the mimicry of pain. The partnership established between these two men had many similarities with other renowned collaborations, such as those of Darwin and Oscar Gustav Rejlander (1813–75), the photographer from whom the British naturalist commissioned some of the most famous pictures included in his work on the expression of emotions.[84] While Rejlander's photographs reproduced some relevant expressions from an evolutionary point of view, with models such as the famous *Ginx's Baby* or even Rejlander himself simulating aston-

---

[80] Verdicchio, *Looters* (cit. n. 79), 87–9; Carlo Brogi, *Il ritratto in fotografia: appunti pratici per chi posa* (Florence, 1895). The Italian expression used by Brogi was *fare una bella figura*, which means the proper manner in which to take and pose for a portrait.

[81] Brogi, *Il ritratto in fotografia* (cit. n. 80), 11–2.

[82] Pancaldi, *Darwin in Italy* (cit. n. 2), 153.

[83] Concerning the political activity of Mantegazza, who was a deputy from 1865 to 1876 and then a senator in the Italian Parliament, see Monica Boni, *L'erotico senatore: Vita e studi di Paolo Mantegazza* (Genoa, 2002).

[84] On the collaboration between Darwin and Rejlander, see Darwin, *Expression of Emotions* (cit. n. 67), 23; Prodger, *Darwin's Camera* (cit. n. 64), 158.

ishment or indignation, Brogi's pictures depicted the physiological experiments that Mantegazza had previously planned in order to shed light on the changing physiognomy of suffering by inflicting painful stimuli on various individuals in the laboratory.[85]

In so doing, Mantegazza was emulating the procedure that Duchenne de Boulogne (1806–75) had developed in his photographic study, in order to analyze the language of the passions in movement, which he praised as a method that was "ingeniously fertile . . . to re-create analogous images to those of moral pain."[86] Like the French physiologist had intended to do, Mantegazza would also attempt to replicate the physiognomy of pain by experimental means, but he did not make use of localized faradization because he thought that it allowed only the replication of the muscle contractions involved in the expression of pain, without reproducing the emotion of pain itself, that is, a real suffering.[87] Reproducing the psychological dimension of pain was only possible through what Mantegazza claimed to be "natural causes," such as being exposed to bright lights or unpleasant noises, ingesting bitter herbs, sniffing ammonium sulfate, or having one's fingers crushed.[88] Thus, in Mantegazza's visual study of the expression of pain, Darwin's evolutionary ideas merged with Duchenne's experimental approach.

The hybrid nature of Mantegazza's *Atlante* was also highlighted because this contribution was situated halfway between the sciences and the arts. The *Atlas of Pain* revealed an eccentric combination of what he called his "photographs on the expression of pain taken by real means" [*le espressioni del dolore prese dal vero*], with reproductions of paintings such as Leonardo da Vinci's *La Medusa* and images representing sculptures such as the famous *Laooconte*.[89] Mantegazza's "experimental photographs" were relatively few in number, only fifteen, in comparison with the profuse reproduction of works of art representing dramatic scenes featuring the heroines of the ancient world, the martyrs of Christianity, as well as paintings illustrating the crucifixion of Jesus, which he briefly referred to as "expressions of pain."[90]

Mantegazza's distinction tells us that he was keenly aware of the differences existing between these two kinds of sources. While the works of art were only intended to move our feelings, the experimental photographs became scientific facts that could lead to the discovery of new laws on the expression of emotions. As he wrote in the introduction of the *Atlante*:

> Under my direction, the famous Florence photographer Brogi has been able to reproduce with rare skills some very fleeting emotions of the human face under direct sunlight, and thanks to them, I have found some very simple laws that govern the expression of pain. . . . By reproducing a man's face under the effects of physical pain with photography, you will find that the many and varied expressions that you are collecting correspond perfectly to certain groups of the expression of moral pains: so here is a new method of investigation that attempts to replace neither the inspirations coming from painting or sculpture nor those

[85] Lorraine Daston and Peter Galison, "The Image of Objectivity," *Representations* 40 (1992): 81–128. See also Daston and Galison, *Objectivity* (New York, 2007), 125.

[86] Paolo Mantegazza, "Dell'espressione del dolore," *Arch. Antropol. Etn.* 4 (1874): 5.

[87] Guillaume-Benjamin Duchenne, *Mécanisme de la physionomie humaine ou analyse électrophysiologique de l'expression de passions applicable a la pratique des arts plastiques* (Paris, 1862). See also François Delaporte, *The Anatomy of the Passions*, ed. Todd Meyers, trans. Susan Emanuel (Stanford, Calif., 2008), 49.

[88] Mantegazza, *Atlante* (cit. n. 7), 2.

[89] Ibid., 16.

[90] Ibid., 5–14.

coming from dramatic art but rather helps them to provide a rich multitude of facts, drawn from the inexhaustible source of nature.[91]

Mantegazza's photographs were transformed into scientific facts by introducing some protocols, such as systematically shooting full frontal and profile pictures that allowed him to measure gestures according to the relative cranial proportions established for each ethnological group.[92] Shooting analogous pain photographs with the same oval frame and on a neutral-colored background was also used as a rhetorical strategy to homogenize pictures and present them as data that could provide scientific evidence for the law that he had previously enunciated.[93]

The use of these protocols together with other instruments, such as pendulums for determining facial angles, would become the characteristic approach of the anthropological school flourishing in late nineteenth-century Florence, where Mantegazza also founded the Museum of Anthropology and Ethnology in 1869. The main concern of this Florentine school was to develop anthropometry, a science that measured man in his own habitat and compared his cultural expressions to those observed in other living societies.[94]

Indeed, one of the most famous scientists who applied anthropometry to the newest science of criminology was Lombroso, Mantegazza's controversial colleague, who would also praise the virtues of the camera in identifying socially deviant groups such as malefactors, prostitutes, and the insane. In this respect, Mantegazza's *Atlante* helped establish photography as a fairly common practice within the late nineteenth-century Italian positivist school of thought. The proponents of photography thought it could be used to capture the distinctiveness of humankind by exploring the expression of emotions in normal as well as pathological states.

However, Mantegazza's *Atlas of Pain* had not only a scientific but also an ideological purpose. The fact that he organized the experimental photographs that he took with Brogi along with images belonging to the art of ancient Greek culture, Christianity, and the Renaissance, which were preserved in Italian museums and, notably in Florence, was intended to shape the identity of the new Italian citizen by reviewing the value of his inherited artistic tradition. Moreover, he consciously made a comparison with other physiognomies from distant cultures, such as Polynesians, Jews, Japanese, and what he called "the Negroes."[95]

Unfortunately, Mantegazza's anthropological vision represented one of the most extreme examples of late nineteenth-century Western colonialism when, for instance, he asserted that "the expression of the white man is higher than that of the Negro, and the latter higher than that of the ape."[96] This type of judgment, which Mantegazza formulated as the "Morphological Tree of the Human Race," would also orient the series of photographs that he made with Brogi, when they reproduced the different ways in which Italians and black people expressed their suffering.[97]

---

[91] Ibid., 1.

[92] Piccardi, "L'espressione del dolore" (cit. n. 41), 90.

[93] On the role of photographs as scientific data, see Prodger, *Darwin's Camera* (cit. n. 64), xxiii. See also Cosimo Chiarelli, "Fissare le emozioni: Mantegazza e Darwin e la fotografia delle espressioni," *Arch. Antropol. Etn.* 128 (1998): 125–56.

[94] Monica Zavattaro, Maria Gloria Roselli, and Paolo Chiozzi, *Obiettivo Uomo* (Florence, 2010), 5. See also Verdicchio, *Looters* (cit. n. 79), 89.

[95] On the aesthetic inferiority of other races, see Mantegazza, *Physiognomy* (cit. n. 32), 75.

[96] Ibid., 87.

[97] Ibid., 312.

As will be discussed in the final sections of this article, Mantegazza's photographs known as *Espressioni del dolore prese dal vero* immortalized the experiments carried out on various individuals while he inflicted painful stimuli, illustrating what the performativity of pain means. Photographing pain was not only the best way to guarantee the self-control of the scientist, by mediating between his feelings and the suffering of the subject posing for the camera but was also used for culturally constructing the expression of pain in the white man in contrast to the black one.[98]

WHAT DOES PAIN LOOK LIKE?

In the aforementioned letter, which Mantegazza sent to Darwin on 23 December 1872, he also added some commentaries about the observations that he had made during his travels in America, which seemed to have been especially useful throughout the conception of his *Atlas of Pain*:

> Injuries to self-esteem, when they cannot be exhibited externally, oblige man to an absolute immobility of the facial muscles, and one swallows saliva, exactly as when tasting very bitter substances. The physiognomy of a man with aloes in his mouth, as captured by photography, is identical to the expression that a man takes on involuntarily when he is humiliated. I have found other striking analogies between physical pain and pain of a higher order as you will see in my album of photographs taken from nature.[99]

Mantegazza's observation of pain in other cultures during his travels in South America led him to establish some analogies, such as the common linguistic expressions used in Spanish, Italian, and French "to express the humiliation which follows an offense."[100] All these terms—the Spanish *trajar*, the Italian *ingoiare*, and the French *avaler*—indicated that injuries to self-esteem were described through references to gustatory pain, such as those related to disgust, which literarily means "something offensive to the taste."[101] As he commented to Darwin in his letter, this kind of anthropological observation provided him with inspiration for the formulation of his new law on the expression of pain, "which explains many obscure facts of human expression and of the highest psychology."[102]

The organization of the experimental photographs included in his *Atlas of Pain* was intended, indeed, to corroborate his law, which stated that the expression of specific pains of the senses displayed "the artifices of defense, as well as other laws of sympathy that connect each sense with a given region of the brain, and, in consequence, of feeling and thought."[103] Therefore, he classified pictures following the tripartite schema that he had previously applied to the production of pain in his laboratory, a choice that also enabled him to examine physical and moral pain as two entangled aspects of this emotion, which appeared gradually throughout the *Atlante*.

According to his previous classification of painful expressions, plate number one brings together four photographs representing the suffering of specific senses. The first image depicts gustatory pain in a well-dressed, middle-aged Italian man after

---

[98] White, "Darwin's Emotions" (cit. n. 12), 820.

[99] Burkhardt et al., *Correspondence of Charles Darwin* (cit. n. 65), 646–7.

[100] Ibid., 647. An account of Mantegazza's travel to South America can be found in Paolo Mantegazza, *Rio de la Plata e Tenerife, viaggi e studi di Paolo Mantegazza* (Milan, 1867).

[101] Darwin, *Expression of Emotions* (cit. n. 67), 255.

[102] Mantegazza, *Physiognomy* (cit. n. 32), 128.

[103] Ibid.

he has tasted the bark of *Quassia amara*, a particularly bitter Latin American plant.[104] As indicated by the caption situated below the picture, Mantegazza interpreted "the expression of gustatory pain, and especially that produced by a bitter taste," as having many similitudes "to that of the dumb anguish of self-esteem."[105]

The second photograph portrays olfactory pain by showing a more violent contraction of the facial muscles in the same subject after smelling ammonium sulfate, which we can imagine as the content of the glass that he is holding with his right hand in the picture.[106] Drawing once again on the analogies between the expression of physical and moral pain, Mantegazza interpreted the expression of olfactory pain as being intimately related to those of contempt and disdain, because we react in a similar way to "a vile thing or an infamous man."[107] In accordance with Mantegazza's instructions, which were inspired by Darwin's evolutionary theory, Brogi captured the gesture of the Italian man closing his nostrils as the particular expression of a man's offended dignity.[108]

Alternatively, the third photograph represents Mantegazza himself suffering from an acoustic pain produced by the unpleasant sound of glass being scratched with nails.[109] Thus, the scientist appears in a dramatic stance, revealing not only the movement of facial muscles but also the tension of his whole gesture. He interpreted his expression of acoustic pain performed by himself as being analogous to those manifesting moral injuries. Because "hearing is the sense most intimately and closely associated with feeling," wrote Mantegazza, "the expression of the specific sense of hearing is identical to that of the most cruel wounding of our affections."[110]

Finally, the expression of visual pain was reproduced in Brogi's studio by exposing the same person who expressed gustatory and olfactory pain in the first and the second photographs to a bright light.[111] This time, however, the man frowns slightly, suggesting that his eyes are suffering as he contracts "the muscles which are in anatomical and physiological relations with the orbicular of the eyelids."[112] In Mantegazza's view, his expression was closer to the physiognomy of someone portraying intellectual pains "of the most elevated nature," such as the reaction resulting from aesthetic disgust provoked by the contemplation of an ugly sculpture or painting.[113]

Mantegazza presented these analogies between the pain of the senses and moral pains as the result of long scientific considerations. However, the photograph included in the *Atlante*, in which he appeared representing acoustic pain, was classified in Darwin's photographic collection as depicting visual pain, according to the legend written in French by Mantegazza himself (see fig. 1). This was proof that photography always involves interpretation by an observer, as the Italian scientist actually changed his mind sometime between February 1872, when he sent these photographs to Darwin, and 1876, when the *Atlante* was published.

In spite of Mantegazza's enthusiastic defense of photography as a method that was exempt from any subjective interpretation, he was uncertain about the meaning of the

[104] Mantegazza, *Atlante* (cit. n. 7), 2.
[105] Mantegazza, *Physiognomy* (cit. n. 32), 130.
[106] Mantegazza, *Atlante* (cit. n. 7), 2.
[107] Mantegazza, *Physiognomy* (cit. n. 32), 130.
[108] Ibid. On the expression of contempt, see Darwin, *Expression of Emotions* (cit. n. 67), 255.
[109] Mantegazza, *Atlante* (cit. n. 7), 2.
[110] Mantegazza, *Physiognomy* (cit. n. 32), 129.
[111] Mantegazza, *Atlante* (cit. n. 7), 2.
[112] Mantegazza, *Physiognomy* (cit. n. 32), 129.
[113] Ibid.

Figure 1. Photograph by Carlo Brogi depicting specific visual pain. Expression produced by sunshine. Darwin's photographic collection. Reproduced with the permission of Cambridge University Library.

pictures that he used to illustrate the physiognomy of pain. He therefore consciously introduced captions to control what he wanted the viewer to observe.[114] Furthermore, we know that Mantegazza carefully selected the pictures, which he took with Brogi, even discarding seven of them because he thought that they did not properly represent

[114] Mantegazza, *Atlante* (cit. n. 7), 2.

the expression of pain.[115] All these decisions show that Mantegazza had consciously defined what pain should look like throughout the photographs that he included in the *Atlante*.

Plate number two was intended to capture the expression of pain in black people (see fig. 2).[116] Mantegazza repeated the same experiments on sensory pain, allowing Brogi to take similar profile pictures in order to show the variation in the expressions of pain of different individuals. By examining the expression of pain in the black man, Mantegazza was attempting to demonstrate that even if all humans expressed emotions in the same way according to Darwin's theory of evolution, there were also racial differences due to "the prominence of the facial muscles," revealing the "grotesque and simian expressions" of "the Negroes and the Negritoes."[117] As he explained in his *Physiology of Pain*,

> With the Negro, the expression of pain is distorted, hard, tumultuous, very bestial, as the muscles of the face are not contracted in isolation or in small groups; all of them contract and relax together, marking only the major traits and the most characteristic of emotion.[118]

The reiteration of pain gestures represented in Mantegazza's *Atlas of Pain* were in this way performative, as it was intended to give emotional meaning to the physiognomy of the black man when compared to the white man.[119] The captions that accompanied the pictures were also aimed at defining what the word "Negro" meant by repeating this term, which had a long history in European science with strong colonialist implications.[120] What Mantegazza considered the more distorted reactions exhibited by the black man while he experienced different types of sensory pain were deliberately highlighted through these captions in order to provoke certain disgust in the white observer, who would recognize in another man's skin color a kind of social transgression against what were considered the civilized norms of expression in late nineteenth-century Europe.

As Sara Ahmed has pointed out, the feeling of disgust is linked to "a history, whereby the mobility of white European bodies involves the transformation of native bodies into a knowledge, property and commodity."[121] Darwin had already explained the evolutionary meaning of disgust in relation to his experience at Tierra del Fuego, when he "felt utter disgust" at his food while "being touched by a naked savage, though his hands did not appear dirty."[122] Mantegazza also stressed the importance of disgust in his conception of the *Atlante*, when he wrote in his letter to Darwin about his observations on the expression of gustatory pain in other cultures. In this manner, disgust helped Mante-

---

[115] Today these photographs are archived at Fondo Mantegazza, Museo di Storia Naturale di Firenze, *Espressioni del dolore prese dal vero*, 7883, 7885, 7886, 7887, 7888, and 7889. On these photographs, see also Cossimo Chiarelli, "Mantegazza e la fotografia: una antología de immagini," in *Paolo Mantegazza e l'Evoluzionismo in Italia*, ed. Cossimo Chiarelli and Walter Pasini (Florence, 2010), 95–120.

[116] Mantegazza, *Atlante* (cit. n. 7), 3.

[117] Prodger, *Darwin's Camera* (cit. n. 64), 214; Mantegazza, *Physiognomy* (cit. n. 32), 232.

[118] Mantegazza, *Fisiologia del dolore* (cit. n. 7), 320.

[119] On repetition of the expression of emotions as a technique to produce social norms, see Sara Ahmed, *The Cultural Politics of Emotion* (New York, 2004), 12.

[120] On the meaning of the term "Negro" and its colonialist implications, see Cristina Malcolmson, *Studies of Skin Color in the Early Royal Society* (Farnham, 2013), 3.

[121] Ahmed, *Cultural Politics* (cit. n. 119), 82.

[122] Darwin, *Expression of Emotions* (cit. n. 67), 255.

Figure 2. Photograph by Carlo Brogi depicting specific gustative pain. Expression in a black person produced by *Quassia amara*. Darwin's photographic collection. Reproduced with the permission of Cambridge University Library.

gazza not only to discover new analogies between physical and moral pain but also to differentiate Europeans from other native cultures having a different "ethnic sensibility."[123]

---

[123] Mantegazza, *Fisiologia del dolore* (cit. n. 7), 186. The feeling of disgust was also a central component of the working definition of pain that Mantegazza had proposed in his physiological work, like "an alteration in the sensitivity that results in extreme disgust in those who experience it."

Thus, by creating a parallel between the repulsive reaction provoked by some food that could potentially cause disease and contamination with a community of persons, which did not fit with the Western social norms of expression, disgust became a way of picturing what pain should look like in Mantegazza's *Atlas of Pain*. His negative interpretation about the brutality represented by the expressions of the Negro made sure that the potential spectator would differentiate himself from primitive people, establishing an irremediable identification with the higher moral values embodied by the white man that were associated with the portraits representing well-dressed and educated Italian men.

It was not just a coincidence that, in plate number three, Mantegazza identified "diverse expressions of pain obtained . . . by crushing intensely the fingers of their two hands" that were "analogous to those situated in the highest of the hierarchy of moral pain" with the refined physiognomy of a white man like himself, who expressed emotions in a more sophisticated way as a result of the civilization process.[124] He was convinced that "children, the insane and savages could not feel" the delicate sufferings of melancholy, as well as those corresponding to other intellectual pains, such as ennui and hypochondria, the very disease from which Mantegazza is said to have suffered.[125]

By showing how pain shaped "other bodies," such as those of black people, Mantegazza was trying to define the expressions through which Italians should express their suffering in a proper manner, according to what he called "the Aesthetic Tree of the Human Race," which considered Europeans not only the apogee of human progress but also the embodiment of the beauty of civilization. From Mantegazza's colonialist point of view, pain became a kind of emotional practice that he mobilized with the help of Brogi's camera in order to shape national identities, such as those representing the recently created Italian state.[126]

THE PERFORMATIVITY OF PAIN

From a history-of-emotions perspective, Mantegazza's science is an excellent example with which to illustrate what the performativity of pain means and what this emotion does when shaping bodies according to late nineteenth-century gender, race, class, or human-animal distinctions. Rather than regarding pain as an abstract concept, Mantegazza preferred to analyze carefully its influence on physiological processes such as breathing and nutrition, as well as on the changing physiognomy of the white man in comparison to his black counterpart. When vivisection was conducted, for instance, pain appeared to result from repeatedly applying a set of scientific practices that were an integral part of the so-called practices of experimentation and observation, as well as those aimed at creating an unemotional scientific self.

Drawing on recent work on social theory, cultural studies, and medical humanities, the performativity of pain has been explained in two complementary ways throughout this article, each corresponding, respectively, to the physiological and psycholog-

---

[124] Mantegazza, *Atlante* (cit. n. 7), 4.
[125] Mantegazza, *Fisiologia del dolore* (cit. n. 7), 251.
[126] On the role of new evolving media techniques aimed at mobilizing emotions, see Anja Laukötter, "How Films Entered the Classroom: The Sciences and the Emotional Education of Youth through Health Education Films in the United States and Germany, 1910–30," in this volume.

ical research conducted by Mantegazza in the laboratory, as well as in the photographic studio of his colleague Carlo Brogi.[127] Mantegazza's dual education as a physician and an anthropologist, one of the first to hold a chair in this discipline in Europe, allowed him to understand pain as a phenomenon that was biologically conditioned, but also culturally constructed and socially situated. In Mantegazza's science, experience and expression do not work as dichotomies that describe the inner and the outer reality of emotions, but rather as reciprocal processes through which we may shed light on the meaning of pain. This correspondence is ultimately explained through his positivist approach, which led him to understand psychological pain as almost an epiphenomenon of physical pain, whose multiple emotional manifestations—ranging from suffering to pleasure or fear—resulted in an irremediably "subjective experience, a fact of consciousness."[128]

Throughout his physiological research, Mantegazza understood pain as a phenomenon that could be reproduced, observed, and controlled in the laboratory through the use of controversial experimental practices, such as vivisection. As described in his *Physiology of Pain*, Mantegazza's experiments were aimed at examining the vital changes undergone by animals while they endured various painful stimuli, such as having numerous sharp nails driven through their feet and limbs. Vivisection brought into question not only the moral ethos of the physiologist but also his objective vision, because of his inability to control his own feelings when scrutinizing another being's suffering. Even if the physiologist was trained to remain impassive toward animals' blood and viscera, Mantegazza recognized that he was irremediably touched when observing the agonies of other living beings.

Therefore, Mantegazza did not hesitate to develop some strategies in order to silence his own natural feelings toward others' suffering within his experimental research. For instance, he proposed using animals with few similarities to humans for vivisection purposes. He was convinced that lower species were likely to feel less suffering than highly organized beings and, above all, because their sorrows were so markedly different from ours, that they could not move the tenderhearted feelings of the physiologist. While the pain of animals only deserved our pity as humans, the future elimination of suffering in mankind became the overriding goal of the modern physiologist, who justified his profession in utilitarian terms as the best manner to progress toward creating a world "in which only pleasure would measure" human action.[129]

This type of disquisition on the sensibility of certain living beings shows us that the late nineteenth-century debate about vivisection implied profound philosophical questions, such as what is considered a human, and which types of beings seem closer to animals, such as children, women, or black people. Apparently innocent metaphysi-

---

[127] On performativity, see Ahmed, *Cultural Politics* (cit. n. 119), 4; Monique Scheer, "Emotions" (cit. n. 8), 193–210; Alberti, *Medicine, Emotion and Disease* (cit. n. 9), xviii; Michelle Z. Rosaldo, "Toward an Anthropology of Self and Feeling," in *Culture Theory: Essays on Mind, Self and Emotion*, ed. Richard A. Shweder and Robert A. LeVine (Cambridge, 1995), 137–57; William M. Reddy, *The Navigation of Feeling: A Framework for the History of Emotions* (Cambridge, 2001); Jo Labanyi, "Doing Things: Emotions, Affect and Materiality," *J. Spanish Cultur. Stud.* 11 (2010): 223–33; Ian Burkitt, *Emotions and Social Relations* (Los Angeles, 2014); Jan Plamper, *The History of Emotions: An Introduction* (Oxford, 2015), 265–70.

[128] Mantegazza, *Fisiologia del dolore* (cit. n. 7), 12. On Mantegazza's physiological psychology, see Giovanni Landucci, *Il Darwinismo a Firenze: Tra scienza e ideologia (1860–1900)* (Florence, 1977), 181.

[129] Mantegazza, *Fisiologia del dolore* (cit. n. 7), 18.

cal distinctions between the different levels of sensibility perceived in organisms were used to account for a politics of pain that justified power relations and excluded certain minorities from the realm of suffering, such as women and the other races in Mantegazza's case, or the mentally ill and the socially degenerated in Lombroso's criminology. As an experienced anthropologist, Mantegazza was fully aware of the capacity that pain had for constructing a socially ordered world according to gender, class, and race differences: a perfect world like ours, in which certain pains matter, while others, such as those of immigrants attempting to arrive in Europe, do not.

The performativity of pain has also been discussed in relation to Mantegazza's physiognomic work, which included, notably, his hilarious visual study on the expression of pain, the *Atlante della espressione del dolore*. While his Darwinian-oriented investigation of expression stated that the primary functions of the mimicry of pain were defense and sympathy, Mantegazza's *Atlas of Pain* introduced the camera as a scientific instrument that enabled the examination of the fleeting action of emotions in human physiognomy. In the *Atlante*, photography became not only a technique of observation that allowed the scientist to distance himself from the emotional reaction that he could potentially feel when studying other bodies in pain but also a useful method for deducing some fundamental laws revealing the analogies existing between the expression of physical and moral suffering, such as those related to gustatory pain and the humiliation of self-esteem.

Although Mantegazza claimed in numerous passages of his work that his "experimental photographs" were revealing nature as it was, the fact is that they were reproducing previously planned physiological experiments aimed at confirming his ideas about the common expressions perceived between pain related to the senses and other higher pains, such as those concerning feelings and intellect.[130] In this respect, Mantegazza's photographs were constructing what pain should look like, by means of describing certain experimental conditions, such as the intensity of light or the reproduction of violent noises, as well as the introduction of some anthropometrical protocols such as systematically taking full and profile pictures. Thus, photography was not just a mirror of reality but also a strategy that helped Mantegazza to give emotional meaning to certain bodies, such as those of the white man, while marginalizing those of other individuals, including those he referred to as "the Negroes" and "the Negritoes."[131]

The performativity of pain became notably apparent throughout the organization of the captions that Mantegazza included in the *Atlante*, as they were arranged in order to show the different ways through which white and dark-complexioned physiognomies expressed their suffering. The repetition of shots reproducing physical pain related to the senses in "civilized" bodies in contrast to the "primitive" ones was intended to reinforce our identification with the expression of the white man, by means of provoking our disgust toward the suffering articulated by other physiognomies, which were portrayed as brutal and, therefore, violating the European cultural norms of feeling.[132] Thus, the performativity of pain allowed Mantegazza to shape not only physical but also political bodies by showing what the expression of suffering should look like in late nineteenth-century Western culture and, notably, when Italians posed for the camera to immortalize their "beautiful profile."[133]

[130] Mantegazza, *Atlante* (cit. n. 7), 1.
[131] Mantegazza, *Physiognomy* (cit. n. 32), 232.
[132] On the performativity of disgust, see Ahmed, *Cultural Politics* (cit. n. 119), 23–8.
[133] Verdicchio, *Looters* (cit. n. 79), 87.

On the whole, Mantegazza's science of emotions revealed to what extent pain became a practice that was socially controlled by a minority of late nineteenth-century physiologists who were often involved in politics. These Italian scientists were fully convinced that by studying emotions, they would improve health conditions and, thereby, increase the degree of civilization of the population. After having achieved the political unification of the country, the main concern of the Italian intelligentsia was to define the social agent that should embody the national identity of the recently created state.[134] This is why late nineteenth-century experimental scientists such as the criminologist Cesare Lombroso, the physiologist Angelo Mosso, and Mantegazza showed a special interest in studying emotions in both normal and pathological states, as they conceived science as an extension of politics, which was focused on making Italians modern citizens.

From a Bourdieuian approach to emotion history, Manteggaza was reenacting the style of thinking that flourished during the last decades of the nineteenth century within the Italian scientific community, combining the influence of social Darwinism with the emergence of a distinctive Italian positivist school.[135] As possibly one of the most emblematic personalities representing this style of thinking, Mantegazza believed that the progress of society could be achieved only through the government of emotions such as pain, an energy that science would gradually eliminate through the invention of more effective anesthetics. His efforts to popularize this modern conception of pain might be regarded, therefore, as a central political strategy aimed at secularizing late nineteenth-century Italian society against the strong influence that the Catholic Church exerted upon spiritual matters.

Beyond Italy, "the warfare between science and theology generated by the debates over evolution" also reveals some relevant connections between Mantegazza's science and broader emotional cultures, such as the late Victorian shift to what Peter Stearns has called a "cool emotionology," which was most noted for control and discipline with regard to emotions.[136] As the social arena in which we can probably best perceive the internalization of this cool and restrained emotionology, late nineteenth-century science revealed a particular commitment to the experimental study of emotions in order to control them outside the laboratory, in the broader context of Western societies.[137] From this integrated history-of-science and history-of-emotions perspective, Mantegazza's science represents a crucial step in the battle against human pain. Indeed, Mantegazza had already envisioned victory in this battle when he described in his science fiction book, *The Year 3000*, a world in which conflicts have been abolished, economic barriers and cultural differences have disappeared, and "physical suffering no longer exists."[138]

---

[134] Stewart-Steinberg, *The Pinocchio Effect* (cit. n. 3), 1.

[135] The foundational manifesto of this Italian positivist school was written by Salvatore Tommasi, *Il naturalismo moderno: Discorso inaugurale pronunziato il di 15 Novembre 1866 nell'Università di Napoli* (Naples, 1866). See also Pancaldi, *Darwin in Italy* (cit. n. 2), 159.

[136] Peter Stearns, *American Cool: Constructing a Twentieth-Century Emotional Style* (New York, 1994), 149.

[137] Dror, Hitzer, Laukötter, and León-Sanz, "Introduction" (cit. n. 5), 36.

[138] Mantegazza, *The Year 3000* (cit. n. 1), 76.

Tempering Madness:
Emil Kraepelin's Research on Affective Disorders

by Eric J. Engstrom*

This essay examines some of the research practices and strategies that the German psychiatrist Emil Kraepelin (1856–1926) deployed in his efforts to account for the significance of emotions in psychiatric illnesses. After briefly surveying Kraepelin's understanding of emotions and providing some historical context for his work in the late nineteenth century, it examines three different approaches that he took to studying emotions. First, it discusses his work in experimental psychology and his use of so-called artificial insanity to study affective disorders. It then turns to his clinical research, exploring his particular interest in the course and outcome of psychiatric disorders and then showing how those concerns related to his nosological delineation of manic depressive illness. Finally, it considers briefly how he attempted to expand his "clinical gaze," turning it outward onto larger, nonhospitalized populations in an attempt to study subclinical forms of affect or temperaments. The article argues that the inadequacies and limitations of his own experimental and clinical research practices contributed to his evolving understanding of affective disorders. In particular, they led him to expand and differentiate his understanding of manic-depressive illness so as to take greater account of premorbid symptoms or temperaments.

INTRODUCTION

The German psychiatrist Emil Kraepelin (1856–1926) is probably best known today for having distinguished between dementia praecox (or what we have come to call schizophrenia) on the one hand and manic-depressive illness on the other. This basic dichotomy of the so-called endogenous psychoses has been enormously influential in the history of twentieth-century psychiatry. Over the past few decades, Kraepelin's legacy has been appropriated by neo-Kraepelinian psychiatrists who have been eager to enshrine him as their historical lodestar—as a new, post-Freudian "father" of clinical psychiatry, of the Diagnostic and Statistical Manual of Mental Disorders (DSM), and of applied quantitative methods. They have repeatedly evoked his name in support of efforts to strengthen the biomedical model of mental disorders, to maintain psychiatry's status within the mental health professions, and to reinforce its position as a research-based medical specialty.[1] Indeed, what some have described as a "rev-

---

* Department of History, Humboldt University, Berlin, Germany; engstroe@geschichte.hu-berlin.de.
[1] See Rick Mayes and Allan V. Horowitz, "DSM-III and the Revolution in the Classification of Mental Illness," *J. Hist. Behav. Sci.* 41 (2005): 249–67; Lucille Parkinson McCarthy and Joan Page Gerring, "Revising Psychiatry's Charter Document DSM-IV," *Written Comm.* 11 (1994): 147–92.

olutionary"[2] paradigm shift toward biological psychiatry in the late twentieth century has gone hand in hand with recourse to Kraepelin's work. Kraepelin's legacy as a nosologist has manifested itself in numerous historical accounts of the convoluted paths that led to his system of classification. These accounts have often taken the form of stories about the "lumping" and "splitting" of psychiatric symptoms, about the serendipitous causes and salubrious effects of his work.

With regard to affective disorders, one of the striking characteristics of this historiography involves the encroachment of two anachronisms upon interpretations of Kraepelin's work. On the one hand, there is a tendency to locate bipolar disorders in the works of early modern and even ancient writers. For example, one recent authoritative account presumes that the ancient Greek physician Aretaeus von Cappadocia described "bipolar disorders" in the first century CE.[3] And on the other hand, today's neo-Kraepelinian researchers are often eager to find evidence in Kraepelin's work that supports their own findings.[4] For a number of reasons, such narrow historiographic perspectives are problematic. For one, they provide no resources with which to resist the reification of Kraepelin's nosology: they remain silent in the face of efforts to chart seemingly timeless lines of continuity stretching from antiquity to DSM-V. Furthermore, they underscore the impoverishment of our historical imagination and of our knowledge about the specific processes that led to the construction of Kraepelin's nosology.

To point to these anachronisms is not to suggest that Kraepelin's work was without precedent, emerging ex nihilo onto the historical stage of psychiatric nosology. Nor is it to discount his influence on the work of late twentieth-century psychiatrists. It is, however, to note that his nosology is often situated in closer proximity to Hippocrates and DSM-III than to his own research practices in Heidelberg or Munich in the years leading up to World War I. Indeed, the very practices and situative contexts from which Kraepelin's manic-depressive forms emerged are usually ignored entirely or simply noted in passing.

And so this article shifts attention away from the larger architecture of his nosology to focus instead on some of the research practices and strategies that he deployed in his efforts to account for the significance of emotions in psychiatric illnesses. I will be fashioning a narrative that relates affective symptoms as Kraepelin understood them to his diagnostic and research practices, and in particular showing how those practices—especially their inadequacies and limitations—contributed to his evolving understanding of manic-depressive illness. After briefly surveying Kraepelin's understanding of emotions and providing some historical context to his work in the late nineteenth century, I will examine three different approaches that he took to studying emotions. First, I will discuss his work in experimental psychology and his use of so-called artificial insanity to study affective disorders. From there I will turn to his clin-

---

[2] W. M. Compton and S. B. Guze, "The Neo-Kraepelinian Revolution in Psychiatric Diagnosis," *Eur. Arch. Psychiat. Clin. Neurosci.* 245 (1995): 196–201; Gerald L. Klerman, "Psychiatric Diagnostic Categories: Issues of Validity and Measurement," *J. Health Soc. Behav.* 30 (1989): 26–40, on 31.

[3] Andreas Marneros and Jules Angst, "Bipolar Disorders: Roots and Evolution," in *Bipolar Disorders: 100 Years after Manic-Depressive Insanity*, ed. Andreas Marneros and Jules Angst (Dordrecht, 2000), 1–35, on 6–7.

[4] See, e.g., A. Jablensky, H. Hugler, M. von Cranach, and K. Kalinov, "Kraepelin Revisited: A Reassessment and Statistical Analysis of Dementia Praecox and Manic-Depressive Insanity in 1908," *Psychol. Med.* 23 (1993): 843–58.

ical research, exploring his particular interest in the course and outcome of psychiatric disorders and then showing how it related to his nosological delineation of manic depressive illness. I will then briefly consider how he attempted to expand his "clinical gaze," turning it outward onto larger, nonhospitalized populations in an attempt to study subclinical forms of affect and temperaments. Finally, I will conclude with some general remarks about the challenges and pitfalls facing the history of emotions at its interface with the history of science, in particular with the history of the so-called psy disciplines of psychology and psychiatry.

KRAEPELIN'S VIEWS ON EMOTIONS

Kraepelin's views on emotions mirrored those of his academic mentor and longtime friend, the experimental psychologist Wilhelm Wundt (1832–1920).[5] In his widely read *Vorlesungen über die Menschen- und Thierseele* (1863–64), Wundt interpreted emotions as sensations that either lacked an external cause or were so intense as to profoundly alter the sensory apparatus and/or physiology. Like other sensations, these sensory emotions could but did not need to be conscious or to have an impact on cognition. In a mental act of apperception, however, the multitude of bodily sensations or feelings came to consciousness as one *Gemeingefühl*. On this physiological foundation, Wundt then built a hierarchy of more complex emotions and moods [*Affekte, Stimmungen*], which in turn gave rise to drives and volition. Crucially, in the process of distinguishing between external stimuli and internal sensations or feelings, Wundt posited a role for emotions in the development of consciousness and, in particular, of a sense of self. Unlike other sensations or ideas, the common distinguishing feature of all different kinds of emotions was their subjectivity, that is, their reference to the self as distinct from the outer world.[6] As Claudia Wassmann has shown, Wundt's wider understanding of consciousness was predicated on this reflexive work of the nervous system and the brain. And for this essentially self-reflexive characteristic of emotions, Wundt drew decisively on the widespread contemporary physiological model of the reflex arc.

Kraepelin also grounded emotions in sensory physiology. Like Wundt, he argued that "every impression of the senses that rises to the level of consciousness creates within us, in addition to the perception itself, a peculiar change in our mental condition that we call an emotion."[7] And again, like Wundt, he understood emotions as a direct characterization of "the position of the self with respect to the perceptions of the external world."[8] He effectively welded emotions to sensory perceptions—be

---

[5] On Wundt's theory of emotions and its historical contexts, see Claudia Wassmann, "Physiological Optics, Cognition and Emotion: A Novel Look at the Early Work of Wilhelm Wundt," *J. Hist. Med. Allied Sci.* 64 (2009): 213–49; and most recently, Wassmann, "'Picturesque Incisiveness': Explaining the Celebrity of James's Theory of Emotion," *J. Hist. Behav. Sci.* 50 (2014): 166–88. According to Wassmann, Wundt developed the "first natural scientific conception of emotion" ("Physiological Optics," 213).

[6] As Wassmann puts it, "The sole difference between an emotion and an idea was that an emotion was interpreted as referring to the Self, whereas an idea was interpreted as referring to the external world. An emotion contained an activity of the subject, whereas an idea was directed to the view of objects." Wassmann, "Physiological Optics" (cit. n. 5), 232.

[7] Emil Kraepelin, *Psychiatrie: Ein Lehrbuch für Studierende und Ärzte*, 8th ed., 4 vols. (Leipzig, 1909–15), 1:338.

[8] Emil Kraepelin, *Psychiatrie: Ein Lehrbuch für Studierende und Ärzte*, 6th ed., 2 vols. (Leipzig, 1899), 1:185. See also Kraepelin, *Psychiatrie*, 8th ed. (cit. n. 7), 1:338.

they conscious or not—and to the psychological correlates of those perceptions, that is, to mental images or *Vorstellungen*. Grounded in sensory physiology, mental images were therefore inherently laden with affect. As far as Kraepelin was concerned, "mood and mental image [*Stimmung* and *Vorstellung*] were simply different expressions of one and the same process."[9] To account for the dynamics of emotional life, Kraepelin also drew directly on Wundt's distinction of three specific affective polarities: pleasure and displeasure [*Lust/Unlust*], excitation and relaxation [*Erregung/ Beruhigung*], tension and resolution [*Spannung/Lösung*].[10] And again like his mentor, he also posited a distinction between lower sensory emotions and higher general emotions (logical, ethical, artistic, and religious emotions).[11]

Kraepelin never seems to have deviated markedly from this basic Wundtian model of emotions. Nor was he much interested in the content (subjective or otherwise) of his patients' emotions. As Paul Hoff has written, his interests went instead to the formal, that is, clinical manifestations of emotions [*Gemütsbewegungen*].[12] But with respect to affective disorders, it is worth noting that emotions per se were not pathogenic. Far more important was a personal disposition and the manner in which one emotionally managed life experiences. It was a lack or failure of a certain mental ability to "compensate for emotional dangers" and to attain emotional equanimity that was pathogenic.[13] Presuming the existence of higher- and lower-order emotions, Kraepelin believed that it was not simply willpower or reason—themselves imbued with emotional content by virtue of their origins in sensory physiology—but also the higher-order, encultured emotions that could control deeper passions and mood swings.

HISTORICAL CONTEXTS: UNITARY PSYCHOSIS AND NEUROPSYCHIATRY

To understand some of the implications of these ideas for psychopathology, two historical contexts are especially noteworthy. The first concerns the notion of "unitary psychosis" [*Einheitspsychose*] that dominated much mid-nineteenth-century German thinking about madness.[14] The theory posited the existence of a single mental illness that evolved through different stages, beginning with melancholy and then proceeding through states of mania and more severe delusional/psychotic conditions [*Wahnsinn/Verücktheit*] and culminating in the complete dissolution of the mental personality (dementia). This model of mental illness situated affective disorders in the early or "primary" stages of insanity, and nineteenth-century asylum psychiatrists—so-

---

[9] Kraepelin, *Psychiatrie*, 8th ed. (cit. n. 7), 1:361.

[10] See ibid., 1:338. See also Wassmann, "Physiological Optics" (cit. n. 5). Throughout his textbook—although not systematically—Kraepelin attributed various combinations of these polarities to specific mental conditions.

[11] See Kraepelin, *Psychiatrie*, 8th ed. (cit. n. 7), 1:341.

[12] Paul Hoff, "Der Affekt in den Psychosekonzepten von Emil Kraepelin und Eugen Bleuler," *Psychiat. Psychotherapie* 6 (2010): 98–103, on 101.

[13] Kraepelin, *Psychiatrie*, 8th ed. (cit. n. 7), 1:123–4.

[14] See Josef Vliegen, *Die Einheitspsychose: Geschichte und Problem*, Forum der Psychiatrie, n.s., 9 (Stuttgart, 1980); Ulrich Trenckmann, *Mit Leib und Seele: Ein Wegweiser durch die Konzepte der Psychiatrie* (Bonn, 1988), 121–61. According to Werner Janzarik, the term—if not the concept—dates from the late nineteenth century and was used pejoratively to describe an outdated nosological idea. Janzarik, "Forschungsrichtungen und Lehrmeinungen in der Psychiatrie: Geschichte, Gegenwart, forensische Bedeutung," in *Handbuch der forensischen Psychiatrie*, ed. H. Göppinger (Berlin, 1972), 1:588–662, on 596.

called alienists—believed that these early affective stages were more amenable to medical intervention.

In terms of psychiatry's institutional development, the concept of unitary psychosis was a stroke of professional genius because it effectively undergirded what can best be described as a dogma of early and rapid institutionalization. Alienists spared no opportunity to stress that illnesses, if they were identified in their early affective stages, had a better chance of being cured and were less likely to evolve into chronic conditions. Changes in mood were, in many respects, the early harbingers that warned of potentially more devastating mental incapacity.[15] Throughout much of the nineteenth and early twentieth centuries, the specter of further decline into debilitating chronic conditions drove widespread efforts to have affectively deviant patients institutionalized as early as possible.

Nevertheless, by the late 1860s, this model of unitary psychosis was becoming increasingly untenable. A growing number of studies had revealed internal contradictions and suggested the existence of several different kinds of mental illness that did not necessarily evolve from melancholy. Emil Kraepelin's research agenda and his efforts to classify psychiatric disorders as discrete pathological entities can be interpreted as a response to the collapse of the unitary psychosis model and the nosological limbo in which it left German practitioners.[16]

The second important historical context concerns the preeminence of neuropsychiatry in the 1870s and 1880s.[17] During these decades, psychiatric practitioners held out great hope that pathological anatomy and physiology would provide a somatically grounded explanation of mental illness. No one did more to encourage this belief than Wilhelm Griesinger (1817–68). His dictum that mental illness was brain disease inspired an entire generation of academically trained laboratory scientists such as Theodor Meynert (1833–92), Carl Westphal (1833–90), and Karl Wernicke (1848–1905). For these cerebral pathologists, the cause of mental illness lay in physical changes in the anatomic structure and physiology of the brain and could, by implication, be localized. They harbored deep skepticism toward clinical empiricism and began relocating their science away from the mental asylums and into university laboratories. But their high hopes of fusing mind and brain, of anchoring the psyche in neurological processes, were soon dashed for lack of reliable evidence and therapeutic applicability. Undaunted, however, they advanced a range of psychosomatic theories by extrapolating from their laboratory results to the clinical symptoms of madness.

An example of this kind of research can be seen in Theodor Meynert's article, "Über die Gefühle," in 1882.[18] Drawing on his observation of frogs, professors, and frustrated students of botany, Meynert grounded "feelings" in cerebral physiology, linking them with vasomotor innervation, blood flow, and the play of mental "asso-

---

[15] "Emotions are the most sensitive signs of all inner changes. In mental illness therefore, it is usually precisely patients' emotional accent [Gefühlsbetonung], their emotional disposition [Gemütsleben] that initially manifests the most obvious disruptions." Kraepelin, Psychiatrie, 8th ed. (cit. n. 7), 1:338.

[16] On this period, see Werner Janzarik, "Die klinische Psychopathologie zwischen Griesinger und Kraepelin im Querschnitt des Jahres 1878," in Psychopathologie als Grundlagenwissenschaft, ed. Janzarik (Stuttgart, 1979), 51–61; W. de Boor, Psychiatrische Systematik: Ihre Entwicklung in Deutschland seit Kahlbaum (Berlin, 1954), 10–19.

[17] See Wolfram Schmitt, "Das Modell der Naturwissenschaft in der Psychiatrie im Übergang vom 19. zum 20. Jahrhundert," Ber. Wissenschaftsgesch. 6 (1983): 89–101.

[18] Theodor Meynert, "Ueber die Gefühle," Jahrb. Psychiat. 3 (1882): 165–76.

ciations." Emotions were ultimately a form of subjective awareness of the "sum of physiological processes" and the expression of a sense of the "nutritional state of the cerebral cortex."[19] He interpreted the free and uninterrupted "play" of the "brain mechanism" and of "associations" with pleasurable feelings, the interruption and inhibition of that play with painful feelings.[20]

Kraepelin was decidedly skeptical of Meynert's efforts to link emotional—and, more generally, all mental—states directly with cerebral processes and structures.[21] He considered these linkages to be premature and highly speculative. He had long criticized Meynert and other neuropathologists for being more interested in the brains of rabbits and dogs than in the condition of the psychiatric patients on their hospital wards. In Kraepelin's view, the failings of early nineteenth-century romantic medicine had driven many later neuropathologists like Meynert to the opposite extreme, prompting them to adopt positions of "naive materialism." As a result, large swaths of their patho-anatomic research on the brain had become only peripherally significant to psychiatry proper.

In his own research, Kraepelin was trying to distance himself from this tradition of cerebral pathology by drawing on Wundt's experimental methodology to rehabilitate a psychological dimension to psychiatric research—a dimension that he believed had gone missing in a decidedly neuroanatomic era. Accordingly, his explanations of mental processes were much more somatically abstinent and psychologically informed than those of other contemporary neuropathologists. Kraepelin was not so much dismissing neurophysiology outright as underscoring the need to study mental processes without recourse to dubious linkages between patho-anatomic and clinical evidence. Adopting the psychophysical parallelism of his mentor Wundt,[22] Kraepelin sought to sever those linkages and to argue that psychological experimentation represented a more promising strategy for studying the mind. He argued that because Wundt had transformed psychology into a natural science, psychiatrists could now embrace it unreservedly and thus move the study of psychological processes to the fore of psychiatric research.

EXPERIMENTAL PSYCHOLOGY

Kraepelin employed many of the same experimental methods that he had observed as a student in Wundt's laboratory in Leipzig. At the core of his laboratory work in the 1880s stood the measurement of basic psychological reaction times. In countless

---

[19] Ibid., 173.

[20] Kraepelin, *Psychiatrie*, 8th ed. (cit. n. 7), 3:1370–1.

[21] See esp. Kraepelin's inaugural lecture in Dorpat: Emil Kraepelin, *Die Richtungen der psychiatrischen Forschung: Vortrag, gehalten bei der Übernahme des Lehramtes an der kaiserlichen Universität Dorpat* (Leipzig, 1887). But even decades later, Kraepelin and his associates were still debunking Meynert's views. See Kraepelin, *Psychiatrie*, 8th ed. (cit. n. 7), 3:1370–1; Wilhelm Weygandt, *Über die Mischzustände des manisch-depressiven Irreseins: Ein Beitrag zur klinischen Psychiatrie* (Munich, 1899), 61. This distinction again underscores the fact that Kraepelin was not quite the "brain psychiatrist" that some neo-Kraepelinian disciples have made him out to be. On the neo-Kraepelinians' misappropriation of Kraepelin, see Eric J. Engstrom and Kenneth S. Kendler, "Emil Kraepelin: Icon and Reality," *Amer. J. Psychiat.* 172 (2015): 1190–6.

[22] On the broader context of Wundt's psychophysical parallelism and the function it had in different discourses, see Mai Wegener, "Der psychophysische Parallelismus: Zu einer Diskursfigur im Feld der wissenschaftlichen Umbrüche des ausgehenden 19. Jahrhunderts," *Z. Gesch. Wiss. Tech. Med.* 17 (2009): 277–316.

stimulus-response experiments, he sought to quantify various mental processes. Slow response times and false starts, for example, could provide important clues about a nervous constitution or a disorder of sensory or neural functions. He sometimes concatenated experiments in an elaborate sequence designed to measure fatigue, attention span, memory, and so on. Experimental subjects were called upon to add numbers, memorize random syllables, or estimate intervals of time and physical stimuli.

The aim of this research was to develop a "quantitative individual psychology [*messende Individualpsychologie*]"[23] capable of grasping the basic mental characteristics of an individual.[24] In Kraepelin's mind, it should have been possible to use a battery of psychological tests to establish the "status psychicus"[25] of his mentally ill patients, in the same fashion that general medicine used chemical and physical tests.[26] Those tests, he hoped, would allow a "rapid psychological mapping [*Kennzeichnung*] of the individual."[27]

This research agenda had a number of professional advantages. By mimicking the rigor of the natural sciences, the psychological experiment was a bid to legitimate psychiatry's disciplinary practices and to advance claims of parity alongside other branches of medical science. Furthermore, in institutional terms, an effective battery of diagnostic tests was of great use in steering hospital admissions and in managing the distribution of patients within a larger system of institutional care. And finally, as a diagnostic tool, Kraepelin's *messende Individualpsychologie* had the potential to speed up diagnostic procedures and thereby regulate the flow of patients through his institution and optimize the conditions under which he could pursue his clinical research.

For the most part, Wundtian psychological methods were deployed to study cognitive function. But could they also be used to study affective disorders? Kraepelin believed they could, and to do so he adapted a technique described as "artificial insanity." Generating artificial insanity involved the inducement of mild symptoms of mental illness through physical or mental exhaustion or through the ingestion of various "poisons" such as tea, coffee, tobacco, alcohol, morphine, or sedatives like bromide. The ensuing states of mind were thought to mimic or even reproduce the emotional effects of mental illness.[28] Experiments conducted before and after the onset of artificial insanity provided Kraepelin with the quantitative evidence he sought in his study of affective disorders. Such psychotropic agents could, in his words,

[23] Emil Kraepelin, "Der psychologische Versuch in der Psychiatrie," *Psychol. Arbeit.* 1 (1895): 1–91, on 43. Kraepelin also expressed his agenda in terms of a "Physiologie der Seele" and a "Zergliederung krankhafter Seelenzustände." Ibid., 27, 28.

[24] Kraepelin used terms such as *persönlichen Grundeigenschaften* or *psychische Grundeigenschaften*. Ibid., 41–65; and Kraepelin, *Psychiatrie*, 6th ed. (cit. n. 8), 1:281.

[25] Emil Kraepelin, "Experimentelle Studien über Associationen," *Allg. Z. Psychiat.* 40 (1884): 829–31, on 829. Elsewhere Kraepelin speaks of the "status praesens." Cf. Kraepelin, "Der psychologische Versuch" (cit. n. 23), 65–79.

[26] Kraepelin, "Der psychologische Versuch" (cit. n. 23), 8.

[27] Ibid., 69. Kraepelin went so far as to outline in detail a five-day sequence of experiments that would test and evaluate the basic properties of personality.

[28] In his textbook *Psychiatrie* (cit. n. 7), vol. 1, Kraepelin explicitly stressed the relationship between changes in mood [*Stimmungswechsel*] and intoxication [*Erfahrungen des Rausches*] and underscored the importance of experimental methods in elucidating that relationship (343). And with reference to the pleasure [*Lustgefühl*] produced by alcohol, Kraepelin cited his psychological experiments as evidence of an affinity with manic agitation (357). Significantly, however, and in keeping with Wundtian assumptions about psychophysical parallelism, Kraepelin offered no explanation of the link between psyche and soma.

decisively alter the entire emotional disposition [*Gemütslage*]. We are unable to measure any of these effects on emotional life. But we can easily establish the type and size of the changes in the mental processes associated with them. And so perhaps it will be possible to use the range of these changes as a measure of the intensity of emotional effects. Furthermore, from the effects of the poisons we might conceivably derive conclusions about the unique characteristics of the existing emotional condition.[29]

So while Kraepelin could not directly quantify changes in emotional states, he sought to measure them indirectly through their manifestation in motor function, impulse, and cognitive ability.

It is important to underscore that, throughout his career, Kraepelin never abandoned the hope of applying experimental methods to the study of emotions.[30] For instance, in several of his last publications on occupational psychology in the early 1920s, he reiterated the important role that emotions played in workers' motivation and enthusiasm [*Arbeitsfreudigkeit*] and hoped that by establishing quantitative techniques to measure fluctuations in their productivity he could shed light on various "emotional processes."[31]

Nevertheless, Kraepelin also recognized the limitations of his own experimental methods. Indeed, various factors—including intractable experimental subjects, the limitations of the instruments and their use, cognitive bias, and so on—all conspired to temper his expectations about what those methods could achieve. Kraepelin recognized that experimental methods alone were fundamentally inadequate to the task of scientifically grasping affective symptoms.[32] Among other things, this inadequacy of experimental methods to fully map patients' psychological profiles drove him to expand the scope of his psychiatric research agenda.

CLINICAL RESEARCH

One of the psychological experiment's most significant limitations was the inability to capture more than an immediate snapshot of patients' mental processes. Therefore, in the 1890s, Kraepelin turned his attention to a clinical research project that would allow him to study not only the immediately visible and/or experimentally measurable symptoms at hand but also the entire course and outcome of his patients' illnesses. Indeed, the classification of psychiatric disorders on the basis of their course and outcome became one of the hallmarks of his clinical method and nosology. Kraepelin believed that specific mental diseases exhibited a common course and that by studying the evolution of patients' illnesses over time he could distinguish between different types of insanity. Furthermore, he believed that specific mental diseases

[29] Kraepelin, *Psychiatrie*, 6th ed. (cit. n. 8), 1:280.

[30] See Emil Kraepelin, "Ziele und Wege der psychiatrischen Forschung," *Z. Gesam. Neurol. Psychiat.* 42 (1918): 169–205, esp. 191–2.

[31] See Emil Kraepelin, "Arbeitspsychologische Ausblicke," *Psychol. Arbeit.* 8 (1925): 431–50, on 446–7; Kraepelin, "Gedanken über die Arbeitskurve," *Psychol. Arbeit.* 8 (1925): 533–47, on 546.

[32] See also Kraepelin, *Psychiatrie*, 8th ed. (cit. n. 7), 1:338–9: "But the assessment of these symptoms runs into certain difficulties because, far less so than in the realm of cognition, we lack a reliable guiding principle that could help us to determine exactly the gradual deviations from healthy behavior." And by 1913, Kraepelin was openly skeptical about his earlier ambition of establishing patients' status praesens: "There is, and I believe can be no psychiatrist who would be able, based only on clinical symptoms [*Zustandsbilde*], to recognize whether any given manic episode belongs to one or the other group of diseases [*Gruppe von Erkrankungsformen*]. Ibid., 3:1374.

had specific outcomes, and so he sought to develop a "science of prognostics"[33] with "prognostic rules"[34] that would help demarcate psychiatric disorders.

In practice, the development of a clinical research agenda focusing on course and outcome involved the deployment of a rigorous regime of clinical inscription that could track and document patients' symptoms over extended periods of time.[35] This regime incorporated an elaborate system of cards and administrative protocols that organized patient documentation and observation in his hospital. It allowed him to intensify his observation of patients and to expand the clinical window onto their institutional lives.

Among the most important clinical research tools that Kraepelin deployed in order to help him chart disease course were his so-called diagnostic cards [*Zählkarten*].[36] Kraepelin introduced these cards into the work routine on the surveillance ward of the psychiatric clinic in Heidelberg in the early 1890s. These cards were essential tools for the construction of his nosology and reflected his emphasis on clinical observation and longitudinal analysis. The cards included excerpts of the patient history and outlined the "essential characteristics of the clinical picture."[37] Although structurally the cards made no explicit provision for affective symptoms, they enabled Kraepelin and his colleagues to register those symptoms and their changes over time. The cards documented the observations of hospital physicians, patients' own statements, and sometimes the claims of third parties, such as the patients' family members. The cards were supposed to avoid technical jargon and give as objective—that is, empirically grounded—an account of the clinical facts as possible. Whenever a patient was admitted to the clinic, a new card would be prepared on which a diagnosis and course of the disease could be recorded. Even after the patient had left the clinic, the cards would continue to be updated until a final outcome could be determined. Keeping track of patients after they left the clinic was crucial to the success of Kraepelin's research project.

It should be emphasized that Kraepelin's clinical research agenda was not solely concerned with nosology. It was also fundamentally about developing reliable diagnostic techniques that, in turn, would lead the way toward greater prognostic certainty in day-to-day clinical practice. And prognostic certainty had important ramifications for clinical practice insofar as it was a crucial administrative precondition for the release of patients from his overcrowded psychiatric hospital. That Kraepelin came to place such nosological importance in the prognosis of psychiatric disorders, was in part a derivative of its great institutional utility. Not surprisingly, therefore, his clinical research agenda met with fierce administrative and professional resistance. Medical officials complained vociferously about his eager and sometimes wayward prog-

---

[33] Emil Kraepelin, "Ziele und Wege der klinischen Psychiatrie," *Allg. Z. Psychiat. Psych.-Gerichtliche Med.* 53 (1897): 840–8, on 844.

[34] Kraepelin, *Psychiatrie*, 8th ed. (cit. n. 7), 3:1348.

[35] For more details about this regime, see Eric J. Engstrom, "Die Ökonomie klinischer Inskription: Zu diagnostischen und nosologischen Schreibpraktiken in der Psychiatrie," in *Psychographien*, ed. Cornelius Borck and Armin Schäfer (Zurich, 2005), 219–40. For his clearest articulation of this inscription regime, see Emil Kraepelin, "Die Erforschung psychischer Krankheitsformen," *Z. Gesam. Neurol. Psychiat.* 51 (1919): 224–46.

[36] On the cards and their nosological significance, see Matthias Weber and Eric J. Engstrom, "Kraepelin"s Diagnostic Cards: The Confluence of Empirical Research and Preconceived Categories," *Hist. Psychiat.* 8 (1997): 375–85.

[37] Emil Kraepelin, *Lebenserinnerungen* (Berlin, 1983), 142.

nostications, which disrupted the wider system of hospital care and compromised the health of patients. And professional colleagues suggested that—contrary to fundamental principles of general pathology—Kraepelin had essentially raised the practical institutional necessity of rapidly establishing a prognosis to an instrument of patient diagnosis and disease classification.[38]

MANIC-DEPRESSIVE ILLNESS

Doubtless one of the most influential and lasting achievements of this clinical research agenda was Kraepelin's description of manic-depressive illness.[39] And in his architecture of the illness, its prognosis and course took on particular significance. Famously, he argued that the prognosis of manic-depressive illness was generally favorable: unlike patients suffering from schizophrenia—or what Kraepelin then termed "dementia praecox"—whose conditions were likely to degenerate into dementia, manic depressives were seen as far more likely to recover from their illnesses. Kraepelin used the favorable outcomes of these patients to argue—with nosological hindsight—that this "uniform prognosis" represented a "common bond uniting all forms of manic depression."[40]

That Kraepelin came to group such apparently incompatible conditions as mania and melancholy together was also a consequence of the "psychological observation" that the course of both disorders was characterized by mood swings.[41] The key to understanding the disorder lay in this variability of the clinical picture. What made Kraepelin's description so remarkable was that he used this variability as an argument not against, but rather for, the constitution of both manic and depressive symptoms as a single nosological entity.

In arguing the case for this unified view, Kraepelin focused on so-called mixed states [*Mischzustände*] that combined cognitive, affective, and psychomotor symptoms.[42] Specifically, mixed states were conditions that exhibited either manic symptoms (flight of ideas, euphoria, hyperactivity) or depressive symptoms (inhibition of

---

[38] On these issues, see Eric J. Engstrom, *Clinical Psychiatry in Imperial Germany: A History of Psychiatric Practice* (Ithaca, N.Y., 2004), 121–46.

[39] For a recent analysis of the nosological development of Kraepelin's concept of manic depressive illness, see Katharina Trede, Paola Salvatore, Christopher Baethge, Angela Gerhard, Carlo Maggini, and Ross J. Baldessarini, "Manic-Depressive Illness: Evolution in Kraepelin's Textbook, 1883–1926," *Harvard Rev. Psychiat.* 13 (2005): 155–78. More generally—and thoroughly—on the evolution of Kraepelin's nosological categories, see Paul Hoff, *Emil Kraepelin und die Psychiatrie als klinische Wissenschaft: Ein Beitrag zum Selbstverständnis psychiatrischer Forschung* (Berlin, 1994). It must be noted that Kraepelin's clinical approach by no means obviated experimental research on manic-depressive illness, as his own work and that of his students suggests. Cf. Weygandt, *Über die Mischzustände* (cit. n. 21), 61.

[40] Kraepelin, *Psychiatrie*, 8th ed. (cit. n. 7), 3:1184–5.

[41] Emil Kraepelin, *Psychiatrie: Ein Lehrbuch für Studierende und Aerzte*, 5th ed. (Leipzig, 1896), 597; Kraepelin, "Die klinische Stellung der Melancholie," *Monatsschr. Psychiat. Neurol.* 6 (1899): 325–35, on 325.

[42] Kraepelin first articulated the notion of mixed states in the sixth edition of his textbook in 1899. Specifically, mixed states were delineated by comparing the relationship in the oscillating course of manic and depressive forms in three areas: (1) *Denkstörung* (cognitive disorder), (2) *Verstimmung* (mood disorder), and (3) *Willenstörung* (volitional disorder). See the diagram of curves in Kraepelin, *Psychiatrie*, 8th ed. (cit. n. 7), 3:1287. See also the study of Kraepelin's student in Heidelberg, Weygandt, *Über die Mischzustände* (cit. n. 21), 2–3.

thought, depressed mood, weak volition or abulia) but that were also characterized by isolated symptoms typical of their respective opposite pole. Although their overlapping symptoms prompted him to elaborate numerous subforms of mixed states, he interpreted this as evidence for their nosological unity. Furthermore, Kraepelin argued that these mixed states represented not simply deviations from either mania or melancholy but rather the transition between the two.[43] It was the careful examination of subtle shifts in the course of his patients' illnesses that had initially led him to isolate these mixed states and to conclude that they were indeed intermediate, transitional forms between more basic manic and depressive conditions.[44] So studying disease course over time became crucial in identifying manic-depressive illness as a clinical entity. Looking simply at isolated clinical states was not enough; one had to investigate the basic forms over time with an eye toward transitions passing through these mixed states.[45]

Kraepelin's classification evoked considerable protest and led to long and drawn-out debates about the frequency of acute episodes and about what counted as being "periodic" or "cyclical" in manic-depressive illness.[46] Critics took issue with his very liberal interpretation of periodicity because it allowed, in some cases, years to pass between bipolar episodes. They also disputed the assumption that mixed states represented transitions between bipolar conditions rather than simply episodes of unipolar ones. Other critics complained that the notion of *Mischzustände* effectively opened the floodgates on manic-depressive illness, allowing it to be filled with all manner of affective conditions.[47] Kraepelin was certainly aware of these criticisms, as well as the limitations of his own clinical method, and he never tired of stressing that, in many respects, the demarcation of individual symptoms was "artificial and arbitrary."[48] Nevertheless, in his nosology, much of the affective spectrum of psychiatric disorders began to gravitate toward manic-depressive illness. From the late 1890s, as German Berrios has suggested, Kraepelin's concept of manic-depressive illness became such an "omnibus" and "over-inclusive" category that much of its long

---

[43] Kraepelin, *Psychiatrie*, 8th ed. (cit. n. 7), 3:1288–95.

[44] Ibid., 1301: "The mixed states discussed here are most usually temporary manifestations within the disease course."

[45] Ibid., 1301–2: "We are justified in interpreting the clinical cases as mixed states and symptoms of manic-depressive illness only on the basis of their course [and] their transition between the respective [manic and depressed] states." See also Trede et al., "Manic-Depressive Illness" (cit. n. 39), 174.

[46] See, e.g., the deliberations in Kraepelin, *Psychiatrie*, 8th ed. (cit. n. 7), 3:1321–46. For a more recent perspective on these issues, see Stephen Tyrer, "What Does History Teach Us about Factors Associated with Relapse in Bipolar Affective Disorder?" *J. Psychopharmacol.* 20 (2006): 4–11.

[47] According to his successor in Munich, Oswald Bumke (1877–1950), Kraepelin's mixed states contributed to manic-depressive illness "encompassing most everything that did not causally result in dementia." Bumke went on to emphasize that efforts to delineate it based on studies of heredity had demonstrated that any method based strictly on "statistical techniques was useless." Furthermore, the very anatomic and hereditary unboundedness of manic-depressive forms suggested not only a more normatively determined concept of disease but also an "infinite variety that is impossible to categorize, a variety from which one can instead only extract specific types, just as one uses certain nonmorbid temperaments as types or landmarks for orientation." Bumke, "Hoffnungen und Sorgen der klinischen Psychiatrie," *Klin. Wochenschr.* 5 (1926): 1905–8, on 1905–6.

[48] Kraepelin, *Psychiatrie*, 8th ed. (cit. n. 7), 3:1237. Kraepelin arrived at this conclusion not only because of the "occurrence of fluid transitions between all of the different clinical pictures, but also because, over a very short period of time, a single case can undergo enormous changes. Accordingly, the nosology presented in this book can only be considered as a rough attempt at sorting the great variety of clinical evidence."

twentieth-century legacy represented "no more than an analysis of the fragmenta-
tion of the Kraepelinian notion."[49]

Nowhere was the gravitational pull of manic-depressive illness on affective symptoms greater than in the case of melancholy. Whereas in mid-nineteenth-century psychiatric parlance "melancholy" had described the initial depressed mood at the outset
of potentially still curable mental illness, by 1900 Kraepelin had radically reinterpreted the term to mean simply a "depressive mental disorder of old age" [*depressive
Geistesstörung des Rückbildungsalters*], that is, to encompass only those forms of depressed mood exhibiting a poor prognosis and declining into dementia.[50] Furthermore, the scope of neurasthenia, which by the turn of the century had become a
fashionable ailment, was radically circumscribed: Kraepelin used it only to describe
chronic exhaustion of the nervous system due to overwork.[51] Not even that most affectively laden diagnosis of hysteria remained entirely immune. Certainly, in 1913
Kraepelin continued to understand hysteria as characterized chiefly by emotional reactions and the "discharge of affective tension" [*Entladung gemütlicher Spannungen*],
which he interpreted as an expression of "instinctual defense mechanisms" [*triebartige
Verteidigungsmaßregeln*].[52] But he rejected a strictly affective explanation of hysteria.
Indeed, he played down emotional reactions in order to play up the importance of the
will in governing primal emotions and drives. To his mind, the prominence of affective
symptoms in hysteria was symptomatic of the absence or impoverished development of
higher, civilized volitional faculties—in other words, hysteria was as much a disorder of
the will as of affect.

By 1913, Kraepelin was himself hinting that his clinically derived conception of
manic-depressive illness had drawn together too many other cases that exhibited periodic mood swings.[53] Furthermore, in the years leading up to World War I, he came
to conclude that disease course did not, in fact, suffice as justification for the clinical
unity of manic-depressive illness. Years of clinical research had led him to conclude
that there was no evidence of

> anything even approximating the well-defined course that, on the basis of certain indi-
> vidual observations, one had once assumed. It is precisely this [clinical] experience that
> has made untenable any attempt to demarcate and group disorders based on specific var-
> iations in their course.[54]

So heterogeneous, so divergent had the symptoms of his patients' illnesses become,
that even his finely tuned clinical technologies now failed to discover any specific
course common to all forms of manic-depressive illness.

[49] German Berrios, "Mood Disorders," in *A History of Clinical Psychiatry: The Origin and History
of Psychiatric Disorders*, ed. Berrios and Roy Porter (London, 1995), 385–408, on 387. For specifics
on the nosological inflation of manic depressive forms, see also Hoff, *Emil Kraepelin* (cit. n. 39).

[50] Kraepelin, "Die klinische Stellung der Melancholie" (cit. n. 41), 331. See also Kraepelin's introduction to Georges L. Dreyfus, *Die Melancholie: Ein Zustandsbild des manisch-depressiven Irreseins*
(Jena, 1907), v–vi. For a contemporary critique of Kraepelin's expansion of manic-depressive forms
at the expense of melancholy, see Alfred Hoche, "Die Melancholiefrage," *Zentralblatt Nervenheilkunde
Psychiat.* 31 (1908): 193–203.

[51] Cf. Emil Kraepelin, "Die Diagnose der Neurasthenie," *München. Med. Wochenschr.* 49 (1902):
1641–4, esp. 1643–4.

[52] Emil Kraepelin, "Über Hysterie," *Z. Gesam. Neurol. Psychiat.* 18 (1913): 261–79.

[53] "Doubtless because of all of these additions, the scope of manic-depressive illness has grown very
substantially." Kraepelin, *Psychiatrie*, 8th ed. (cit. n. 7), 3:1382.

[54] Ibid., 1336–7, 1376–7.

TEMPERAMENTS AND CLINICAL EPIDEMIOLOGY

If course could no longer be relied upon to hold manic-depressive illness together as a nosological construct, then Kraepelin surmised that perhaps the hereditary predispositions of the temperaments could. Kraepelin's interest in mixed states had helped to refocus his research on—or at least to acquire a greater appreciation of—the interludes [*Zwischenzeiten*] between acute manic and depressive episodes of the illness.[55] In these interludes he found that the vast majority of patients exhibited no morbid/clinically relevant symptoms at all. They did, however, show slight but persistent mental aberrations [*leichte Störungen des psychischen Allgemeinbefindens*].[56] Kraepelin described these aberrations as *Grundzustände*[57] or temperaments. Temperaments represented

> at times certain preliminary stages that don't evolve into more serious disorders, at times the slightest hint of more explicit psychoses that barely cross the threshold of morbidity, and at times simple, inconspicuous permanent conditions that form a foundation on which we otherwise observe more intense, acute conditions. And so from this perspective we can consider manic, depressive, irritable, and cyclothymic predispositions. We encounter them as personal peculiarities of manic-depressive patients either prior to an acute attack or during the symptom-free intervals of their suffering. But often enough we also encounter them in cases that never evolve into an actual illness.[58]

The aberrations that typified these temperaments were by and large emotional [*Besonderheiten des Gemütslebens*], and they were not specific to manic-depressive patients. Nor did they necessarily evolve into manic depression, although he associated them with its various clinical subgroups.[59] They derived their clinical relevance from the observation that they were often found in patients' families. And accordingly, Kraepelin believed them to be hereditary, concluding that there existed a "certain predisposition" that could be considered an initial or prodromal stage in the development of manic-depressive illness.

Today, contemporary psychiatrists have taken enormous interest in these subclinical predispositions, and, like Kraepelin, many see them as making up the foundation of affective disorders.[60] They have understood Kraepelin's integration of temper-

---

[55] See ibid., 1337–44.

[56] Ibid., 1303: "Between episodes, the vast majority of manic-depressives, especially those with rare episodes, manifest no deviation from normal healthy behavior."

[57] See ibid., 1303–19. On Kraepelin's views about temperament, see also U. Ehrt, P. Brieger, and A. Marneros, "Temperament und affektive Erkrankungen—geschichtliche Grundlagen einer aktuellen Diskussion," *Fortschr. Neurol. Psychiat.* 71 (2003): 323–31, esp. 325–6; H. S. Akiskal, "Dysthymia and Cyclothymia in Psychiatric Practice a Century after Kraepelin," *J. Affect. Disorders* 62 (2001): 17–31.

[58] Kraepelin, *Psychiatrie*, 8th ed. (cit. n. 7), quoted in Ehrt, Brieger, and Marneros, "Temperament" (cit. n. 57), 325–6.

[59] Kraepelin subdivided these *Grundzustände* into (1) *depressive Veranlagung* [*konstiutionelle Verstimmung*], (2) *manische Veranlagung* [*konstitutionelle Erregung*], (3) *reizbare Veranlagung*, as well as (4) a mixture of the first two [*zyklothyme Veranlagung*]. On the considerable difficulties of distinguishing between these subclinical groups, see Kraepelin, *Psychiatrie*, 8th ed. (cit. n. 7), 3:1321.

[60] Current debate in psychiatry turns on the question of whether there is a continuum between subthreshold (temperament, affective dispositions) and major depressions. Hagop Akiskal's research is concerned with the transition from gloominess to dysthymia and from moodiness to cyclothymia, and especially the implications of these conditions for early intervention strategies. Like Kraepelin, Akiskal argues that "temperamental dysregulation" underlies affective disorders. Akiskal, "Dysthy-

aments into his interpretation of manic-depressive illness as a decisive expansion of the affective spectrum of psychopathology and as confirmation of the continuities between morbid and premorbid symptoms.[61] And indeed, in colonizing those interludes with temperaments, he effectively undergirded vast swaths of his nosology with subclinical, affective types. His conceptualization of temperaments was at once a symptom and a cause of a more general ballooning of manic-depressive forms of mental illness within his nosology.

But one of the difficulties faced by today's clinicians is that the "symptoms" of these aberrations tend not to show up on the wards of psychiatric hospitals.[62] And this was precisely the problem that, in his day, confounded Kraepelin as well. Neither ward visits nor elaborate, institutionally rooted inscription regimes could adequately capture the most subtle, prodromal signs of these temperaments.[63]

In studying these emotional aberrations and individual predispositions—and much more—Kraepelin therefore took his psychiatric science beyond the walls of the hospital and into the daily lives of the general populace. What remained beyond the grasp of both his experimental and clinical methodologies would now be rendered visible using new epidemiological strategies. In an article written in 1908, "On the Question of Degeneration [*Zur Entartungsfrage*]," he sketched his vision of a gigantic, proto-epidemiological research project in which he called for "extensive, careful, decades-long studies" across entire geographic regions by "specially trained commissions."[64] His aim was to "gather knowledge by means of expert analysis of individual cases— knowledge that we can never acquire through regular, large-scale population statistics."[65] Kraepelin was advocating a massive project in psychiatric epidemiology that would project the technologies that he had developed in his clinic research outward and facilitate widespread "clinical" observation of the general populace:

> Regions of sufficient size, at least a large city or county, must be studied systematically, frequently, and with the utmost accuracy, so that we can discover in the greatest possible detail not only the current state of things, but also any changes that occur. These studies must be undertaken by specially trained commissions comprised of doctors and statisticians whose attention is devoted solely to the task of investigating the question of degeneration. Beyond the number and fertility of marriages, the rates of illness and mortality, the life expectancy and military fitness, consideration would also need to be given to

---

mia and Cyclothymia" (cit. n. 57), 28. For an assessment of current research on affective disorders that draws on Kraepelin's work, see also Olga Zivanovic and Aleksandra Nedic, "Kraepelin's Concept of Manic-Depressive Insanity: One Hundred Years Later," *J. Affect. Disorders* 137 (2012): 15–24.

[61] According to Ehrt, Brieger, and Marneros, Kraepelin moved the *Grundzustände* inside the category of manic-depressive illness and thereby achieved a "paradigm change" in terms of an "expansion of the affective spectrum." Ehrt, Brieger, and Marneros, "Temperament" (cit. n. 57), 325.

[62] Indeed, Kraepelin's nosology has been explicitly criticized for its very narrow empirical foundation. As German Berrios has noted, "the Kraepelinian synthesis had been based on the description of asylum cases, and had left out a large group of disorders composed of protracted griefs, dysphorias, minor depressions, anxiety disorders, and neurasthenias." Berrios, "Mood Disorders" (cit. n. 49), 401. See also Friederike Fritze, U. Ehrt, and P. Brieger, "Zum Konzept der Hyperthymie: Historische Entwicklung und aktuelle Aspekte," *Fortschr. Neurol. Psychiat.* 70 (2002): 117–25, esp. 124.

[63] Kraepelin, *Psychiatrie*, 8th ed. (cit. n. 7), 3:1319–20, 1384–5.

[64] Emil Kraepelin, "Zur Entartungsfrage," *Zentralblatt Nervenheilkunde Psychiat.* 31 (1908): 745–51. A good portion of the other research conducted at Kraepelin's psychiatric clinic in Munich was likewise informed by questions about degeneration. See Wolfgang Burgmair, Eric J. Engstrom, and Matthias M. Weber, eds., *Emil Kraepelin: Kraepelin in München I, 1903–1914* (Munich, 2006), 52.

[65] On this point, see also Felicity Callard, "The Intimate Geographies of Panic Disorder: Parsing Anxiety through Psychopharmacological Dissection," in this volume.

rates of crime, prostitution, drunkenness, and syphilis, as well as to occurrences of mental illness, idiocy, psychopathy, epilepsy, and the transmission of these disorders to progeny. Across a limited region we can, in all of these and perhaps other respects, gather knowledge by means of expert analysis of individual cases—knowledge that we can never acquire through regular, large-scale population statistics. Yet this knowledge forms the essential scientific foundation that will enable us to clarify the degeneration question, to assess the nature and size of the danger, and then to determine the preventative measures to be taken.[66]

Kraepelin's ambitious, albeit dystopian, dream of complete psychiatric observation was never realized. But ever resourceful and not known for procrastination, he was already moving to expand the range of his own clinical gaze.[67] As early as 1905, he was actively soliciting the support of his colleagues throughout Bavaria.[68] He petitioned the courts to gain access to defendants who were being held pending psychiatric evaluation. He collected information on the mental states of juvenile delinquents from the files of detention and reeducation facilities. In support of the demographic and genealogical research of Ernst Rüdin (1874–1952), he secretly—so as to avoid alarming the unwitting participants—requisitioned Bavarian school records. And to glean additional information on "military fitness," he exploited the files of Bavarian army recruitment offices. Beyond these local efforts, Kraepelin also called for coordinated research on a global scale: at the Fourth International Congress for the Care of the Mentally Ill in Berlin in 1910 he sponsored (together with Rüdin) a resolution proclaiming "that it was necessary to conduct an elaborate epidemiological study [*Bevölkerungsstatistik*], over many decades, of a limited geographic region in order to collect the requisite evidence to resolve the question of whether mental and nervous diseases were increasing and what mitigating social factors were involved."[69]

The enormous diversity of sources from which Kraepelin hoped to glean evidence about subclinical affective predispositions underscores the fact that his proto-epidemiological research project relied significantly on other nonpsychiatric institutions and their respective inscription regimes. In other words, it was embedded in a network of social relationships and administrative jurisdictions, and its implementation demanded heightened cooperation between psychiatrists and various other professional groups and public officials. The result of these initiatives was a "massive influx of clinical observation material" that demanded that clinical work be conducted "on a grand scale."[70] In his eagerness to expand the range of clinical data at his disposal, Kraepelin appears to have had few qualms about drawing on the observations of officials not trained in psychiatry. Indeed, his solicitation of evidence that could never have satisfied his own exacting standards of clinical observation suggests internal tensions within his own work and contrasts sharply, for example, with his insistence on gathering knowledge "by means of expert analysis of individual cases."

[66] Kraepelin, "Zur Entartungsfrage" (cit. n. 64), 751.
[67] For more detail on these efforts, see Eric J. Engstrom, "On the Question of Degeneration," *Hist. Psychiat.* 18 (2007): 389–404.
[68] Emil Kraepelin, "Fragestellungen der klinischen Psychiatrie," *Centralblatt Nervenheilkunde Psychiat.* 28 (1905): 573–90, esp. 589.
[69] [Johannes] Bresler, "Bericht über den IV: Internationalen Kongress zur Fürsorge für Geisteskranke," *Psychiat.-Neurol. Wochenschr.* 12 (1910–1): 305; Alois Alzheimer, "Ist die Einrichtung einer psychiatrischen Abteilung im Reichsgesundheitsamt erstrebenswert?" *Z. Gesam. Neurol. Psychiat. Origin.* 6 (1911): 242–6, on 244.
[70] Kraepelin, *Lebenserinnerungen* (cit. n. 37), 141–2.

CONCLUSION

To conclude, let me be clear. I am not suggesting that we can easily map Kraepelin's epidemiological research agenda one-to-one onto the architecture of his nosology, let alone derive it from his evolving understanding of manic-depressive illness. But I am suggesting that there was a remarkable convergence between his interest in temperaments and his larger research agenda. We encounter a clinical problem of observing early affective symptoms within the institutional context of academic hospitals that were ill-suited to track and document them. We find Kraepelin trying to exploit the information contained in prison, school, and military records in order to close the empirical gaps in his clinical evidence. We see him advocating for large-scale psychiatric evaluations of the general population with the aim of identifying signs of hereditary disposition. And finally, in 1913 we see him expanding and differentiating the scope of manic-depressive illness in this textbook to take greater account of premorbid symptoms or temperaments.

Furthermore, I am suggesting that the development of Kraepelin's research on affective disorders was driven in part by what he believed to be the limitations or "failings" of his experimental and clinical practices. Not least because the laboratory-bound methods of experimental psychology were poorly equipped to capture patients' affective symptoms and track their course over time, Kraepelin developed his elaborate clinical inscription regime. And in turn, because that hospital-bound inscription regime was poorly equipped to identify subclinical affective states, he promoted a proto-epidemiological research agenda that sought to deploy clinical research tools outside hospital settings and across the general population. Or, to put it another way, Kraepelin came to recognize the significance of temperaments and to incorporate them in his nosology only after finding that neither the protocols of experimental psychology nor the clinical regimes used to document and track psychiatric patients' illnesses were adequate to the task of grasping affective symptoms, delivering reliable diagnoses and prognoses, and demarcating specific disease entities.

If we step back from Kraepelin's narrower research agenda, we see that his growing interest in milder affective states also needs to be interpreted in the context of Germany's evolving network of psychiatric institutions. The widespread construction of urban psychiatric hospitals—and especially the university clinics in which he worked—had been premised on the notion that they would facilitate the hospitalization of patients with acute psychiatric illness. Unlike larger rural asylums, which often became repositories for chronically ill patients, urban hospitals served, in a sense, as frontline crisis intervention centers. In particular, they were designed to treat acute but passing episodes of mental illness that often displayed strong affective symptoms. Kraepelin recognized, however, that these manifestly affective symptoms were usually "irrelevant" in assessing the specific cause of an illness.[71] By contrast, patients' mental dispositions—including but not limited to their emotional temperaments—were of greater causal importance. And the specific pathogenic factors that affected those dispositions manifested themselves "neither only, nor perhaps even critically in the emergence of explicit mental disorders, but rather in countless, more or less vis-

---

[71] Kraepelin, "Ziele und Wege der psychiatrischen Forschung" (cit. n. 30), 175.

ible phenomena of daily life."[72] Efforts to identify such milder forms of affective deviance reflected not only the stark limitations of institutionally based clinical science
but also a much wider shift toward extramural forms of psychohygienic intervention.

And finally, if we step back still farther, we find that it is worth reflecting on what
this story can contribute to discussions about the relationship between the histories of
emotions and science. Attentive readers will rightly have recognized that my narrative hews more closely to a history of mental illness and psychiatry than to a history
of emotions. In forging that narrative, however, I have raised an important question
that historians of emotions need to address. For if nothing else, my account illustrates
that historians are Johnny-come-lately participants in the community of scholars interested in emotions. Whereas historians have only relatively recently turned their attention to emotions, many other researchers—and indeed, entire disciplines—have
long since taken an abiding interest in them. In sharpening their analytic and termi-
nological arsenal, historians will therefore surely benefit from paying special attention to the narrower perspectives and developments of these older and richer scientific communities.

In particular, ambitious historians aspiring to a "grand narrative of the interrelationships between the emotions and science"[73] will need to grapple with the crucial and
distinguishing characteristics of disciplines like psychology and psychiatry. Certainly,
as the editors of this volume suggest in their introduction, studying the (often hypo-)
emotional styles or communities of scientists or measuring the historical influence of
their theories can yield insightful results. But unlike most other sciences, and as several of the contributions to this volume demonstrate, the so-called psy disciplines can
take emotions themselves as the objects of their inquiry and indeed sometimes consider those emotions to be constitutive of their own science. Although I have not attempted to do so here, writing the histories of these "sciences of the emotions" as histories of emotion will therefore demand significant methodological sophistication,
simply because their respective historical and scientific objects overlap in nontrivial
ways. A history of emotions that fails to recognize or that takes no notice of this overlap will be hobbled from the outset.

Be that as it may, the approach I have taken in this article can also serve as a
reminder of some of the pitfalls awaiting historians of emotions along the interface
with the history of science, and especially with the history of medical science, for
I have been interested neither in a hagiographic uncovering of the origins of neo-
Kraepelinian psychiatry nor in critically debunking those origins. Instead, I have narrated an account that seeks to evade two all-too-common historiographic tropes: on
the one hand, a naive presentist approach that traces only those topics and trajectories
that culminate (and hence resonate) in contemporary (neuroscientific or other) discourses about emotions, and on the other hand, a slightly less naive approach that
uses historical evidence simply to critique contemporary scientific theory and practice, not least by illustrating how the future always tends to outwit scientists' ephemeral certitudes. Both of these approaches unnecessarily circumscribe and shortchange

---

[72] Ibid., 194.
[73] Otniel E. Dror, Bettina Hitzer, Anja Laukötter, and Pilar León-Sanz, "An Introduction to *History of Science and the Emotions*," in this volume.

the potential of the history of emotions because they are overinvested in either legit-
imizing or critiquing modern science. Historians of emotions can do better than suc-
cumb to these two symbiotic historiographic pitfalls. And they are bound to do better
if instead they pay closer attention to the specific dynamics, contexts, and contingen-
cies of their historical evidence.

How Films Entered the Classroom:

The Sciences and the Emotional Education of Youth through Health Education Films in the United States and Germany, 1910–30

by Anja Laukötter*

ABSTRACT

This essay focuses on health education films in Germany and the United States in the first decades of the twentieth century, illustrating how these films developed their potential as a teaching tool capable of shaping the emotions and changing the behavior of audiences. The essay argues that the films' educational goals were inspired by certain contemporary ideas on the relation between perception, cognition, and emotions. In concentrating on youth as a target audience, it traces the way in which the sciences of psychology and pedagogy discovered the significance of emotions to this specific age group's learning process. The essay discusses the deployment of both general and specially created films in the classroom as a new educational practice, arguing that these films can be read as a negotiation of the modern human subject and its emotions.

UNSETTLING EMOTIONS

In 1924, in a theater in Munich, representatives of the Munich police department, together with local authorities in medicine, politics, and education, previewed a new health education film, *Die Geschlechtskrankheiten* (Venereal diseases). In a letter written shortly after the screening, they informed the Department of the Interior about their concordant judgments. To them, it was obvious that

* Max Planck Institute for Human Development, Berlin, Lentzeallee 94, 14195 Berlin, Germany; laukoetter@mpib-berlin.mpg.de.

Earlier versions of this text have been discussed with Alexa Geisthövel, Uffa Jensen, and my three coeditors, Otniel E. Dror, Bettina Hitzer, and Pilar León-Sanz. I am very grateful for their hints and advice. I would like to thank the anonymous reviewers and the editor of *Osiris* for their comments and suggestions, which helped improve the essay. I also thank Kate Davison for her great support with the copyediting of this article.

this film—through its opening scene, the amalgamation of disgusting, terrifying images, the many, often superfluous close-up shots of male and female genitals—can be expected not merely to overexcite the fantasy of the young, but to also have a harmful effect on the moral and hygienic development of youth. It is unbearable that such a film should be shown to young people of any age.[1]

Although the censorship board had given approval for this film on venereal disease to be shown to young people, on the condition that screenings be limited to single-gender audience groups, these experts were concerned that it would have counterproductive effects on the educational development of young viewers by (over)stimulating their fantasies and emotions of fear and disgust. Such apprehensions represent one side of a wider discourse on the status of health education films on human development in the first decades of the twentieth century, especially in the United States and Germany.[2]

While Scott Curtis has convincingly illustrated how training films were efficiently used as a pedagogical tool for medical students at the beginning of the twentieth century, this essay will focus on educational films produced for the wider public.[3] More specifically, it will focus on their educational function, which was ultimately geared toward shaping the emotions of young viewers and thereby inducing them to change their behavior.[4] Thus, in focusing on health education films in the United States and Germany, it will illustrate how this new medium not only was regarded as an "emotional engineering technique" but was even intentionally developed as such. This essay will not focus on how audiences felt while watching these films but on how their emotions were conceptualized. In this context, reflections by film theorists and psychologists such as Hugo Münsterberg on the potential for film to configure emotions are of immense interest, particularly with regard to how they inspired ideas about the relationship between perception, cognition, and (moving) emotions. The essay will then outline how young people as a specific audience category and the significance of their emotions for the learning process were discovered by the sciences of psychology and pedagogy. And yet the emotional development of youth was conceptualized

[1] Letter from the Police Department of Munich to the State Ministry of the Interior, State Ministry of the Interior/MInn 66561, Hauptstaatsarchiv, Munich, Germany (hereafter HSTA). The medical authorities were Professor Dr. v. Zumbusch, Dr. Uhl, Dr. Tyroll, Dr. Seiffert, and Dr. Seyderer; the political authorities were Dr. Orth (State Ministry for Education and Culture) and Mr. Eichner (State Ministry of the Interior); and the educational authority was Mr. Schäfer (director, Youth Welfare Service and representative of the school board).

[2] Sylvia Kesper-Biermann, "Kommunikation, Austausch, Transfer: Bildungsräume im 19. Jahrhundert," in Transnationale Bildungsräume: Wissenstransfers im Schnittfeld von Kultur, Politik und Religion, ed. Esther Möller and Johannes Wischmeyer (Göttingen, 2013), 21–42. On the international development of educational films, see Christian Bonah and Anja Laukötter, "Moving Pictures and Medicine in the First Half of the 20th Century: Some Notes on International Historical Developments and the Potential of Medical Film Research," in "Film and Science," special issue, Gesnerus 66 (2009): 121–45.

[3] Scott Curtis, "Dissecting the Medical Training Film," in Beyond the Screen: Institutions, Networks and Publics of Early Cinema, ed. Marta Braun, Charles Keil, Rob King, Paul S. Moore, and Louis Pelletier (New Barnet, 2012), 161–7.

[4] The educational function of visual representations of scientific knowledge and their connection to emotions is not merely a phenomenon of the twentieth century. Anne Secord, e.g., has shown how in early nineteenth-century botany the ability of pictures to stimulate pleasure was fiercely debated in relation to whether they could encourage new audiences for the science. See Secord, "Botany on a Plate: Pleasure and the Power of Pictures in Promoting Early Nineteenth-Century Scientific Knowledge," Isis 93 (2002): 28–57.

as being endangered in the crisis of the "educational machinery."[5] Against this backdrop, the essay will examine the deployment of films in the classroom as a new educational practice and show how film can be read not only as the "problematic embodiment of cultural modernity"[6] or the "aesthetic reality of modernity,"[7] but also as a negotiation of the modern human subject and his or her emotions, a negotiation that was strongly shaped by the sciences.

THE EMERGENCE OF HEALTH EDUCATION FILMS: PRESENTING FACTS AND ADDRESSING EMOTIONS

Following its invention at the end of the nineteenth century, the technique of cinematography quickly found application in the field of medicine in both Europe and the United States. Together with the "truthful wonder instruments" of microcinematography, slow- and fast-motion cinematography became important tools for research and training within the discipline.[8] It was assumed that with the technology of film, reality could be seen and described in novel ways.[9] Commentators held the belief that through films, the last "secrets of nature" could be decoded.[10] The X-ray films made by John Macintyre just before the turn of the twentieth century seemed to provide a new insight into the body. As film technology developed, it was used to study movement disorders in neurological patients with the benefit of unlimited repetition.[11] This intervention in time and even the idea of a new form of "serial looking" was highly appreciated, as it freed scientists from the constraints imposed by "case material" and the pitfalls of human error in direct observation.[12] Film enabled the creation of a new way of looking, a "scientific gaze," and thus a new way of specifying and obtaining knowledge.[13] It was regarded as a tool to reveal new truths—a tool whose evidence was based on the logic of the visible.[14]

Beyond becoming an instrument of high relevance in scientific research, the medium of film evolved into an important part of health campaigns to educate the wider

[5] John W. Slaughter, *The Adolescent* (New York, 1911), 23.

[6] Mitchell G. Ash, *Gestalt Psychology in German Culture, 1890–1967: Holism and the Quest for Objectivity* (Cambridge, 1995), 299.

[7] Jörg Schweinitz, *Prolog vor dem Film: Nachdenken über ein neues Medium* (Leipzig, 1992), 5.

[8] R 86 Nr. 2595, Bundesarchiv, Berlin, Germany; Georg Michael, "Medizin und Kinematographie," *Tägliche Rundschau* (Berlin), 23 October 1926; Curtis, "Dissecting" (cit. n. 3), 161–7.

[9] "Microkinematography," *Nature* 88 (1911): 213–5; R 86 Nr. 2595, Bundesarchiv (cit. n. 8); Erich Bulhud, "Wunderwerkzeuge," *Stadt-Anzeiger* (Cologne), 4 January 1927.

[10] "Roentgen Cinematography," *Brit. Med. J.*, 19 November 1910, 1645.

[11] Robert Kutner, "Die Bedeutung der Kinematographie für medizinische Forschung und Unterricht sowie für volkshygienische Belehrung," *Z. Ärztl. Fortbild.* 8 (1911): 249–51, on 250; Ute Holl, "Neuropathologie als filmische Inszenierung," in *Konstruierte Sichtbarkeiten: Wissenschafts- und Technikbilder seit der Frühen Neuzeit*, ed. Martina Hessler (Munich, 2006), 217–40, on 221–2.

[12] Walter G. Chase, "The Use of the Biograph in Medicine," *Boston Med. Surg. J.* 63 (1905): 571–3. He used film to study epileptic seizures.

[13] Hannah Landecker, "Microcinematography and the History of Science and Film," *Isis* 97 (2006): 121–32, on 123; Kirsten Ostherr, *Medical Visions: Producing the Patient through Film, Television and Imaging Technologies* (New York, 2013), 3–27.

[14] Monika Dommann, "'Sehen ist sicherer denn Fühlen': Die Radiographie als Repräsentationstechnologie (1895–1935)," in *Körper macht Geschichte—Geschichte macht Körper: Körpergeschichte als Sozialgeschichte*, ed. Bielefelder Graduiertenkolleg Sozialgeschichte (Bielefeld, 1999), 299–320. On the close connection between visual representation and the emergence of objectivity in the nineteenth century, see Lorraine Daston and Peter Galison, *Objectivity* (New York, 2007), and other works by the same authors.

public on disease at the beginning of the twentieth century. What can generally be defined as health education films were produced with the aim of communicating messages on health topics that would appeal to a popular audience: to provide information on the transmission, prevention, and treatment of illnesses such as sexually transmitted diseases, tuberculosis, smallpox, or malaria.[15] Furthermore, they could be used to promote prophylactic or curative medical practices such as vaccination or radiation therapy, and to advise viewers to give up unhealthy habits such as drinking alcohol. Primarily screened in public venues such as theaters, lecture halls, museums, and schools, they sometimes also accompanied feature films in the main movie theaters. As in the letter from the Munich police quoted at the beginning of this essay, most authorities recommended that they be preceded by a lengthy, detailed talk by a qualified doctor.[16]

Throughout the twentieth century, a large number of these films were produced in Europe and the United States with the shared aim of educating the public on how to lead a healthy lifestyle and working to reduce socially undesirable behaviors. Ideas and methods for how to achieve these goals were not only widely discussed but were also implemented in the films in various ways. As such, most of the films on these various topics were hybrids, in that they combined informative and fictional elements.

A common format that appears in several of these films involved the explication of necessary information about the disease by a physician. This science-based knowledge was presented in different forms, including microscopic views of the agents, statistics highlighting the spread of the disease, diagrams of the mortality rate, and maps of the disease's distribution. Insights into the body or images of degenerated (sexual) organs were often shown in close-ups or through animation techniques.[17] The use of such varied methods for representing knowledge linked these films with the scientific realm and served to highlight the power and validity of the films' evidence.

The fact-based parts of the films are often interwoven with fictional dramatizations of individuals' encounters with the disease. Differing in length and dramatic content, these scenes aimed to illustrate how the presented facts might play out in real life and often encompassed a demonstration of the appropriate reaction by individuals in the event of diagnosis (i.e., the reaction that would lead to a successful life), contrasted with the wrong reaction (which would lead to death). For example, the anti-syphilis film *Feind im Blut* (Enemy in the blood; 1930), which successfully reached a large number of people in Germany, France, and Switzerland as well as the United States,

---

[15] Various terms have been used in contemporary literature to designate this genre, such as public health films, institutional films, propaganda films, educational films, teaching films, and others. Even more historical stakeholders involved in the production and screening of these films used varied terminology.

[16] Such lectures were quite common in the early twentieth century. This practice served as a technique of rhetorical reworking in order to adjust or orient the visual material shown to the public. See Anja Laukötter, "Listen and Watch: The Practice of Lecturing a Film and the Epistemological Status of Sex Education Films in Germany," in "Screening Sex Hygiene Films in the First Half of the 20th Century," special issue, *Gesnerus* 72 (2015): 56–76.

[17] Kirsten Ostherr, "Cinema as Universal Language of Health Education: Translating Science in 'Unhooking the Hookworm' (1920)," in *The Educated Eye: Visual Culture and Pedagogy in the Life Sciences*, ed. Nancy Andersen and Michael R. Dietrich (Hanover, N.H., 2012), 121–40, on 122. On the varied application of animation in health education films, see Anja Laukötter, "Wissen als Animation: Zur Transformation der Anschaulichkeit im Gesundheitsaufklärungsfilm," in "Animationen," special issue, *Montage AV* 2 (2013): 79–96.

used a variety of visual representation techniques to explain how the disease was spread, what it looked like, and how it could be prevented.[18] This factual part was linked with three fictional scenes depicting the respective destinies of a student and his friend, a male worker, and a married woman, all of whom (as well as the worker's baby) become infected (see figs. 1, 2).

With this "vivifying technology" of "living pictures," the makers of these films hoped that audiences' emotions would be stimulated.[19] For example, Jean-Benoît Lévy, a French film director who was well known in Germany and the United States, reflected that in what he called a "propaganda film" (by which he meant an educational film), "which is intended for a large public, the action must have an emotional basis and consist of pre-eminently human features, which in our opinion can alone grab the masses through their appeal to the heart and feelings."[20] By stimulating the emotions and at the same time providing scientifically supported information, these films would encourage the audience to follow "the right road."[21]

Proponents of these films emphasized that this conversion into action had already been successful. For example, Thomas C. Edwards, of the National Health Council, asserted in 1926 that the prenatal care film *A Two-Family Stork* led to a great increase in the number of visits to maternity clinics.[22] He also claimed that the screening of *One Scar or Many*, a film about smallpox, prompted "1600 requests for vaccination," and that *New Ways for Old*, on immunizing children against diphtheria, changed parents' attitudes about bringing their children in for treatment.[23] Statistical data were also provided for other productions, such as the German film *Falsche Scham* (False shame; 1925) about venereal disease: according to one contemporary commentator, after the film's release, "people who sought advice in clinics, hospitals and outreach clinics increased by up to 100 percent."[24] However, such assertions that the viewing of films followed a clear stimulus-reaction model were more the exception than the rule. Furthermore, the appropriate emotional reactions to these films (especially but not only the ones about sex education) were controversial—both in theory and in practice.

[18] Martin S. Pernick, *The Black Stork: Eugenics and the Death of "Defective" Babies in American Medicine and Motion Pictures since 1915* (New York, 1996), 163; Robert Eberwein, *Sex Ed.: Film, Video, and the Framework of Desire* (New Brunswick, N.J., 1999).

[19] Oliver Gaycken, "The Cinema of the Future: Visions of the Medium as Modern Educator, 1895–1910," in *Learning with the Lights Off: Educational Film in the United States*, ed. Devin Orgeron, Marsha Orgeron, and Dan Streible (Oxford, 2012), 67–89, on 70.

[20] Jean-Benoît Lévy, "Views of a Film-Maker," *Int. Rev. Educ. Cinematogr.* 2 (1930): 977–81, on 977. Lévy differentiates here between scientific films and educational films (which he calls "propaganda films"). While scientific films only have to "emphasize the essential points," "propaganda films" need emotions. Like his colleague Jean Painlevé, he was involved not only in educational film production but also in other scientific and vanguard film projects in France: Pierre Thévenard and Guy Tassel, *Le cinéma scientifique français* (Paris, 1948); Valérie Vignaux, *Jean-Benoît Lévy ou le corps comme utopie* (Paris, 2007); Richard Millet, "Jean Painlevé cinéaste," in *Le cinéma et la science*, ed. Alexis Martinet (Paris, 1994), 86–94.

[21] Gaycken, "Cinema" (cit. n. 19), 70.

[22] Thomas C. Edwards, "Health Pictures and Their Value," in "The Motion Picture in Its Economic and Social Aspects," special issue, *Ann. Amer. Acad. Polit. Soc. Sci.* 128 (1926): 133–8, on 135.

[23] Ibid., 135–7.

[24] Curt Thomalla, "Die Entwicklung des medizinischen und hygienischen Lehr- und Kulturfilms in Deutschland," *Int. Lehrfilmsch.* 1 (1929): 468–94, on 480–1.

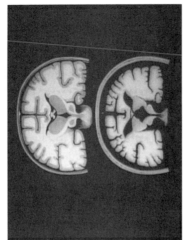

Figure 1. Still images from the film Feind im Blut, *directed by Walter Ruttmann (Praesens-Film, 1930/1931). Left, the woman is presented as a prostitute. Middle, insights into medical research. Right, diagrams of brains: a healthy brain compared to one infected by syphilis. © Praesens-Film AG, Zurich.*

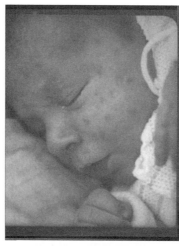

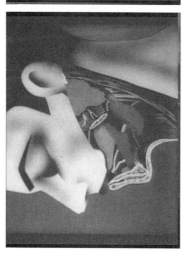

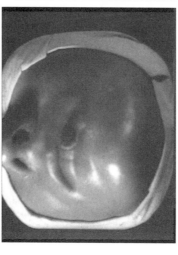

Figure 2. More still images from the film Feind im Blut, directed by Walter Ruttmann (Praesens-Film, 1930/1931). Left, bolus (wax molding) of a female head with syphilitic lips. Middle, drawing of a head from the side: animation shows the process of syphilitic infection. Right, real film image of an infected baby. © Praesens-Film AG, Zurich.

FILM AND EMOTION: CONTEMPORARY CONCEPTIONS

The notion that films and emotions are strongly related was first theorized by the German psychologist and film theorist Hugo Münsterberg in his 1916 work, *The Photoplay*.[25] Münsterberg moved to the United States in 1897, but he maintained his professional contacts in Germany. At the Harvard Laboratory, he continued his studies on visual perception, emotions, and memory and developed what he called the "psychotechnique" in his 1914 work, *Grundzüge der Psychotechnik*.[26] That year, after having been to the movies for the first time, Münsterberg changed his previously low opinion of cinema and its "unhealthy influence" and seductive character, which he believed led to criminality and indulgence in guilty pleasures.[27] He became interested in the exploration of cinematography techniques and was increasingly convinced of the great potential of the medium.[28] From then on, Münsterberg ascribed an exceptional subjectivity to film as a medium because of its "immersive potential."[29] Drawing on a traditional conception of art, Münsterberg developed a belief that, with its specific techniques such as close-ups, flashbacks, and montage rhythms, film could trigger, guide, and objectify the mental processes of individual viewers, from attention, memory, and imagination to active emotional responses.[30] In this sense, for Münsterberg the viewer's mental processes were not free-floating; rather, the film itself shaped the viewer's mind with regard to what to watch and how to do so. In his view, the mind was thus strongly constrained.[31]

According to Münsterberg, the display of emotions was the main aim of film. Only actions that are governed by emotions, expressed in gestures, actions, and facial expressions, created meaning.[32] As the Munich authorities observed in their letter, this expression of emotions could be supported not only by using filmic techniques such as close-ups, but also by showing in a concentrated way specific "numberless shades into the feeling tone,"[33] with the subtle art of the camera or the creation of a specific

[25] Hugo Münsterberg, *The Photoplay: A Psychological Study* (New York, 1916); Jörg Schweinitz, ed., *Hugo Münsterberg—Das Lichtspiel: Eine psychologische Studie* (1916; repr., Vienna, 1996).

[26] Jörg Schweinitz, "Psychotechnik, idealistische Ästhetik und der Film als mentalstrukturierter Wahrnehmungsraum: Die Filmtheorie von Hugo Münsterberg," in Schweinitz, *Das Lichtspiel* (cit. n. 25), 9–26, on 9; Hugo Münsterberg, *Grundzüge der Psychotechnik* (Leipzig, 1914). On the relationship between Münsterberg's work, psychotechniques, and vocational psychology, see Jeremy Blatter, "Screening the Psychological Laboratory: Hugo Münsterberg, Psychotechnics, and the Cinema," *Sci. Context* 28 (2015): 53–76. For the use of Münsterberg's psychotechniques in order to improve the efficiency of work, see Andreas Killen, "Weimar Psychotechnics between Americanism and Fascism," *Osiris* 22 (2007): 48–71.

[27] Hugo Münsterberg, *Psychology: General and Applied* (New York, 1914), 454.

[28] Schweinitz, "Psychotechnik" (cit. n. 26), 13.

[29] Jörg Schweinitz, "The Aesthetic Idealist as Efficiency Engineer: Hugo Münsterberg's Theories of Perception, Psychotechnics and Cinema," in *Film 1900: Technology, Perception, Culture*, ed. Annemone Ligensa (New Barnet, 2009), 77–86, on 79.

[30] Matthew Hale, *Human Science and Social Order: Hugo Münsterberg and the Origins of Applied Psychology* (Philadelphia, 1980), 146.

[31] Ibid.

[32] Münsterberg, *The Photoplay* (cit. n. 25), 113–4.

[33] Ibid., 113. Similar thoughts about the connection of close-ups and emotions described in a musical language (within the silent era) can be found in the theoretical reflections of Walter Bloem. To him, film in general was the "art of emotions" because it affected the primary human instincts: to feel and to see. Moreover, Bloem described close-ups as intensifying the display of emotions, especially if they showed emotions customarily expressed very "silently." Bloem, *Die Seele des Lichtspiels: Ein Bekenntnis zum Film* (Leipzig, 1922), 32.

environment, such as an "emotionalizing of nature."[34] As far as audience reception was concerned, Münsterberg distinguished two levels of audience emotions. On one level, the emotions displayed by the actors are transmitted to the audience: "The visual perception of the various forms of expression of these emotions fuses in our [the audience's] mind with the conscious awareness of the emotion expressed; we feel as if we were directly seeing and observing the emotion itself."[35] On the second level, the audience reacts to the film with an "independent affective life."[36]

In terms of educational films, health officials hoped that the first level of emotions would be achieved. In viewing the fear of infection, shame about immoral life choices, or disgust toward diseased body parts projected on screen, the audience would be reminded of their knowledge about these emotions and thus be led to imagine that they themselves felt the emotion displayed. The underlying assumption of this process was that it would motivate a change in behavior. Scott Curtis has shown that there was a "mimetic presumption" to "copy the movements" displayed by the actors in training films, while in the educational films discussed in this essay there was a mimetic presumption to copy displayed emotions.[37]

But this task was not easy, as several examples illustrate. The film *Wege zu Kraft und Schönheit* (Path to strength and beauty; 1925), the greatest success of the German film company UFA, illustrates how difficult it was to create the right emotions. The film was distributed in Europe and the United States but attracted criticism and vociferous protests in many countries. In the Belgian city of Leuven, for example, massive student protests supported by church authorities had to be broken up by the police.[38] For the most part, these strong reactions were provoked by the multitude of naked female bodies displayed in the film. Another example of such films provoking the wrong emotions was *Frauennot-Frauenglück* (Misery of women—happiness of women; 1929/1930)—a film that stirred controversy, demonstrations, and denunciations in Germany, France, and Switzerland. In July 1931, the production company Präsens-Film-AG filed a lawsuit against the Munich police department for prohibiting the film from being shown. The police justified their decision by claiming that the film would jeopardize the health of those who watched it.[39] Because the film showed childbirth, a cesarean section, and the deadly consequences of an abortion, it was assumed that these scenes would create a "fear of childbirth." More concretely, they cited members of the audience who "felt poorly" and a woman who "screamed hysterically" upon seeing the film.[40] In other words, if these films showed a reality that became too real, it was feared that the film would induce shock or even phobias and thus violate the emotions of the audience, which would in turn have counterproductive effects on their behavior.

---

[34] Münsterberg, *The Photoplay* (cit. n. 25), 120. Before Béla Balázs, Sergei M. Eisenstein, and Wsewolod I. Pudovkin had extensively explored the effects of film technique, such as close-ups and montage, Münsterberg had theorized them as emotional.

[35] Ibid., 123–4.

[36] Münsterberg, *The Photoplay* (cit. n. 25), 125.

[37] Curtis, "Dissecting" (cit. n. 3), 165.

[38] "Die Vorgänge in Löwen," *Film-Kurier* 54 (4 March 1927).

[39] Letter from the State Ministry of the Interior to the State Ministry of Finance, 24 November 1931, State Ministry of the Interior/MInn 72683, HSTA.

[40] Of course, the discussions on *entsittlichende* (demoralizing) films hit their peak in times of new social movements, as, e.g., the *Nacktkultur-Bewegung* (nudism movement). Police Department of Munich, Folder 2: 7417 and 2: 7402, HSTA.

Emotions were thus clearly at stake. They were the subject of controversial debates as to how, why, and to what extent emotions were configured by health education films. In the following section, I will demonstrate how this presumption concerning educational film as an emotional engineering technique played out in the creation of a certain audience, namely, youth.[41]

THE CREATION OF SPECIFIC EDUCATIONAL FILM AUDIENCES

Who was the audience for these educational films, or, more precisely, who were the audiences? According to contemporaries, health education films were regarded as the essential medium of urban mass culture.[42] Because of their "universal language" and their ability to address "simple instincts,"[43] these films could reach increasingly broad, even global, audiences and had "extended the mental horizon of great masses."[44] Beyond enthusiasm about the fact that these films could address everyone regardless of age, class, and race,[45] from very early on specific audiences were identified and targeted. They were thus conceptualized as target audience groups whose emotions needed to be guided and shaped in specific ways, albeit for different reasons. The medium of film was well suited to this task.

Soldiers formed one of these target audience groups. Far from home, they were thought of as being morally weak and as such potentially attracted to paying for sex. During World War I, health education films—particularly about sexually transmitted diseases, such as the transnationally distributed American film *Fit to Fight* (1918)—were produced with the intention of being screened in military camps. These films aimed to prevent soldiers from contracting infections and diseases and in doing so to ensure the strength of the armed forces as a whole.

Another group was not categorized by occupation or moral shortcomings but rather by geography: rural people. Far removed from the city theaters and cinemas, those living in rural areas were "the most difficult people to reach with any teaching message" but also the "most susceptible to the emotional appeal" of films.[46] No reasons were given as to why people in rural areas would be more affected by films than those in urban areas, but it was a popular assumption.[47] Thus, so-called health vehicles fully

---

[41] On the use of a newly evolving technique to address emotions, see Dolores Martín Moruno, "Pain as Practice in Paolo Mantegazza's Science of Emotions," in this volume. Whereas Martín Moruno shows how photography was applied to create an emotional distance, my essay shows how films were designed to engage the emotions.

[42] Ash, *Gestalt Psychology* (cit. n. 6), 299.

[43] F. von Welsch, "Psychologisches in der Kinematographie," *Kinema* 3 (24 May 1913): 1–5, on 2.

[44] Ernest L. Crandall, "Possibilities of the Cinema in Education," in "The Motion Picture" (cit. n. 22), 109–15, on 110. The comparison of motion pictures with words, including the notion that moving pictures have a universal language, persisted even decades later. See, e.g., Adolf Nichtenhauser, "A History of Medical Film," unpublished manuscript, Nichtenhauser Papers, 1954, National Library of Medicine, Washington, D.C. Nichtenhauser wrote the first extensive history of educational and scientific films from an international perspective. I thank David Cantor for pointing me toward Nichtenhauser's work.

[45] Charlotte Perkins Gilman, "Public Library Motion Pictures," in "The Motion Picture" (cit. n. 22), 143–5, on 144.

[46] Edwards, "Health Pictures" (cit. n. 22), 136.

[47] Curt Thomalla, "Der Film im Dienste der ländlichen Gesundheitspflege in Deutschland," *Int. Lehrfilmsch.* 2 (1930): 577–84.

fitted out with projection equipment were used to bring health education films from urban venues to the countryside.[48]

The third specific audience was defined by age: youth. Debates and publications about film stressed that young people were particularly susceptible to negative influences because of their inexperience in life. In other words, they were in need of special protection and guidance.[49] This notion was supported by the establishment of pediatrics as a subdivision of medicine at the turn of the twentieth century. From then on, the physical and mental development of infants, children, and adolescents was monitored by health experts—with an emphasis on disease prevention as the primary means of maintaining health. The view that it was important for children to be wholesome in body and mind was reinforced after World War I, when children came to be regarded as replacements for men lost at war. Moreover, the hygiene and health of young people became a significant issue not only in medicine but also for mothers who wanted to be successful in their child-rearing.

The following part of this essay will focus on this last group—youth—as one of the target audiences. I will illustrate why the physical and mental development of youth attracted renewed attention and why such an important role was ascribed to emotions in the learning process. Moreover, I will show how and why health education films for children (and mothers) gained a certain relevance—not only among health educators and teachers but also among pedagogues and psychologists.

THE CONCEPTION OF YOUTH AND THEIR EMOTIONS
AND THE CRISIS OF THE "EDUCATIONAL MACHINERY"[50]

Even before Sigmund Freud developed his widely popular notion that early life was fundamental to the subsequent maturing process, American childhood had attracted the attention of the newly emerging discipline of psychology.[51] As Ann Hulbert has observed, efforts to observe and structure children's lives turned "human development into a laboratory and its study into a science."[52] Although differing in their approach, some of the first psychologists, such as L. Emmett Holt and G. Stanley Hall, paved the way for establishing the study of child development.[53] At the turn of the twentieth century, researchers began conducting physiological experiments with children not only in the United States but also in Germany. In Berkeley, California, children's behavior and perception were explored in a simulated classroom; in Germany, Ernst Meumann investigated children's aesthetic experiences, perceptions, and feelings regarding art.[54] The invention of the intelligence test for children by child psychologist

[48] *Educational Screen*, March 1926, 158. Such "health vehicles" were used not only in the United States and Germany, but also in France, Russia, Great Britain, and other countries. At least in the United States and Germany, it seems that no films were produced specifically for this rural audience.

[49] Alois Funk, *Film und Jugend: Eine Untersuchung über die psychischen Wirkungen des Films im Leben der Jugendlichen* (Munich, 1934).

[50] Slaughter, *The Adolescent* (cit. n. 5).

[51] Ellen Herman, "Psychologism and the Child," in *The Modern Social Science*, ed. Theodore M. Porter and Dorothy Ross, *Cambridge History of Science*, vol. 7 (Cambridge, 2003), 649–62, on 655.

[52] See Ann Hulbert, *Raising America: Experts, Parents, and a Century of Advice about Children* (New York, 2003), 38–9.

[53] Ibid. Holt focused on parental power and stressed its importance ("parent-centered approach"), while Hall concentrated on the natural development of children ("child-centered approach").

[54] Till Kössler, "Die Ordnung der Gefühle: Frühe Kinderpsychologie und das Problem kindlicher Emotionen, 1880–1930," in *Rationalisierung des Gefühls: Zum Verhältnis von Wissenschaft und Emotionen 1880–1930*, ed. Daniel Morat and Uffa Jensen (Munich, 2008), 189–210, on 191.

Lewis Terman in 1916 and the subsequent popularity of this social technology in assessing pupils' intelligence in both the United States and Germany underlined a new perception of children and adolescents and their learning processes.[55] This technique of "mind-measuring" and its extended use called into question what, when, and how children learned in schools and created anew the possibility of "school failure."[56] Furthermore, the market for advice literature on children's education and the importance of their proper development grew.[57] This normative literature varied greatly in the way it advised parents on various issues: in the prescribed virtues that were to be implemented in children's upbringing; in the educational goals that were identified for parents; and in whether emotions played a crucial role in education, which emotion(s) were emphasized, and to what extent.[58] This advice literature, together with research within the scientific fields of pedagogy, pediatric medicine, and psychology, highlighted the fact that the process of managing children's development and their education was problematic.

In this overall process of reevaluating childhood, the emotions of children attracted great attention. According to John Willis Slaughter, professor of sociology at the Rice Institute (and one of Hall's close colleagues), adolescence was not merely defined through emotional development but was almost synonymous with it. In his words: "The central fact of adolescence is emotional change."[59] The idea of transforming emotions inspired research such as Hall's study on fear and even Meumann's study on the development of aesthetic feelings.[60] With the advent of Watson's behaviorist approach, which emphasized the formative power of the environment, emotions came to be viewed as being subject to conditioning and thus highly predictable.[61] The emotional development of young people was not constructed as a natural process but rather a sensitive phase of life that could be shaped. Simple guidance was not sufficient, however—the right guidance was essential. In other words, in offering to regiment young people's emotions, the sciences of pedagogy and psychology promised "conscious control."[62] Normal emotional development was defined through the criterion

---

[55] Lewis M. Terman, *The Measurement of Intelligence: An Explanation of and a Complete Guide for the Use of the Stanford Revision and Extension of the Binet-Simon Intelligence Scale* (Boston, 1916); Alexa Geisthövel, *Intelligenz und Rasse: Franz Boas psychologischer Antirassismus zwischen Amerika und Deutschland, 1920–1942* (Bielefeld, 2013), 30, 131; Leila Zenderland, *Measuring Minds: Henry Herbert Goddard and the Origins of American Intelligence Testing* (Cambridge, 1998).

[56] Zenderland, *Measuring Minds* (cit. n. 55), 115.

[57] The importance of advice literature for the creation of emotional norms and emotional cultures has been exhaustively analyzed by Peter N. Stearns. See, e.g., Stearns, "Girls, Boys, and Emotions: Redefinitions and Historical Change," *J. Amer. Hist.* 80 (1993): 36–74. See also Uffa Jensen, "Mrs Gaskell's Anxiety," in *Learning How to Feel: Children's Literature and the History of Emotional Socialization, 1870–1970,* by Ute Frevert, Pascal Eitler, Stephanie Olsen, Uffa Jensen, Margrit Pernau, Daniel Brückenhaus, Magdalena Beljan, et al. (Oxford, 2014), 21–39. German authors of child-rearing advice literature in the early twentieth century included Julian Borchert, Adele Schreiber, Heinricht Lhotzky, Clara Heitfuß, Nikolaus Faßbinder, and Adolf Mathias. Their work was widely read in Germany, and some of it reached an international audience. In the United States, Alice M. Birney, Ernest and Glydy Groves, John Anderson, and Marian Faegre, among others, were highly influential.

[58] For an example of the historical change in the way certain emotions were evaluated in advice literature throughout the nineteenth and twentieth centuries, see Peter N. Stearns and Timothy Haggerty, "The Role of Fear: Transitions in American Emotional Standards for Children, 1850–1950," *Amer. Hist. Rev.* 96 (1991): 63–94.

[59] Slaughter, *The Adolescent* (cit. n. 5), 23.

[60] Kössler, "Die Ordnung der Gefühle" (cit. n. 54), 199.

[61] Hulbert, *Raising America* (cit. n. 52), 125.

[62] Arnold Gesell, *Infancy and Human Growth* (New York, 1928), 408–10.

of being subject to such control; on the other hand, it provided a platform where emotional aberrations could come to light. Some conceived of emotions as being strongly linked to modes of behavior, and therefore a "valuable means of diagnosis" for teachers.[63] To others, including Hall and other early childhood researchers, a child's emotional development was linked to his or her intellectual development as an adolescent. They claimed that emotions motivated and structured a child's intellectual growth.[64] Not only was thinking itself regarded as being affective, but a child's interest in learning processes was also understood to be based on feelings of pleasure. In turn, a positive influence on young people's emotions would support the learning processes.[65] The creation of a stimulating learning environment thus came to be of great interest and importance, and it was against this backdrop that the use of motion pictures as a new teaching technique in the classroom also gained huge popularity.[66]

VISUAL EDUCATION: FILMS IN SCHOOL

In 1915, the early film critic Vachel Lindsay envisioned that textbooks in schools would, in the future, "be illustrated by standardized films." To him it was obvious that film would "be in the public school to stay."[67] Two years later, Münsterberg described educational films as "great transmitters of knowledge" and as emotional training tools for young people.[68] He believed that educational films could depict knowledge in a lively way and highlighted the fact that the camera enabled children to observe things that were practically impossible to see by any other means. At the same time, he attributed even greater importance to film in its function as an emotional educator and also, therefore, as a moral educator, "for the young mind has to learn more than facts":

> No lesson is more needed than that of wholesome emotion and pure feeling and sentiment. The moral education is still more important than the intellectual; what an inspiration and genuine cultivation of the heart may be won in those thrilling hours in front of the screen. Noble ambition for highest life work, sympathy with the sufferer, distaste for the unclean, contempt for the dishonest, may be developed: a high-minded patriotism may be kindled; enthusiasm for everything great and up-building may be stirred in the mind throughout those most formative years of youth.[69]

Drawing on his work *Grundzüge der Psychotechnik*, Münsterberg had already argued that "seelische Wirkungen" (mental effects) were necessary in order to increase the

[63] "Eye and Ear Mindedness," in *A Cyclopedia of Education*, ed. Paul Monroe (New York, 1911), 2:564–5. This concept was influenced by the James-Lange theory, in which bodily reactions preceded the emotional reaction.

[64] G. Stanley Hall, *Adolescence: Its Psychology and Its Relations to Physiology, Anthropology, Sociology, Sex, Crime, Religion and Education*, vols. 1, 2 (New York, 1919); Dorothy Ross and G. Stanley Hall, *The Psychologist as Prophet* (Chicago, 1972), 325.

[65] Kössler, "Die Ordnung der Gefühle" (cit. n. 54), 208.

[66] On the role of emotions in teaching processes, see Rafael Mandressi, "Affected Doctors: Dead Bodies and Affective and Professional Cultures in Early Modern European Anatomy," in this volume.

[67] Nicholas V. Lindsay, *The Art of the Moving Picture* (New York, 1915), 253.

[68] Although he agreed with proponents of film censorship for children that "weak" and "nervous" children in particular were threatened by the "unreal, exaggerated emotionalism of most melodramatic photoplays" and that criminal films might stimulate criminal acts by children, he was against prohibiting young people from viewing films. Hugo Münsterberg, "Gefahren für die Kindheit im Kino [1917]," in Schweinitz, *Das Lichtspiel* (cit. n. 25), 117–22, on 120; Münsterberg, "Peril to Childhood in the Movies," *Mother's Magazine* 12 (1917): 109–10, 158–9, on 110.

[69] Münsterberg, "Gefahren für die Kindheit," 121; Münsterberg, "Peril to Childhood," 158 (both cit. n. 68).

"seelischen Sinn" (mental sense) for hygiene and a healthy lifestyle in order to combat disease.[70] To Münsterberg, educational films seemed to be an important instrument for social engineering and an effective emotional "psychotechnique" for young people. But how were these visions put into practice?

From the late 1910s onward, new publications such as the *Educational Film Magazine* (1919) and *The Screen* (1920), as well as organizations like the American Education Motion Picture Association (1919) and the Society for Visual Education (1920) were established in the United States.[71] A similar development in this new era of visual education, which for some even constituted a "movement,"[72] can be observed in Germany, where in 1919 the Zentralinstitut für Erziehung und Unterricht vom Preußischen Kultusministerium (Prussian Ministry of Culture's Central Institute for Education and Teaching) established a new institution called Bildstelle (Picture Center).[73] Led by a former professor of geography, Felix Lampe, the Bildstelle tested the adequacy of film for teaching purposes and tried to investigate the needs of schools for films and their cinematic educational material requirements.[74] In September 1924, two smaller German associations involved in promoting the use of educational films for youth (one led by the teacher and historian Walther Günther) merged into one: the Bildspielbund deutscher Städte (Moving Pictures Association of German Cities). Because of high costs and teachers' lack of interest in attending screenings at public film theaters with their school groups, the association strongly advocated that films should be screened in schools, either in gyms or in classrooms for specific age groups.[75] Lists of films highlight the fact that in the 1920s, health education films for young people were distributed and shown in theaters, youth-related institutions such as the YMCA, and most of all in schools (fig. 3).[76] Some experts even postulated that the screening of films during lessons should be part of the daily routine. Arguments were formulated by drawing on different practices in schools in other countries, for example, Great Britain and the United States, in order to underscore how common it was to screen films in these educational contexts.[77] Thus, demands to establish a projection unit in each school became quite frequent, but efforts to meet them were constrained by the

---

[70] Münsterberg, *Grundzüge der Psychotechnik* (cit. n. 26), 1, 291–2.

[71] For an overview, see Devin Orgeron, Marsha Orgeron, and Dan Streible, "A History of Learning with the Lights Off," in Orgeron, Orgeron, and Streible, *Learning with the Lights Off* (cit. n. 19), 15–66, on 27–8.

[72] George E. Stone, "Visual Education: A Retrospective, an Analysis and a Solution," *Educational Screen*, June 1926, 329–37, 348, on 330.

[73] It is also worth mentioning similar international developments, such as the efforts of the League of Nations to start the International Institute of Educational Cinematography in 1928, including the establishment of the *International Review of Educational Cinematography*. An important step in efforts to create films for an international public was the founding of the United Nations Film Board by the French educational film director Jean-Benoît Lévy in 1947. On the importance of the World Health Organization and their production of educational films, see Kirsten Ostherr, *Cinematic Prophylaxis: Globalization and Contagion in the Discourse of World Health* (Durham, N.C., 2005). On the international development of educational films, see also Bonah and Laukötter, "Moving Pictures" (cit. n. 2).

[74] Ursula Keitz, "Wissen als Film: Zur Entwicklung des Lehr- und Unterrichtsfilms," in *Geschichte des dokumentarischen Films in Deutschland*, ed. Klaus Kreimeier, A. Ehmann, and Jeanpaul Goergen, vol. 2, *Weimarer Republik 1918–1933* (Stuttgart, 2005), 120–150, on 126–7.

[75] Ibid., 121–2.

[76] Nelson L. Greene, "Motion Pictures in the Classroom," in "The Motion Picture" (cit. n. 22), 122–30.

[77] *Kinema* 22 (1919): 4.

Figure 3. *Distribution headquarters for films and slides. See William M. Gregory, "A Teacher's Training Course in Visual Aids," Educational* Screen, *February 1925, 88–90, 119, on 88.*

high costs of acquisition.[78] In Germany, so-called *Schulkinogemeinden* (school theater associations) were created. These groups of teachers helped facilitate the exchange of film material between schools in order to reduce costs.[79]

The potential of using film as an interactive tool in teaching (and thus to reject the opinion that film was an immersive medium that created passive recipients) was especially emphasized by reform pedagogues in Germany and the United States in the late 1920s, as film technology became more accessible.[80] Arthur Rarig, an author for *Educational Screen*, reported on the extraordinary initiative of one local health committee in producing an educational film together with a class of high school students called "What Price Folly?" The initiative not only had the purpose of demonstrating that film screening had become a common practice in the classroom but also aimed to illustrate to what extent film could be part of the daily teaching practice.[81]

The educational philosophy that guided this process of institutionalization stressed that film provided a specific and incomparable *Anschauung* (outlook) for young people in line with the ideas of the Swiss educational reformer Johann Heinrich Pestalozzi.[82] His ideas on concrete perception had not only inspired educational reforms in Europe, especially in Germany, but had also influenced the progressive education movement in the United States since the late nineteenth century.[83] The importance of

[78] "Jugend und Kino," *Kinema* 28 (1913): 2–3.

[79] These *Schulkinogemeinden* were promoted by the Bildstelle. See Keitz, "Wissen als Film" (cit. n. 74), 128; Edgar Beyfuß, "Schule und Film," in *Das Kulturfilmbuch*, ed. Edgar Beyfuß and Arthur Kossowsky (Berlin, 1924), 64–71, on 68.

[80] Keitz, "Wissen als Film" (cit. n. 74), 139–40. This idea goes back to the work of education reformer Georg Kerschensteiner and his concept of *Arbeitsschulen* (vocational schools). See Kerschensteiner, *Begriff der Arbeitsschule* (Munich, 1912).

[81] Arthur Rarig, "How a High School Produced an Educational Movie," *Educational Screen*, December 1928, 269, 276.

[82] F. von Welsch, "Psychologisches in der Kinematographie," *Kinema* 23 (1913): 1–4, on 1. See also Beyfuß, "Schule und Film," (cit. n. 79), 64–71, on 66.

[83] Gaycken, "Cinema" (cit. n. 19), 70–1.

the sense of vision in education was repeatedly stressed, as film seemed to provide this form of outlook on the real thing and as such favored concrete versus abstract learning processes.[84] As a result, the notion that vision was the main sense associated with the acquisition of practical knowledge was established. Moreover, the eye, as Hall had described earlier, was seen as the sense closest to the mind.[85] According to Nelson L. Greene, editor of *Educational Screen*, educational films in particular had the ability to "let the mind work directly upon the realities of the physical world through facsimiles presented to the eye" (besides other visual aids such as maps, diagrams, posters, and cartoons).[86] Felix Lampe, believing that the learning processes of young people were strongly influenced by affects, made similar arguments. To him, educational films were so important and successful because this "form of teaching is based on intuition [*Veranschaulichung*], avoiding logical-discursive thoughts and abstract concepts, turning toward associative thought processes with a heavy emphasis on feelings [*Gefühlsbetonung*]."[87] He believed that by means of educational films, movie theaters could become schools and thus combat the strong rationalistic orientations of contemporary education. What he termed the "intellectualism" [*Intellektualismus*] that dominated the classroom could be counteracted only by actively involving feelings [*Gefühlsbeteiligung*].[88] To others as well, this so-called intellectualism in education was too unilateral, causing "half-education" [*Halbbildung*] and "emotional stupefaction" [*Gefühlsverblödung*].[89] Only with the use of films could the "imitative instincts" and "virtues" described by Ernest C. Crandall, such as honesty, fair play, and clean living, be enforced and the formation of relevant habits be encouraged.[90] As Walther Günther explained, educational films could enhance young people's ability to see, feel, and act for themselves.[91]

Although these enthusiastic endorsements were in line with contemporary pedagogic and psychological approaches vis-à-vis affective learning processes, critics expressed concern. They warned that such films might trigger adverse (emotional) reactions that would compromise or limit the learning process and that they might therefore be a potential threat to young people. They denigrated educational films as mere entertainment, which served only to distract and encourage "habits of divided attention." Münsterberg's colleague, the psychiatrist Robert Gaupp, asserted that they were just as senseless as they were tiresome and boring for children.[92] As we saw with the Munich police and their colleagues quoted at the beginning of the essay, still others stressed the negative effects of evoking emotions such as disgust and fear on the health of those who experienced them.

---

[84] Eef Masson, *Watch and Learn: Rhetorical Devices in Classroom Films after 1940* (Amsterdam, 2012), 37–40.

[85] Hall, *Adolescence* (cit. n. 64), 32–3.

[86] Greene, "Motion Pictures" (cit. n. 76), 122–3.

[87] Felix Lampe, *Der Film in Schule und Leben* (Berlin, 1924), 26.

[88] Ibid., 11.

[89] D. U. Lang, "Will es endlich tagen?" *Kinema* 6 (1919): 1–3, on 2. Lang went even further, blaming this "intellectualism" for the outbreak of World War I.

[90] Crandall, "Possibilities" (cit. n. 44), 111; Ernest L. Crandall, "Habit Formation as Affected by the Motion Picture," *Educational Screen*, March 1925, 140–1.

[91] Walther Günther, "Kulturfilm und Jugend," in Beyfuß and Kossowsky, *Das Kulturfilmbuch* (cit. n. 79), 42–59, on 58.

[92] Anna V. Dorris, "The Possibilities of Mass Instruction with Motion Pictures," *Educational Screen*, October 1927, 361–3, on 362; Robert Gaupp, "Die Gefahren des Kino," *Süddeutsche Monatsheft*, 1911/12, 64–9, on 64; Schweinitz, "Psychotechnik" (cit. n. 26), 11.

It is unclear whether it was this type of criticism that motivated the educational committee of the Commonwealth Fund to provide the University of Chicago with $10,000 in order to conduct several experiments concerning educational films.[93] In the general summary, the head of the study, educational psychologist Frank N. Freeman, explained that these experiments had been designed "to determine how fast the schools ought to go in adopting the present forms of visual education and in what direction the development of visual education should proceed."[94] Several experiments on the use of motion pictures in the classroom were conducted to establish potential shortcomings and ways to improve the new tool.[95] Freeman then reported on several experimental settings to compare the "effectiveness" of this new medium to that of other teaching tools such as textbooks, oral instructions, reading, and pictorial presentations.[96] The particular efficacy of film was defined as its ability "to give information or to teach how to perform an activity."[97] The tests therefore sought to identify which parts of educational films achieved their aims.[98] Other aspects of the filmic dispositive were also tested: one study by Freeman and his colleagues attempted to establish whether motion pictures should be accompanied by verbal advice. They tested twenty-six fifth-grade groups (altogether 865 children) in Joliet, Illinois, public schools by screening eight health education films: *Knowing Gnome, The Romance of the White Bottle, The Four M's, Milk, the Perfect Food, The Story of the Orange, Mrs. Brown and the High Cost of Living, Guard Your Mouth, the Gateway to Health,* and *Toothache.*[99] Interestingly, in a second experiment, only two films—*Knowing Gnome* and

[93] Frank N. Freeman, *Visual Education: A Comparative Study of Motion Pictures and Other Methods of Instruction* (Chicago, 1924), preface.

[94] Ibid., 4.

[95] This experiment was one of the first within a classroom, but not the first overall. The first experiments on the emotional and informational impact of a venereal disease film on its audience were conducted and the results reported by the experimental psychologists Karl S. Lashley and John B. Watson, *A Psychological Study of Motion Pictures in Relation to Venereal Disease Campaigns* (Washington, D.C., 1922). For an analysis of these experiments, see Anja Laukötter, "Measuring Knowledge and Emotions: American Audience Research on Educational Films at the Beginning of the Twentieth Century," in *Communicating Good Health: Movies, Medicine and Cultures of Risk in the Twentieth Century*, ed. Anja Laukötter, Christian Bonah, and David Cantor (Rochester, N.Y., forthcoming).

[96] Frank N. Freeman, "The Methods of Investigation in Visual Education," *Educational Screen*, March 1923, 103–8, on 103–4; Freeman, *Visual Education* (cit. n. 93). See also Orgeron, Orgeron, and Streible, "A History of Learning," (cit. n. 71), 32–4.

[97] Freeman, "Methods of Investigation" (cit. n. 96), 105.

[98] H. Y. McClusky, "Study of the Content of Educational Films," in Freeman, *Visual Education* (cit. n. 93), 377–88.

[99] Carolyn Hoefer and Edna Keith, "An Experimental Comparison of the Methods of Oral and Film Instruction in the Field of Health Education," in Freeman, *Visual Education* (cit. n. 93), 346–76, on 347. From Hoefer and Keith's report, we can see that these films covered a wide range of topics, from advice on nutrition and sleeping to dental care. The film *Knowing Gnome* included a Lilliputian figure with a jelly bag cap, "the gnome," as protagonist (see images after p. 374). In the intertitles, this figure was described in the following way: "Once upon a time high among the mountains, there lived a little Gnome whose duty it was to watch over the children of the earth, whom he believed to be the world's greatest treasure." Then: "From rock to rock he went peering into the world below in search of little children who needed his lesson of Health and Happiness." Thus it was not the physician but the "gnome" who functioned as the expert on health questions for children. Beyond their primary themes, these films had to address the specific emotional demands of children and scientific criteria. As Hoefer and Keith noted, "Difficulty was experienced in obtaining satisfactory film because the following qualifications had to be considered: (1) Adaptability to the interest of children of this age. (2) Emphasis upon the positive rather than the negative phase of health instruction. . . . (3) Scientific accuracy as well as interest. . . . (4) The films had to include health habits which could be measured at the close of the experiment. Not all of the films cited above fulfilled these conditions to a high degree. *The Story of*

Mrs. Brown and the High Cost of Living—were tested on 170 pupils in the seventh and eighth grades of the Washington School in Illinois, but the testing group was divided into two groups based on the previous results of the "Terman's Intelligence Test, Form A."[100] Terman, a student of Hall's, argued that these tests did not measure the "entire mentality of the subject," defined as intellect, emotions, and will, but only intelligence. Thus, emotions and will were not explicitly tested "beyond the extent to which these naturally display themselves in the tests of intelligence."[101] However, in this experiment intelligence not only was linked with cognition of the film's message but was regarded as a determining factor in the process of perceiving the screened films. In this vein, we can also read the experiments with so-called subnormal children, who received lessons on nutrition through relevant health education films, as quite a successful enterprise, as the author Jean Parnes reported.[102] In Germany, one of the most influential proponents of visual education, Walther Günther, argued in 1924 that despite psychotechniques, an extensive mass observation of the effects of film on young people could not be conducted because the methods for assessing the intelligence of the participants were still in the early phases of development.[103]

The experiments, conducted under Freeman's supervision, attracted huge interest and perhaps inspired further projects—such as the one at Lincoln School in New York, which in cooperation with the Department of Education at Yale University experimented with "the dramatic film as an educational device."[104] Experiments were designed with the aim of overcoming the limitations of the small number of schools tested in order to obtain a broader, more comparative scope. Among these experiments were the first "nationwide classroom experiments in visual education," which began in Rochester, New York, and were eventually extended to public schools in twelve US cities.[105]

Aside from the way these experiments were conducted, it is interesting to examine their results. The setting of oral and film instruction in health education might be considered as exemplary. In their short conclusion, researchers Carolyn Hoefer and Edan Keith pointed out that the "probable effect of the film instruction does not warrant the expenditure of the amount of money involved in the rental of such films as are available at the present time." They further cautioned, "Before definite conclusions in regard to the benefits derived from film instruction can be made . . . it is necessary to have films produced which are more adaptable to classroom use."[106] Neither of these findings was new. The question of cost had always been connected with the production of educational films. Moreover, from the early 1920s, increasing demands were raised by health officials in Germany and the United States to produce films exclu-

---

the Orange, for example, which was used to emphasize the value of fruit in the diet, was, perhaps, more of a geography film than a health film, and its application to the daily lives of the children had to be made by the teacher" (351, 347–8).

[100] Ibid., 375.

[101] Terman, *Measurement* (cit. n. 55), 48–9.

[102] Jean Parnes, "The Use of Visual Aids for the Subnormal Child," *Educational Screen*, March 1931, 76–7. See Zenderland, *Measuring Minds* (cit. n. 55), 105–42, where she explores the statistical meaning of "subnormality."

[103] Günther, "Kulturfilm und Jugend" (cit. n. 91), 45.

[104] Nathaniel W. Stephenson, "The Goal of the Motion Picture in Education," in "The Motion Picture" (cit. n. 22), 116–21.

[105] "Classroom Films," *Film Daily* 40, no. 9 (11 April 1927).

[106] Hoefer and Keith, "An Experimental Comparison" (cit. n. 99), 56–61, 376.

sively for young audiences.[107] Until then, it was common practice to show the same educational films to adults and children, films such as *How Life Begins*[108] or the films in the series *Science of Life*.[109] Indications that this practice changed can be found a few years later, and this change contributed to the emergence of a subgenre of educational films for the classroom.[110] However, in the present context, it is more significant that neither the intention of the experiment (to give insights into the efficacy of film as an educational medium) nor the implicit assumptions regarding the structure of the experiment (to correlate intelligence with *Anschauung*) were addressed in the study's conclusion. In other words, these psychologists problematized and researched the medium and its impact on learning without producing any results.

If we take into account all of the experiments mentioned, it is interesting to consider not only their subject matter but also what was excluded. As discussed earlier, in psychology, pedagogy, medicine, hygiene, and beyond, it was commonly thought that these films would enhance thinking and learning processes. However, their impact on the emotions was not investigated in every study, even though William Hays, the powerful director of the Motion Picture Producers and Distributors of America, claimed that motion pictures had the "most direct route alike to the emotions and the intelligence."[111] It was several years later that the so-called Payne Fund Studies—a series of thirteen psychological studies on children in the early 1930s—focused on the emotions of young people in relevant experiments.[112]

CONCLUDING REMARKS

It was the intent of this essay to show how educational films evolved as a teaching tool in Germany and the United States at the beginning of the twentieth century. The essay has argued that the notion of a specific "emotional engineering technique" was associated with these films. As such, they appeared as a new and promising technology to address and emotionally reorganize not only adults but also pupils in classrooms. Despite the skepticism toward educational films during this period, the majority of experts within the fields of education, medicine, and psychology outlined the positive effects of film as an educational tool, in particular with regard to young people's

[107] Keitz, "Wissen als Film" (cit. n. 74), 131.

[108] Undated report that reached the Public Health Service on 15 March 1920, RG 90/130/65/16/4 Pierce File 1918–21, Box 331, National Archives, Washington, D.C. The film *How Life Begins* was shown not only in the United States, where it was adopted by the YMCA, but also in France, New Zealand, Japan, etc. See Stone, "Visual Education" (cit. n. 72), 329–37, 348, on 330.

[109] RG 90 130/65/16/2 Venereal Disease Division, General Records (NC 34 Entry 42), Box 214, National Archives, Washington, D.C.

[110] Thomas C. Edwards emphasized that educational films produced especially for young people were very well made in terms of "foot of scene" (image of a scene) and in the creation of straight and simple titles. Edwards, "Health Pictures" (cit. n. 22), 136.

[111] Williams H. Hays, "See and Hear: A Brief History of Motion Pictures and the Development of Sound," *Kinema*, 6 January 1930 and 14 January 1930. Others used terms such as "stimulation of the imagination" or a specific "film sense" to emphasize that film would enhance ways of thinking as well as the memory and learning processes of young people—even of those of "middle-rate intelligence" [*mittelmäßiger Intelligenz*]. "Kino und Schule," *Kinema* 3 (6 September 1913): 3–4; F. Felix, "Kino und Schule," *Kinema* 4 (11 April 1913): 1–3; Gilman, "Public Library Motion Pictures" (cit. n. 45), 143–5; Günther, "Kulturfilm und Jugend" (cit. n. 91), 51.

[112] See Arthur R. Jarvis, "The Payne Fund Reports: A Discussion on Their Content, Public Reaction and Affect on the Motion Picture Industry, 1930–1940," *J. Pop. Cult.* 25 (1991): 127–40; Brenton J. Malin, "Mediating Emotion Technology, Social Science, and Emotion in the Payne Fund Motion-Picture Studies," *Tech. & Cult.* 50 (2009): 366–90.

emotions and their learning processes. Those experts provided the argumentative background for film's application as a teaching tool in the classroom.

At the same time, the sciences of psychology and pedagogy kept the use of film and its relationship to the learning process open to research. These sciences not only offered to regiment young people's emotional and intellectual development; they also promised a comprehensive "conscious control," creating their own prospective subjects of research.

This hitherto underexplored relationship between the emergence of a new educational medium and different scientific fields is revealed from the perspective of the history of emotions. It is precisely this perspective that underlies Hannah Landecker's claim that science is an important part of the history of cinema just as cinema is an important part of the history of science.[113] This case study has shown that emotions can be regarded as a common mediator bridging the history of educational films and the history of sciences, from psychology and pedagogy to medicine and hygiene.

As a final word, it is only possible to hint very briefly at a few effects of the described scenario. On a social practice level, one can say that the interest in (and disputes about) this new teaching medium for youth as well as its deployment in classrooms was sustained by the fact that film was not an ephemeral medium; rather, its use in this context became a common practice that, to varying degrees, lasted until the late twentieth century in the United States and both East and West Germany.[114] On a conceptual level it seems obvious that these early reflections on the perception of films and its relation to behavior and emotions did not end but continued and inspired later influential concepts such as Rudolf Arnheim's "descriptive thinking" [anschauliches Denken] in which visual perception and its cognitive processing were explored.[115] Furthermore, some contemporary studies in cognitive sciences imply that these early thoughts experienced a renaissance. Thus it seems that there is an emerging field of contemporary interdisciplinary work on the link between film theory and theory from within the cognitive sciences.[116] These works translate into certain experimental settings in which, for example, the "emotional effect of films" and more precisely the role of sadness and enjoyment in aesthetic experience are explored.[117]

This essay begins and ends with questions on the relevance of negative emotions in film watching. However, whether and to what extent we can identify continuities and changes in these practices and concepts within the broader time frames of the twentieth century remains to be explored in future work.

---

[113] Landecker, "Microcinematography" (cit. n. 13), 132.

[114] Orgeron, Orgeron, and Streible, "A History of Learning" (cit. n. 71). A coherent overview for the German case, especially a comparative study of its development in East and West Germany, is still lacking.

[115] Rudolf Arnheim, Film als Kunst (Berlin, 1932); Arnheim, Anschauliches Denken: Zur Einheit von Bild und Begriff (Cologne, 1972); Johannes v. Moltke and Jörg Schweinitz, "Für Rudolf Arnheim," Montage AV 9 (2000): 5–17.

[116] Some of the first to explore these connections were the film theorists David Bordwell, Noell Carroll, and Torben Grodal in the 1990s. See, e.g., Torben Grodal, Moving Pictures: A New Theory of Film Genres, Feelings, and Cognition (Oxford, 1997). For current interdisciplinary approaches, including those of neuroscientists, see Julian Hanich, Valentin Wagner, Mira Shah, Thomas Jacobsen, and Winfried Menninghaus, "Why We Like to Watch Sad Films: The Pleasure of Being Moved in Aesthetic Experiences," Psychol. Aesthet. Creativ. Arts 8 (2014): 130–43; Arthur P. Shimamura, Psychocinematics: Exploring Cognition at the Movies (Oxford, 2013).

[117] Hanich et al., "Why We Like to Watch Sad Films" (cit. n. 116), 133, 140. In this study, participants watched short scenes of a "saddening scenario" and then had to fill out questionnaires regarding their affective responses to the scenes. The authors argue that negative emotions contribute to the "pleasurable feeling of being moved."

NEW EMOTIONS—NEW KNOWLEDGE—NEW SUBJECTIVITIES

The Intimate Geographies
of Panic Disorder:

Parsing Anxiety through
Psychopharmacological Dissection

by Felicity Callard*

ABSTRACT

The category of panic disorder was significantly indebted to early psychopharmacological experiments (in the late 1950s and early 1960s) by the psychiatrist Donald
Klein, in collaboration with Max Fink. Klein's technique of "psychopharmacological dissection" underpinned his transformation of clinical accounts of anxiety and
was central in effecting the shift from agoraphobic anxiety (with its spatial imaginary of city squares and streets) to panic. This technique disaggregated the previously unitary affect of anxiety—as advanced in psychoanalytic accounts—into
two physiological and phenomenological kinds. "Psychopharmacological dissection" depended on particular modes of clinical observation to assess drug action
and to interpret patient behavior. The "intimate geographies" out of which panic
disorder emerged comprised both the socio-spatial dynamics of observation on
the psychiatric ward and Klein's use of John Bowlby's model of separation anxiety—as it played out between the dyad of infant and mother—to interpret his adult
patients' affectively disordered behavior. This essay, in offering a historical geography of mid-twentieth-century anxiety and panic, emphasizes the importance of
socio-spatial setting in understanding how clinical and scientific experimentation
opens up new ways in which affects can be expressed, shaped, observed, and understood.

MARY J.'S PANIC

In 1981, in the popular psychology magazine *Psychology Today*, psychiatrists Paul
Wender and Donald Klein heralded the promise of biological psychiatry by emphasizing that revelations about the centrality of biological malfunctions in mental illness

* Department of Geography, Durham University, Lower Mountjoy, South Road, Durham, DH1
3LE, United Kingdom; felicity.callard@durham.ac.uk.
I am grateful to Donald Klein for his generosity in allowing me to interview him at length about his
psychopharmacological research in 2011. I thank the European Neuroscience and Society Network
(funded by the European Science Foundation) for a Short Visit Grant, which funded my trip to the
United States, and the New York University Child Study Center, which kindly hosted me for the duration of this grant. I gratefully acknowledge the Max-Planck-Institut für Wissenschaftsgeschichte and
thank its librarians: while there as a research fellow in the summer of 2012, I conducted additional

would show many of the tenets of psychodynamic theory to be "irrelevant or even misleading."[1] They illuminated their argument with brief case histories. One was of a twenty-three-year-old agoraphobic woman whom they called "Mary J." She was an unmarried buyer for a department store who was suffering from debilitating panic attacks: "She would suddenly be overcome by dizziness, a pounding heart, and an inability to catch her breath while walking down the street or riding on public transportation."[2] Wender and Klein noted that physiologically oriented specialists had tended to diagnose her as suffering from nerves or a virus and had been unable to help Mary J. when she stopped using subways and buses in favor of taxis, en-sconced herself in her parents' home, and gave up her job. Mary J., feeling increas-ingly desperate, sought a psychoanalyst. Wender and Klein reported that on the couch:

> she began to suspect that the panics might be related to a love affair. Six weeks before the attacks started, she had been quite upset: her lover had moved to another city. The analyst closed in on that possibility with penetrating questions. Had her sexual adjustment been guilt-free after she had begun the affair? Didn't her fear of being out on the street reflect her unconscious doubts about her sexual self-control—that is, her fear of identification as a streetwalker? Didn't her clinging dependence on her family show her fear of adulthood and her unconscious desire to substitute her father for other men?[3]

Mary J. spent four years "rework[ing] such baroque structures" while her symptoms came and went. In the end, she left the analyst disillusioned and turned to behavior therapy, a newer, shorter, and cheaper form of treatment. Wender and Klein described how she was given instruction in relaxation exercises and desensitization—exposure to public places—and how, despite initial progress, the panic attacks returned with a vengeance. Mary J. and her parents were, by then, reportedly desperate and resigned

---

research for this article. Earlier research for this article was supported by a Charlotte W. Newcombe Fellowship. My research is supported by the Wellcome Trust (103817/Z/14/Z).

[1] Paul Wender and Donald Klein, "The Promise of Biological Psychiatry," *Psychol. Today*, Febru-ary 1981, 25–41, on 25.

[2] Ibid., 31.

[3] Ibid. The reference to "her fear of identification as a streetwalker" alludes to psychoanalytic mod-els of agoraphobia. Freud, in his letter to Wilhelm Fliess of 17 December 1896, wrote, "I actually con-firmed a conjecture I had entertained for some time concerning the mechanism of agoraphobia in women. No doubt you will guess it if you think of 'public' women. It is the repression of the intention to take the first man one meets in the street: envy of prostitution and identification" (Sigmund Freud and Wilhelm Fliess, *The Complete Letters of Sigmund Freud to Wilhelm Fliess, 1887–1904*, trans. Jeffrey Moussaieff Masson [Cambridge, Mass., 1985], 217–8). While Freud himself wrote little spe-cifically on agoraphobia, the associations he made between agoraphobia and sexual fantasies regard-ing the street were taken up by the next generation of psychoanalysts; see, e.g., Walter Schmideberg, "Agoraphobia as a Manifestation of Schizophrenia: The Analysis of a Case," *Psychoanal. Rev.* 38 (1951): 342–52. The figure of a woman who sequesters herself indoors and thereby removes herself from circulation in the public sphere has appeared a number of times in twentieth-century psychiatric and psychological writings on phobias and anxiety. In 1964, e.g., A. H. Roberts brought to visibility the figure of "the housebound housewife" in a retrospective study of married women in London. See Roberts, "Housebound Housewives—A Follow-Up Study of a Phobic Anxiety State," *Brit. J. Psy-chiat.* 110 (1964): 191–7; see also subsequent research on this figure by psychiatrist and behavioral therapist Tom Kraft (e.g., "Sexual Factors in the Development of the Housebound Housewife Syn-drome," *J. Sex Res.* 6 [1970]: 59–63). Understanding the overdetermined way in which gender, class, and the articulation of the public and private spheres are—within modernity—bound up with each other, is, I would argue, central to understanding histories of fear and anxiety—both within and be-yond the domains of science and medicine.

to her becoming a long-term invalid. But Wender and Klein reported that only a year later, she had experienced a profound affective and social transformation: she was living in the city on her own in an apartment, had returned to her job, and was excited about possible marriage plans. (Let us notice, here, how effective therapeutics transforms "bad" affect into "good," such that equanimity and happiness are demonstrated through a turn to normative forms of sociality [returning to a job, excitedly focusing on marriage].) What had happened? The psychopharmacologists revealed that Mary J. had volunteered to join a clinical experiment for the treatment of phobias that involved taking an antidepressant medication. The drug had stopped the panic attacks, and accompanying psychotherapy had "helped her to control her anticipatory anxiety and allowed her to resume normal activities." The symptoms, we are told, did not return when Mary J. stopped taking the medication six months later.

Wender and Klein's vignette of Mary J. was in the service of their own strong commitment to the effectiveness of biological psychiatry—and as such demanded a disparagement of other clinical approaches to agoraphobia, particularly that of psychoanalysis, and a narrative that culminated in the triumphant success of drug therapy. Their article was published a year after the American Psychiatric Association had anointed the new category of panic disorder in the third edition of the *Diagnostic and Statistical Manual of Mental Disorders* (*DSM-III*)—a category that had been brought into being in large part through Donald Klein's psychopharmacological research that stretched back to the late 1950s, and whose triumphant arrival on the stage of American—and subsequently international—psychiatry was, after a long journey, undoubtedly ensured by Robert Spitzer (the chair of the *DSM-III* Task Force) selecting Klein to join the Task Force on Nomenclature and Statistics.[4]

Wender and Klein's article should be read as a celebration not only of drug treatment as cure but of drug treatment as diagnostic dissection tool. The article not only celebrated how psychopathological affect might be successfully treated via drugs but exemplified a more wide-ranging logic, developed in large part by Klein, and assembled from both empirical and conceptual elements, in which the very shape and timbre of that psychopathological affect might be identified and parsed through drugs.[5] In the same year that Wender and Klein mused in print over Mary J.'s panic, Klein, in a chapter intended for a scientific readership, celebrated what he considered his analytical and methodological breakthrough, namely, "the power of the experimental technique of pharmacological dissection whereby one can pierce through the fascinating, confusing web of symptoms and dysfunctions to tease out the major participant variables by attending to specific drug effects."[6] It was a bold claim that came to inaugurate and cement a new approach to the study of anxiety. Klein's pharmacological interventions had, he averred, disinterred panic disorder—a phenomenon, and a very particular manifestation of pathological affect, that had hitherto remained largely ignored by dint of erroneous formulations concerning agoraphobia.

---

[4] American Psychiatric Association, *Diagnostic and Statistical Manual of Mental Disorders*, 3rd ed. (Washington, D.C., 1980). See Hannah S. Decker, *The Making of DSM-III: A Diagnostic Manual's Conquest of American Psychiatry* (New York, 2013), chaps. 4, 5.

[5] Klein was of course operating as part of a much larger collectivity of researchers grappling with how best to conceptualize and describe drug action. Nonetheless, there was something singular as well as compelling about how he formulated his model of "psychopharmacological dissection," which buttressed what would become a new nosological entity.

[6] Donald F. Klein, "Anxiety Reconceptualized," in *Anxiety: New Research and Changing Concepts*, ed. Donald F. Klein and Judith G. Rabkin (New York, 1981), 235–62, on 242.

What might dwelling on this particular moment—the disinterring of an emotion by a clinician and a scientist via his observation and analysis of drug effects—contribute to our understandings of the long history of models and experiences of fear, anxiety, and phobias? Otniel Dror and colleagues, in their introduction to this volume, ask whether "discrete scientific developments structure the expression, experience, visibility, or nature of emotions?"[7] While panic as a phenomenon and a topic of concern in the history of science and medicine stretches much further back than Klein's post–World War II drug experiments, the emergence of panic disorder arguably established a new way through which manifestations of overwhelming, negative affect—experienced by individuals in particular socio-spatial settings—could be understood, narrated, and, indeed, experienced. If agoraphobia as a term consistently posed the question of the agora (Why did it provoke fear? How ought it to be construed?), panic disorder posed questions about the ontology of the affective phenomenon—panic—itself. Klein himself argued that drug action allowed the observation of two ontologically distinct kinds of anxiety (anticipatory anxiety and panic) that had been conflated in earlier models and theorizations of anxiety. As historians of science and of the emotions, we might instead continue to ponder whether the particular socio-spatial arrangement of Klein's psychiatric wards—one that meshed patients' physiologies, pharmacological action, patient and staff behavior, and practices of observation—structured the very way in which particular affects came to be expressed, shaped, and understood.

I will be particularly attentive, here, to the need to understand how geography figures both in the production of psychopathological affects and in clinical and scientific accounts of those psychopathological affects. The socio-spatial assemblage of epistemological, methodological, and observational techniques that underpinned Klein's work of psychopharmacological dissection operated significantly differently from the one that, since the 1870s, had at its center a clinician puzzling over how to understand and interpret the actions as well as the affective distress of a figure attempting to navigate his way through an urban, public landscape that he could not comfortably inhabit or traverse.[8] The neurologist and psychiatrist Carl Friedrich Otto Westphal, who originated the term agoraphobia in the early 1870s, for example, opened his seminal essay by describing how, "for some years patients have repeatedly approached [him] with the peculiar complaint that it is not possible for them to walk across open spaces and through certain streets and that, due to the fear of such paths, they are troubled in their freedom of movement."[9] Westphal's task, and subsequently that of many neurologists and psychoanalysts who saw agoraphobic patients in their consulting rooms and in outpatient clinics in the late nineteenth and early twentieth centuries, was to respond to those patients' phenomenologically rich accounts of their inability to move through

---

[7] Otniel E. Dror, Bettina Hitzer, Anja Laukötter, and Pilar León-Sanz, "An Introduction to *History of Science and the Emotions*," in this volume.

[8] Note that I have used "he" here. Part of the complex history of gender in relation to agoraphobia is addressed in Felicity Callard, "Understanding Agoraphobia: Women, Men, and the Historical Geography of Urban Anxiety," in *Exploring Women's Studies: Looking Forward, Looking Back*, ed. Carol Berkin, Judith Pinch, and Carol Appel (Upper Saddle River, N.J., 2006), 201–17.

[9] Carl Friedrich Otto Westphal and Terry J. Knapp, *Westphal's "Die Agoraphobie" with Commentary: The Beginnings of Agoraphobia*, trans. Michael T. Schumacher (Lanham, Md., 1988), 59. The German original is Westphal, "Die Agoraphobie: Eine Neuropathische Erscheinung," *Arch. Psychiat. Nervenkrankheit.* 3 (1871): 138–61. Westphal's other writing on agoraphobia included "Über Platzfurcht: Briefliche Mitteilungen," *Arch. Psychiat. Nervenkrankheit.* 7 (1877): 377–83; "Nachtrag zu Dem Aufsatze 'Über Agoraphobie,'" *Arch. Psychiat. Nervenkrankheit.* 3 (1872): 219–21.

open squares and down particular streets. Those narratives bound agoraphobic anxiety tightly to particular urban locales: Westphal, for example, memorably recounted features of the Berlin cityscape (including particular squares, as well as the Charlottenburg zoo) that precipitated extensive fear in the three agoraphobic individuals whose case histories he enumerated in his 1871 article.[10] The logic outlined by Klein emerged from a different socio-spatial world. It was at some distance both from Westphal's symbols and markers of nineteenth-century urbanicity and from the sedate consulting rooms of neurologists and psychoanalysts; it entailed different dynamics of clinical observation. If, for Westphal, the figure, gestures, and affective tenor of "the agoraphobe" provoked clinical intrigue by dint of his stalled passage across the public spaces of the agora, Klein's archetypal, phobic-anxiety figure emerged within the claustral spaces of a mental hospital. What would come to define him or her would not be, as in the case of Westphal's patients, an uncomfortable relationship to walls, passageways, and public squares, but panic attacks, helplessness, and the need for a reassuring, parent-like figure. Thus while Wender and Klein's plot regarding Mary J. in *Psychology Today* commenced with the puzzle of her breathlessness on city streets, neither Klein's own practices of observation nor his investigatory frameworks were centrally preoccupied with the textures and socio-spatial specificities of the urban landscape. How Klein's logic came to be articulated and how the framework in which it was housed worked is what I shall track in this article. If histories of the emotions have provided many nuanced accounts of how temporality is construed and mobilized in different models of particular affects, there has perhaps been less explicit attention to spatiality.[11] This essay is intended, then, as a contribution to the historical geography of the emotions: I am particularly interested in how Klein worked with, and characterized in a particular way, his anxious patients' experiences of the inpatient ward, as well as how the diagnosis of panic disorder dispensed with an explicit emphasis on the subject's relationship to space (cf. agoraphobia).

* * *

Wender and Klein's vignette of Mary J. represented in schematic form the trajectory that approaches to agoraphobia, and agoraphobic anxiety, took in the United States in the second half of the twentieth century. That trajectory moved from the hegemony of psychoanalytic or psychodynamic approaches, to the markedly different therapies and etiological arguments of the behaviorists, to the near inescapability of psychopharmacology for the understanding and treatment of anxiety, panic, and phobias.[12] As different therapeutic regimens jostled for preeminence, they were accompanied by different models of phobic anxiety: those models worked with a variety of phenomenologies of psychopathological affect, of the kinds of individuals who were most

---

[10] Westphal, "Die Agoraphobie" (cit. n. 9). See also Felicity Callard, "'The Sensation of Infinite Vastness'; Or, the Emergence of Agoraphobia in the Late 19th Century," *Environ. & Planning D: Soc. & Space* 24 (2006): 873–89.

[11] Though for an account that takes seriously the problems of both temporality and spatiality in providing a finely grained history of how the affective "accidents of everyday existence" were "transformed into the objects of psychiatric epidemiology," see Rhodri Hayward, "Sadness in Camberwell: Imagining Stress and Constructing History in Postwar Britain," in *Stress, Shock, and Adaptation in the Twentieth Century*, ed. David Cantor and Edmund Ramsden (Rochester, N.Y., 2014), 320–42.

[12] Up to the 1950s and 1960s, most clinicians in the United States aligned themselves directly or indirectly either with the work of Freud or with the psychobiological or psychosocial approaches of those such as Adolph Meyer and William Menninger. The late 1950s was a period of great transformation in the American psychiatric establishment. Psychoanalysis was facing attacks from several

susceptible to being gripped by such affect, of narratives about how such affect would manifest on and through the body, and of what precipitated and maintained those manifestations of affect. Those models were subtended by both implicit and explicit claims about the kind of clinical expert, and the kind of observational practices in which he or she was proficient, that were appropriate to identify and then to intervene upon that affect.

I shall be interested in analyzing specifically how Donald Klein's early psychopharmacological research—conducted in collaboration with the psychiatrist and clinical researcher Max Fink—helped to transform the techniques and conceptual apparatus for observing, identifying, and parsing psychopathological manifestations of anxiety.[13] Surprisingly, there has been relatively little fine-grained historical work that has focused specifically on the imbrication of observational and epistemological frameworks in mid-twentieth-century psychiatry, and their centrality in grounding not only particular diagnostic entities and/or symptoms but particular conceptualizations of psychopathological affect.[14] This article intends to make a contribution to that body of literature.

My argument relies on analyses of published documentation. It therefore reckons with, as well as potentially further embeds, the rhetorical power of Klein's and Fink's written arguments—whether in journal articles reporting their empirical findings or in retrospective narrative accounts of the discovery of psychopharmacological dissection. I am aware of the gap that undoubtedly exists between those tidied, published accounts, and the actual heterogeneous practices of observation that would have taken place in the clinical spaces that acted as the crucible for the development of Klein and Fink's conceptual architecture. Nonetheless, given the centrality of the scalpellic logic of psychopharmacological dissection to the emergence of panic disorder, and the surprising dearth of theoretico-historical elaborations of that logic in the history of psychiatry, close consideration of the workings of those published texts is, I argue, justified.[15]

---

quarters: behaviorists accused psychoanalysts of unscientific methods and proposed very different models of fear and anxiety that derived from the early twentieth-century classical conditioning experiments of researchers such as Ivan Pavlov. Prominent among them was Joseph Wolpe, who developed methods of behavioral desensitization to phobic objects. See Wolpe, *Psychotherapy by Reciprocal Inhibition* (Stanford, Calif., 1958).

[13] Other historical and sociological investigations of agoraphobia and anxiety include Anthony Vidler, *Warped Space: Art, Architecture, and Anxiety in Modern Culture* (Cambridge, Mass., 2000); Paul Carter, *Repressed Spaces: The Poetics of Agoraphobia* (London, 2002); Jackie Orr, *Panic Diaries: A Genealogy of Panic Disorder* (Durham, N.C., 2006); Callard, "'The Sensation of Infinite Vastness'" (cit. n. 10); Kathryn Milun, *Pathologies of Modern Space: Empty Space, Urban Anxiety, and the Recovery of the Public Self* (New York, 2007); Shelley Zipora Reuter, *Narrating Social Order: Agoraphobia and the Politics of Classification* (Toronto, 2007).

[14] Let me be clear. There are of course historical, anthropological, and sociological studies that have attended carefully to psychiatric case records to demonstrate how diagnostic categories were made (e.g., Jonathan Metzl, *The Protest Psychosis: How Schizophrenia Became a Black Disease* [Boston, 2009]). However, relatively few have analyzed how particular epistemological frameworks were bound up with certain practices of observation and their eliciting and interpretation of particular data. One example that focuses on late nineteenth-century psychiatric nosology is Kathleen M. Brian, "'Occasionally Heard to Be Answering Voices': Aural Culture and the Ritual of Psychiatric Audition, 1877–1911," *Hist. Psychiat.* 23 (2012): 305–17.

[15] Klein's early experiments with Max Fink are discussed and analyzed in David Healy, *The Antidepressant Era* (Cambridge, Mass., 1997), 191–3; Orr, *Panic Diaries* (cit. n. 13), 170–2, 205–9. Nei-

HOW DO THE DRUGS ACT?

The synthesis of chlorpromazine in 1951, and its arrival on the psychiatric scene in France in 1952, was a key moment in the development of psychopharmacology. It was termed a "major tranquilizer," not least because of its striking effects on the behavior of some psychotic patients. These effects raised intriguing problems for psychiatrists and pharmacological researchers,[16] for it was becoming clear to them that they possessed few methods to assess not only the effectiveness of drug therapies but also the effects of other treatments (such as lobotomy and electroconvulsive therapy) already within the psychiatric therapeutic armamentarium. In the United States, unanswered methodological, epistemological, and ontological questions about drugs and drug action increasingly preoccupied the Committee on Psychiatry of the National Academy of Sciences–National Research Council, the National Institute of Mental Health (NIMH), and the Committee on Research of the American Psychiatric Association. In 1956, the "Conference on the Evaluation of Pharmacotherapy in Mental Illness" was organized by the psychiatrist Jonathan Cole[17] and the neurophysiologist Ralph Gerard in order to break new ground. The conference had grown out of conversations within and across those committees and organizations and was intended to address the nub of the problem, namely: "Do the drugs act? How do the drugs act? What if the drugs act?" The domain of affect would be central to any kind of answer to the first question, since any determination regarding therapeutic benefit for psychiatric disorders would undoubtedly consider potential affective as well as cognitive transformations in the bodies of those ingesting those drugs.

At the time of that conference, the protocols and frameworks surrounding the organization and practice of clinical evaluation in general, and of clinical trials in par-

---

ther systematically addresses the logic underlying the observational practices that underpin Klein's "psychopharmacological dissection." I am sympathetic to Viola Balz's arguments vis-à-vis the dangers of relying on published material by and interviews with psychiatrists and other scientific researchers (Balz, "Terra Incognita: An Historiographic Approach to the First Chlorpromazine Trials Using Patient Records of the Psychiatric University Clinic in Heidelberg," *Hist. Psychiat.* 22 (2011): 182–200). As she rightly notes, this ends up silencing the patient's voice and rendering invisible the complex relation between clinician and patient around which the act and action of drug taking took place. My longer-term aim is to work with the patient records from the Hillside Hospital, to which I have not yet been able to gain access. Here, I focus on the logics mobilized by psychiatric clinicians in their published texts, logics in which the patient's voice is, for multiple reasons (some of which will become clearer later in this article), attenuated if not obliterated. Since the outcome of Klein's logic was wide-ranging and influential, I believe it is worth analyzing.

[16] David Healy provides a comprehensive overview of the arrival of psychopharmacology, including the momentous synthesis of chlorpromazine. Healy, *The Creation of Psychopharmacology* (Cambridge, Mass., 2002); see also Healy, *The Antidepressant Era* (cit. n. 15). Viola Balz and Volker Hess provide a detailed analysis of the challenges raised by chlorpromazine in Germany (Balz and Hess, "Psychopathology and Psychopharmacology: Standardisation from the Bottom Up, Using the Example of Neuroleptics," in *Harmonizing Drugs: Standards in 20th-Century Pharmaceutical History*, ed. Christian Bonah, Christophe Masutti, Anne Rasmussen, and Jonathan Simon [Paris, 2009]). See also Thomas A. Ban, "Fifty Years Chlorpromazine: A Historical Perspective," *Neuropsychiat. Dis. Treat.* 3 (2007): 495–500; Andrew C. Leon, "Evolution of Psychopharmacology Trial Design and Analysis: Six Decades in the Making," *J. Clin. Psychiat.* 72 (2011): 331–40.

[17] On Cole's central role in early psychopharmacological research in the United States, see Martin M. Katz, "Jonathan O. Cole," *Neuropsychopharmacology* 35 (2010): 2647; Nina R. Schooler, "Jonathan O. Cole, MD (1925–2009): Innovator in Clinical Psychopharmacology and of the ECDEU/NCDEU Tradition," *J. Clin. Psychiat.* 72 (2011): 286–7. See also Jonathan O. Cole and David Healy, "Jonathan Cole: The Evaluation of Psychotropic Drugs," in *The Psychopharmacologists: Interviews by Dr. David Healy*, by David Healy, 3 vols. (London, 1996), 1:239–63.

ticular, were still very much in flux. Gerard, for example, in outlining to the audience the scope of the problems that the field of drug testing in human subjects was facing, emphasized how much work there was still to be done in determining "the selection of the experimental and of the control populations, the testing conditions, the criteria for evaluating change, the follow-up procedures, the quantitative judgments and the properties of reporting results."[18] How, he asked, should the timing and dose of the drug be determined? Where should the drug be administered? (In order to emphasize the importance of context for determining how substance and soma interact, Gerard quipped that "alcohol acts different [sic] in the presence of one's boss or one's blond.")[19] How do changes in the ward situation during the experiment affect drug action? Should one select control populations, and if so, on what grounds does one designate them as controls? From whom or from what should one gather and/or elicit information about any changes produced by the drug? Gerard's questions pointed to the complex webs that connected the drug and the patient's body to the various worlds in which she lived and within which the potential action of the drug might be observed and rendered visible. What, exactly, should be held stable for change to be both discerned and measured? How might the emergence or attenuation of particular affects in and through patients' bodies be one conduit through which a claim for the effectiveness of a drug's action might be lodged? And what kinds of practices, housed within which kinds of bodies and drawing on which kinds of observational skills, would be best placed to discern and measure those changes?

Let us zero in on Cole's deliberations over which aspects of the patient needed to be monitored for signs of change, and which technologies and practices of observation needed to be mobilized in order to do so. Cole grappled with the fact that, at that moment, there was neither general agreement nor any unified codes of practice either in relation to describing, or indeed naming, changes in psychopathological behavior or in relation to identifying specific effects that the new drugs helped to bring about. It was on this muddled and muddied terrain—a terrain that featured the patient's body, the psychopharmacological substance, the clinical scientist, other clinical care staff, and the wards in which they were emplaced—that Donald Klein, a few years later, would come to excavate what he argued were ontologically heterogeneous manifestations of anxiety. The crispness of Klein's empirical and conceptual work was underpinned by what he maintained was the near-surgical precision of one particular drug as it functioned as a psychopharmacological dissection tool. As we recognize the boldness of Klein's maneuver, we might do well simultaneously to keep in view the dense and heterogeneous landscape—material, methodological, and epistemological—with which such a maneuver had both to reckon and, ultimately, to dispense.

Cole, in his conference presentation, covered a gamut of tools and techniques that might be used—starting with clinical rating scales, and moving through interview content analysis (of use, he suggested, in fine-grained analyses of progress in psychotherapy, or for following the effects of other therapies by dint of frequent psychother-

[18] Ralph W. Gerard, "Orientation: Analysis of Program," in *Psychopharmacology: Problems in Evaluation: Proceedings of a Conference on The Evaluation of Pharmacotherapy in Mental Illness Sponsored by the National Institute of Mental Health, the National Academy of Sciences–National Research Council, and the American Psychiatric Association, Washington, D.C., September 18–22, 1956*, ed. Jonathan O. Cole and Ralph W. Gerard (Washington, D.C., 1959), 9–19, on 13.
[19] Ibid., 15.

apeutic interviews), psychological tests (which included the Rorschach, as well as personality tests used in clinical psychology), and physiological evaluation (such as estimating autonomic "reactivity" by injecting small doses of sympathetic or para-sympathetic drugs to predict response to somatic therapy).[20] Cole was acutely aware of the problems posed by the contemporaneous scientific push to increase the scale of psychopharmacological experimentation and analysis. For example, he noted that in a study taking place in a large state hospital, a six-item rating scale may be superior to a fifty-item scale and concluded—in a sentence that accurately presaged the incipient arrival of many short psychiatric rating scales in psychiatry—that "*brief* rating scales for judgments of the patient's psychopathology, to be filled out both by the admitting physician and by the nurses and aides, could be of considerable help in delimiting better the types of patients helped and in providing easily usable data amplifying the assigned diagnosis."[21] The conference as a whole was, indeed, filled with scientists' and clinicians' ambitions to push toward larger scales of analysis and evaluation, and with recognition of their need for better technologies to track and capture change across large vistas of clinical experimentation.

We see, then, how central geography was to the challenge of assessing affective and cognitive change. I use the term "geography" to connote both the physical spaces in which and through which drug action might be adjudicated and measured and the various socio-spatial imaginaries mobilized by clinical researchers. Should the evaluative terrain encompass the microgeographies of conversations between patient and psychoanalyst, or the slightly broader circuit between clinician, patient, and the technological device of a printed rating scale? Or should the focus rather be on the dynamic psychosociological topologies of spatial interactions and atmospheres within the ward, or on tracing out a temporally dislocated geography in which social workers are sent on unexpected visits to discharged patients, beyond the reaches of the hospital and of the locus of the treatment itself?[22] Even if clarity could be acquired about the appropriate socio-spatial context in which to evaluate psychopharmacological action, there remained unresolved questions about the accuracy of the behavioral and affectively freighted material that would be acquired. For example, some psychoanalytically oriented researchers worried that assessing therapeutic change by analyzing interview content would result in fixating too much on conscious, verbal communication and occlude analysis of unconscious motivations and nonverbal modes of communication.[23]

[20] Jonathan O. Cole, "The Evaluation of the Effectiveness of Treatment in Psychiatry," in Cole and Gerard, *Psychopharmacology* (cit. n. 18), 92–107.

[21] Ibid., 104; emphasis added. A number of rating scales were developed in the late 1950s and early 1960s; those that became particularly influential in psychiatry include Max E. Hamilton, "The Assessment of Anxiety States by Rating," *Brit. J. Med. Psychol.* 32 (1959): 50–5; Hamilton, "A Rating Scale for Depression," *J. Neurol. Neurosurg. Psychiat.* 23 (1960): 56–62; A. T. Beck et al., "An Inventory for Measuring Depression," *Arch. Gen. Psychiat.* 4 (1961): 561–71; John E. Overall and Donald R. Gorham, "The Brief Psychiatric Rating Scale," *Psychol. Rep.* 10 (1962): 799–812. See also Michael Worboys, "The Hamilton Rating Scale for Depression: The Making of a 'Gold Standard' and the Unmaking of a Chronic Illness, 1960–1980," *Chronic Illness* 9 (2013): 202–19.

[22] Such evaluative methods were used by the psychiatrist Maxwell Jones, pioneer of the concept of the therapeutic community. Jones, *The Therapeutic Community: A New Treatment Method in Psychiatry* (New York, 1953).

[23] See Cole, "Evaluation" (cit. n. 20), 100.

Perhaps it was not surprising, then, that Cole finished his address by noting that, in spite of the clinical and scientific evaluative innovations offered by various technologies, by the use of diverse socio-spatial settings, and by the push toward assessing larger numbers of patients across multiple sites, "The detailed study of the response to treatment of the individual patient under the experienced observer will still, no doubt, provide leads to be tested on a larger group of patients."[24] That statement appears to be a call to order grounded on straightforwardness and simplicity: Find one trained observer who can focus on the individual patient! And, indeed, as we shall see, much of the potency of Donald Klein's early scientific innovations arose precisely from the act of a small number of experienced observers studying the response to treatment of a small number of patients. There is no doubt that Klein's achievement in installing panic disorder as a new nosological entity was significantly dependent on small-scale, intimate geographies and was not born from the large, multisite experiments that were fantasized about at the 1956 drug evaluation conference. But we should not be hoodwinked by the apparent simplicity of Cole's injunction. What constitutes, in practice, the actions and interpretations of any one "experienced observer" in psychiatric research is anything but straightforward.[25]

CREATIVE EXPERIMENTATION IN HILLSIDE HOSPITAL

Historians of US psychiatry have spent much analytical and empirical energy detailing the shift from psychodynamic models (prominent, e.g., in the second edition of the *DSM*) to the classificatory logics of *DSM-III*.[26] Central to that shift toward new classificatory logics was the proliferation in the 1960s and 1970s of psychiatric rating scales in relation to both symptoms and diagnoses. This emphasis on ratings in the service of standardization, in particular as regards the imperative to ensure reliability in relation to psychiatric diagnoses, has had consequences beyond that of directing our focal gaze toward 1980, the watershed year in which *DSM-III* was published. One consequence comprises the tight bonds that have been drawn between the emergence of psychopharmacology, the development of robust clinical protocols, and the overall push toward standardization. This narrative of methodological transformation has made it more difficult, perhaps, to keep in focus some of the most imaginative and influential early psychopharmacological studies, which departed significantly from the driving logic of rating scales with their impetus toward a standardization of symptomatology and an elaboration of target symptoms.[27] My focus here will be on the

---

[24] Ibid., 104.

[25] As Lorraine Daston and Elizabeth Lunbeck have made clear, "Observation is a highly contrived and disciplined form of experience that requires training of the body and mind, material props, techniques of description and visualization, networks of communication and transmission, canons of evidence, and specialized forms of reasoning." See Daston and Lunbeck, "Observation Observed," in *Histories of Scientific Observation*, ed. Daston and Lunbeck (Chicago, 2011), 1–9, on 3. In my essay, I want to draw attention to the authority accruing to a certain kind of clinical observation, and to its importance in foregrounding a particular psychopathological affect. For other examinations of the complex relationship between the clinical and/or scientific observer and that which she observes, see Lunbeck, "Empathy as a Psychoanalytic Mode of Observation: Between Sentiment and Science," and Otniel Dror, "Seeing the Blush: Feeling Emotions," both in Daston and Lunbeck, *Histories*, 255–75, 326–48; Tiffany Watt Smith, *On Flinching: Theatricality and Scientific Looking from Darwin to Shell-Shock* (Oxford, 2014).

[26] Decker, *The Making of DSM-III* (cit. n. 4).

[27] Two important exceptions include scholarship by David Healy and by Viola Balz and colleagues (who emphasize the importance in early German psychopharmacology of long-term clinical observa-

creative experimentation that entangled drugs, bodies, minds, and affects in a far-from-standardized early set of experiments that took place in one small psychiatric hospital in the United States in the late 1950s. I am locating our analytic gaze, in other words, on a historical-geographical site that is significantly different from the perhaps more settled, and perhaps epistemologically less lively, clinical and research landscape that would come to be installed via the logics of the psychiatric randomized controlled trial and, subsequently, the framework of *DSM-III*.

What kinds of scientific and clinical observation of research patients were sanctioned in Klein and Fink's early psychopharmacological experiments, and how did they help consolidate new ontologies of psychopathological anxiety, as well as new kinds of interpretations of anxious bodies?[28] As I emphasized at the start of this article, the affect of anxiety was not front and center in the investigators' field of vision as these experiments commenced. The clinical researchers did not start with a series of questions about how to understand the phenomenology of agoraphobic anxiety; rather, the hinge that shifted the analytical plane and that served to open up the problematic of anxiety was the introduction of the psychopharmacological substances themselves.

Hillside Hospital, where those experiments took place, is located in Glen Oaks, Long Island, New York.[29] In 1954, the hospital established research programs "devoted to an understanding of the mode of action of the psychiatric therapies of the hospital."[30] Hillside Hospital was a Jewish hospital—which, in that period, meant being located beyond the orbit and the sphere of influence of the large, university-affiliated research hospitals—and was, in the late 1950s, largely focused around psychoanalytic therapies. Max Fink, who headed the experimental psychiatry research program, and his colleague Donald Klein (who at that time was a research associate and a mental health career investigator funded through NIMH) were developing new methodologies through which to investigate the mode of action of drug therapies. And if they were preoccupied with exactly the problems enumerated in the 1956 conference on the evaluation of pharmacotherapy—"Do the drugs act? How do the drugs act? What if the drugs act?"—they were particularly interested in figuring out the answer to a fourth question: In relation to which kinds of patients?

---

tion and psychopathological theory, as opposed to a focus on standardized rating scales and target symptoms). Healy, *The Antidepressant Era* (cit. n. 15); Healy, *Creation* (cit. n. 16); Balz, "Terra Incognita" (cit. n. 15); Balz and Hess, "Psychopathology" (cit. n. 16); Viola Balz and Matthias Hoheisel, "East-Side Story: The Standardisation of Psychotropic Drugs at the Charité Psychiatric Clinic, 1955–1970," *Stud. Hist. Phil. Biol. Biomed. Sci.* 42 (2011): 453–66.

[28] See Deborah Blythe Doroshow on Fink's therapeutic practices as regards insulin coma therapy. Doroshow, "Performing a Cure for Schizophrenia: Insulin Coma Therapy on the Wards," *J. Hist. Med. Allied Sci.* 62 (2007): 213–43.

[29] "Hillside Hospital is a 196-bed, open ward, voluntary psychiatric facility for the treatment of patients with early and acute mental disorders whose stay is independent of their ability to pay. All patients are seen in individual psychotherapy, with the expectation that psychotherapy should be given a trial prior to other measures. Somatic therapies are employed by joint decision of the resident therapist and supervising psychiatrist, with the management of medication restricted to the research staff"; Donald F. Klein and Max Fink, "Psychiatric Reaction Patterns to Imipramine," *Am J. Psychiat.* 119 (1962): 432–8, on 432. See also Irving J. Sands, "The First Twenty-Five Years of Hillside Hospital: A Voluntary Psychiatric Hospital," *J. Hillside Hospital* 2 (1953): 199–206; Robert L. Kahn, Max Pollack, and Max Fink, "Social Factors in the Selection of Therapy," *J. Hillside Hospital* 6 (1957): 216–28.

[30] Max Fink, "Experimental Psychiatric Research at Hillside: Review and Prospect," *J. Hillside Hospital* 10 (1961): 159–69.

They therefore created a laboratory within the psychiatric hospital and established a system through which to control the prescription of psychotropic drugs throughout the hospital.[31] They put procedures in place: all prescriptions were dispensed by a psychiatrist—Klein—within the Department of Experimental Psychiatry, who responded to a request made by the therapist of a particular patient and interviewed the patient prior to dispensing the drug. During the period of drug therapy, the patient's response was assessed weekly, and in a variety of ways, from the perspective of various individuals—by the patients themselves, by ward staff, by the patient's therapist, and by the therapist's supervisor. (We shall return shortly to how those perspectives were weighted on the basis of authority and clinical importance.) The dosage and the type of medication could be altered. From October 1958 to October 1959, Klein and Fink treated 120 patients with chlorpromazine, promazine, or prochlorperazine (all phenothiazines) and eighty-seven patients with imipramine. Imipramine was, at that point, a new drug that had emerged through the Swiss psychiatrist Roland Kuhn experimentally examining the effects of a Geigy compound. (This compound was similar in structure to that of chlorpromazine; it did not appear to have much effect on psychotic symptoms but did appear to reduce the depression of patients diagnosed with schizophrenia. Imipramine is now commonly described as the first tricyclic antidepressant.)[32] In an associated study, a total of 215 patients received only imipramine between October 1959 and July 1961; Klein and Fink published two seminal papers relating to those two studies.[33] Subsequently, after the therapeutic success of those early, open experiments, a randomized placebo trial was carried out.[34] In assessing drug action, Klein and Fink paid attention to what they designated "changes in mental status and hospital adjustment," "progress in psychotherapy," and "utilization of hospital facilities." Crucially, the patient's diagnosis was not at stake in the decision over which drug to prescribe, and Klein and Fink also argued that current psychodynamic formulations were of no help in predicting course of illness or treatment. They therefore jettisoned both existing diagnostic classifications and all psychodynamic formulations; instead, they aimed to set aside commonly used frameworks of adjudication the better to attend to the bodies and actions of those patients who had received drugs from the prescribing physician. "Present techniques of evaluating therapies by global improvement scores, imprecise diagnostic classification, and

---

[31] Psychoanalysts at that time were not keen to prescribe drugs, seeing them as disruptive of the transferential relationship between analyst and patient. In many hospitals, one doctor would be designated the "druggist": he or she would prescribe drugs, leaving the remaining psychiatrists free to conduct psychotherapy/psychoanalysis (see Healy, *Antidepressant Era* [cit. n. 15], 191). My account of Klein and Fink's experiments is greatly indebted to their own two journal publications documenting those early studies, as well as Klein's retrospective reflections on these experiments. See Donald F. Klein and Max Fink, "Behavioral Reaction Patterns with Phenothiazine," *Arch. Gen. Psychiat.* 7 (1962): 449–59; Klein and Fink, "Psychiatric Reaction Patterns" (cit. n. 29); Klein, "Anxiety Reconceptualized" (cit. n. 6); Donald Klein and David Healy, "Donald Klein: Reaction Patterns to Psychotropic Drugs and the Discovery of Panic Disorder," in Healy, *The Psychopharmacologists* (cit. n. 17), 1:329–52.

[32] Walter A. Brown and Maria Rosdolsky, "The Clinical Discovery of Imipramine," *Amer. J. Psychiat.* 172 (2015): 426–9.

[33] Klein and Fink, "Psychiatric Reaction Patterns" (cit. n. 29); Klein and Fink, "Behavioral Reaction Patterns" (cit. n. 31).

[34] My focus here is purely on the early experiments that operated outside of the logic of a controlled, placebo trial. The later fixed dosage, double-blind study was reported in Donald F. Klein, "Importance of Psychiatric Diagnosis in Prediction of Clinical Drug Effects," *Arch. Gen. Psychiat.* 16 (1967): 118–26.

target symptoms abstracted from their context were," they emphasized, "felt to be methodologically inadequate."[35] Klein and Fink documented eight distinct patterns of "behavior change" for those treated with the phenothiazines and seven for those treated with imipramine. Their underlying claim was that the interaction of particular patients with particular drugs allowed the identification of distinct "reaction patterns." The "descriptive behavioral typology" that allowed those reaction patterns to become visible was produced through three research psychiatrists reviewing the patients' detailed records and coming to a consensus "concerning the patient's behavioral reaction during the medication period."[36]

For our purposes, what is crucial is the pattern that Klein and Fink identified as relating to "episodic anxiety": it was this pattern that would, in time, become renamed and reimagined as "panic disorder."[37] Those patients grouped under episodic anxiety were variously characterized, before treatment, as experiencing "episodes of felt anxiety and helplessness, associated with fearful clinging and urgent demands for aid,"[38] or as experiencing "the sudden onset of inexplicable 'panic' attacks, accompanied by rapid breathing, palpitations, weakness, and a feeling of impending death."[39] Notably, they defined the "hallmark" of those patients' condition as "expectant fear of lack of support when overwhelmed" (though they also noted that "their condition was often referred to as agoraphobia").[40] With phenothiazine ("major tranquilizer") treatment, these patients' "episodic anxiety" was unaffected: while the tension they experienced might sometimes be reduced, "depressive complaints were not alleviated and phobic limitations on activity continued."[41] With imipramine, in notable contrast, "the 'panic' attacks ceased, . . . [although] the patients were reluctant to change their phobic behavior pattern and required much persuasion, direction and support."[42] The psychiatrists noted, furthermore, in those patients showing a positive reaction to imipramine treatment, "a surprising rise in aggressive self-assertion and rejection in domination"; response to imipramine showed "no special relationship to age or sex."[43]

Episodic anxiety patients provoked particular analytic attention from Klein and Fink "because of the apparently paradoxical nature of their drug response": while the patients were clearly very anxious, the phenothiazines—tranquilizers—strangely had no effect, either on the particular quality of their anxiety or on their "phobic limitations." (These limitations included the behaviors of some patients who, "between episodes [of anxiety] . . . manipulated the staff to enable them to remain within the

[35] Klein and Fink, "Psychiatric Reaction Patterns" (cit. n. 29), 432.

[36] Klein and Fink, "Behavioral Reaction Patterns" (cit. n. 31), 449.

[37] The term "panic disorder" did not emerge until the late 1970s; throughout the 1960s, Klein—while referring to attacks of panic—continued to use anxiety as the dominant nosological term. For example, in 1967, he referred to the "phobic-anxiety reaction"; Klein, "Importance of Psychiatric Diagnosis" (cit. n. 34), 121. In Klein's book on diagnosis, coauthored by John M. Davis and published in the late 1960s, he and Davis defined "panic anxiety" as "the state of being suddenly overwhelmed by fearful sensations," which is "accompanied by massive autonomic responses, both sympathetic and parasympathetic." See Klein and Davis, *Diagnosis and Drug Treatment of Psychiatric Disorders* (Baltimore, 1969), 325.

[38] Klein and Fink, "Behavioral Reaction Patterns" (cit. n. 31), 456.

[39] Klein and Fink, "Psychiatric Reaction Patterns" (cit. n. 29), 435.

[40] Ibid.

[41] Klein and Fink, "Behavioral Reaction Patterns" (cit. n. 31), 456.

[42] Klein and Fink, "Psychiatric Reaction Patterns" (cit. n. 29), 436.

[43] Ibid.

phobically defined safe areas or to have constant attendance by aides.")[44] Why were phenothiazines not effective for those patients, whereas imipramine was effective? If Klein and Fink had dramatically succeeded in rendering visible distinctive reaction patterns among the patients in their cohort, they still needed to explain why imipramine helped patients with episodic anxiety.

PANIC DISORDER'S ORIGIN STORY

Klein offered a number of retrospective reflections on his early experiments at Hillside Hospital, in which he set out, in characteristically vigorous prose, his explanation for the action of imipramine.[45] Notably, Klein chose to exemplify the stakes of his account by turning to clinical observations that were made regarding one particular male patient. Clinical descriptions of this patient's suffering and treatment might, indeed, be classified as the origin story of the nosological category of panic disorder.[46] Through considering this origin tale in some detail, we will be able to discern the relation that was traced between the practice of observation and the phenomenon that was its object—the relation that underpinned Klein's practice of psychopharmacological dissection.

The story begins in Hillside Hospital with the patient's doctor calling Klein. The doctor described his patient as schizophrenic and told Klein that treatment with the phenothiazine chlorpromazine had made the patient worse. Klein was not convinced that the patient was schizophrenic: he was neither delusional nor hallucinating and manifested no thought disorder or restriction of affect. He was, however, Klein emphasized, "hideously anxious, extremely dependent, extremely demanding."[47] Klein described this experimental and clinical situation as one that allowed him to bring into the same terrain a "patient we didn't know what to do with" and "a drug [imipramine] . . . we were unsure what it did." He "mixed them together" through a process that he characterized as "pure empiricism."[48] After a couple of weeks of imipramine treatment, there appeared to be no change in the patient's symptoms and the patient was complaining bitterly of his continuing anguish. After the third or fourth week, however, the nurses in the wards felt that something had altered, though they were unable to put their finger on quite what that was; neither the patient nor his therapist nor the therapist's supervisor believed there to be any change. Finally, one nurse—whom Klein described a number of times as a "good observer"—pointed out that the patient was no longer running to the nurses every few hours wanting help and feeling as though he were dying.[49] After several more weeks, Klein averred that those improve-

---

[44] Donald F. Klein, "Delineation of Two Drug-Responsive Anxiety Syndromes," *Psychopharmacologia* 5 (1964): 397, 398.

[45] Klein's theoretical architecture would become more elaborate over time, though he has not departed from the basic logic of his early accounts.

[46] The story of this "originary" patient is elaborated in Klein, "Anxiety Reconceptualized" (cit. n. 6); Klein and Healy, "Donald Klein" (cit. n. 31), 311. Klein also spoke at some length about this patient in his interview with me (Donald Klein, interview by Felicity Callard, digital recording, The Hamptons, New York, 28–30 July 2011). In "Anxiety Reconceptualized," the one patient becomes a group of patients, although the structure of the narrative is the same. I refer to a singular patient.

[47] Klein and Healy, "Donald Klein" (cit. n. 31), 331.

[48] Ibid.

[49] Klein, in his interview with me, said "one of [the ward staff] who's a good observer said, 'You know, this guy's been coming to the nursing station four times a day for the past nine months saying he's dying'" (Klein, interview by Callard [cit. n. 46]).

ments could not be discounted—even though the patient's own explanation for his behavior was that he had finally realized the nurses could do nothing for him and that he was therefore no longer running to them. Indeed, the good doctor is required to point out changed behavior to the unknowing patient: according to Klein's account, the patient was "stunned" since "he had no idea he had changed his behavior."[50]

Klein described how he and his colleagues were initially puzzled by the strange turn of events. Was the patient primarily depressed rather than anxious or phobic, such that the imipramine, with its antidepressant qualities, was lifting his depression and simultaneously alleviating his anxiety symptoms? Klein noted, however, that "most of the patients [within the episodic anxiety group] neither looked nor acted depressed," and "thoughts of suicide, guilt, and depressive ideas of reference were conspicuously absent."[51]

What Klein characterized as a scene of "pure empiricism" became a scene that—through the interlocking actions of a drug whose action was uncertain and of observers who were not sure what they might be on the lookout for—unfurled two distinct ontologies of psychopathological anxiety. Klein argued that observations of this originary patient allowed him to parse anxiety into two kinds, in contradistinction to the prevailing psychoanalytic model of anxiety.[52] Klein installed "a physiological discontinuity" between what he came to term "paroxysmal anxiety" (which was manifested, e.g., when the patient ran to the nurses) and "chronic anxiety" (from which the patients suffered most of the time).[53] He interpreted imipramine as acting on the paroxysmal anxiety but having no effect on the chronic anxiety because that anxiety was of a different order. Now that Klein had divided anxiety into two phenomenological and physiological kinds, he was able—in subsequent research and publications—to clarify the link between them. The various phobias that beset patients like the "originary" patient, as well as those patients' chronic and anticipatory anxiety, were all directed toward the avoidance of panic attacks:

> In other words, what they feared was having a panic attack, particularly having one while in a helpless situation. We began to understand why such patients would not drive over a bridge or into a tunnel. The simple answer, without resort to psychoanalytic symbolism, was that they realized that once they had committed themselves to a bridge or a tunnel there would be no way to stop, so that if a panic attack occurred, they would be completely helpless and isolated.[54]

Klein emphasized that patients' intense attacks of anxiety came first and were subsequently followed by the patterns of phobic avoidance, general anxiety, and depressed mood. The patients, he claimed, did not realize the difference between the two kinds

---

[50] Klein and Healy, "Donald Klein" (cit. n. 31), 331.

[51] Klein, "Anxiety Reconceptualized" (cit. n. 6), 239.

[52] For a classic psychoanalytic account of anxiety advocated at that moment, see Leo Rangell, "On the Psychoanalytic Theory of Anxiety: A Statement of a Unitary Theory," *J. Amer. Psychoanal. Assoc.* 3 (1955): 389–414.

[53] Klein, "Anxiety Reconceptualized" (cit. n. 6), 239.

[54] Ibid., 240–1. Notably, *DSM-III* carried a specific emphasis on helplessness in the description for Panic Disorder (300.01): "A common complication of this disorder is the development of an anticipatory fear of helplessness or loss of control during a panic attack, so that the individual becomes reluctant to be alone or in public places away from home" (American Psychiatric Association, *Diagnostic and Statistical Manual* [cit. n. 4], 230).

of anxiety because the chronic anxiety submerged the particularity of the panic attacks. (The originary patient's claim, then, that he was no longer running to the nurses
because he realized they could do nothing for him, was, on Klein's account, an erroneous post hoc explanation.) Chronic anxiety remained, Klein explained, because although imipramine brought patients' panic under control, the patients did not know
or believe that this would remain the case. Thus their anticipatory anxiety remained
and kept in place their avoidant mechanisms (the phobic limitations). Klein's formulations would, in time, assist in establishing panic as a central topic for research and
treatment. Klein came to understand panic as a kind of "spontaneous" attack resulting from a dysfunctional somatic mechanism; he argued that imipramine normalized
this dysfunctionality.[55] This, I argue, helped to transform the locus of clinical intervention in cases of panic: consideration of the situations or places in which paroxysmal anxiety had occurred was of secondary interest since the primary question was
how to cure the defective somatic mechanism—which produced the panic—pharma-
cologically.[56]

Klein's formulations—first developed in his articles from the 1960s, though continuing to this day[57]—turned upside down established psychiatric wisdom concerning
the development of paroxysmal anxiety out of chronic anxiety (a formulation that had
loosely followed Freud's understanding of anxiety neurosis).[58] They also shifted the
mise-en-scène of agoraphobia that had been in place since Westphal's first inquiries
into agoraphobia in the 1870s. The scene of Klein's pharmacological dissections—
the hospital ward and a panic-stricken inpatient running to his nurses—moved the
spatial imaginary of the disorder away from the streets and squares that had until then
formed the primary stage for agoraphobic behavior. That Klein's originary panic disorder patient was male rather than female also marked a break with many psychiatric
and psychoanalytic commonplaces concerning women and agoraphobia.[59] Klein's
model replaced the backdrop of public space with the drama of a terrified figure running to be comforted in the closeted space of a hospital ward: at the center of the dis-

[55] See, e.g., Donald F. Klein et al., *Diagnosis and Drug Treatment of Psychiatric Disorders: Adults and Children* (Baltimore, 1980).

[56] Klein, "Anxiety Reconceptualized" (cit. n. 6), 239. Klein, in the latter part of the 1960s, buttressed his own results by citing F. N. Pitts and J. N. McClure, "Lactate Metabolism in Anxiety Neurosis," *New Engl. J. Med.* 277 (1967): 1329–36. Pitts and McClure had run a study showing that in-
travenous lactate infusions bring on a panic attack in those people who suffer from spontaneous panic
attacks but rarely have any effect on normal individuals. For Klein, the lactate-induced panic seemed
to mirror, and provide grounded confirmation of, his model of the spontaneous panic attack. Orr has
argued that Klein's "panic-disordered body is defined by *an absence of relation* to any social reason
for the force or the timing of its terror. Even within its classificatory family of 'anxiety disorders,'
panic disorder stands out as the psychic response to no discernible stimulus" (Orr, *Panic Diaries*
[cit. n. 13], 174; emphasis in the original).

[57] Klein remains a prolific scientific author. In more recent years, he has championed "serendipity"
in psychopharmacology, which he sees as central to the psychopharmcological successes of the 1950s
and 1960s, and which he believes to have been wrongly pushed to the side by the logic of rational drug
development. See, e.g., Donald F. Klein, "The Loss of Serendipity in Psychopharmacology," *J. Amer.
Med. Assoc.* 299 (2008): 1063–5.

[58] Sigmund Freud, "On the Grounds for Detaching a Particular Syndrome from Neurasthenia under
the Description 'Anxiety Neurosis,'" in *The Standard Edition of the Complete Psychological Works
of Sigmund Freud (1893–1899)*, vol. 3, ed. James Strachey (London, 1895), 85–115.

[59] Many post–World War II approaches to agoraphobia associated it with particular kinds of femininity; see, e.g., D. Buglass et al., "A Study of Agoraphobic Housewives," *Psychol. Med.* 7 (1977):
73–86.

order lay not a problem in negotiating public spaces of exchange and sociality but a problem of dependency and need for a substitute mother figure. The tumult of the city receded; a small-scale, intimate parent-child drama took its place.

* * *

In the remainder of this article I want to consider in greater detail the practices of clinical observation—as well as what I shall call their "intimate geographies"—that accompanied those early, creative experiments by Klein and Fink at Hillside Hospital. These are, I believe, central to understanding how those experiments helped focus attention on a psychopathological manifestation of affect that would, in time, allow the inauguration of the new nosological category of panic disorder. My interest lies in understanding how Klein and Fink responded to the challenges posed in the 1956 conference on the evaluation of pharmacotherapy, and how their experiments mobilized particular formulations of psychopathological affect. The originary scene that Klein described under a rubric of "pure empiricism" drew together a complex network of material objects (e.g., the drug imipramine), socio-spatial settings (spaces of psychotherapeutic consultation vs. the regular space of the ward), discursive elements (the speech of patients, therapists, ward staff), bodily movements (patients running, or not running, to their nurses), and changes in affective rhythms and demeanors (e.g., increases in patients' "aggressive self-assertion" upon taking imipramine). For psychopathological anxiety to be transformed from one into two ontologically distinct kinds, which elements within this network were prioritized and valorized, and which, ultimately, were ignored? The 1956 conference had set out multiple ways of traversing and mapping a dense and heterogeneous landscape so as to determine whether, how, and with what consequences drugs might "act." But how widely did Klein and Fink's map extend? How did it end up validating some elements within that landscape and occluding others? And is the concept of "purity" (namely, Klein's claim of "pure empiricism") apposite in characterizing that scientific and clinical scene?

OBSERVING BEHAVIOR

Central to Klein and Fink's framework for adjudicating drug action was their notion of a "behavioral reaction pattern."[60] How did they conceptualize behavior, and what role did affect play? Notably, changes in affect were one of the five criteria—alongside changes in symptoms, patterns of communication, and participation in psychotherapy and social activity—they used to divide patients into groups.[61] But what was meant by affect? The researchers set great store on "gain[ing] a *broad image* of the patient's behavior": not only did they shy away from the enumeration of "simple lists of traits and symptoms," but they also deemed batteries of psychological, psychiatric, and behavioral indices to be of little use in assisting in the carving out of relevant patient subpopulations.[62] Of the eight behavior change clusters that Klein and Fink

---

[60] Klein and Fink's collaborative work followed earlier work by Fink in which he had examined behavioral patterns to explore the effects of convulsive therapy. See Max Fink, "A Unified Theory of the Action of Physiodynamic Therapies," *J. Hillside Hospital* 6 (1957): 197–206; Fink, R. L. Kahn, and M. A. Green, "Psychological Factors Affecting Individual Differences in Behavioral Response to Convulsive Therapy," *J. Nerv. Mental Disease* 128 (1959): 243–8. Fink would go on to become one of the world's leading electroconvulsive therapy researchers.

[61] Klein and Fink, "Behavioral Reaction Patterns" (cit. n. 31), 449, 457.

[62] Ibid.; emphasis added.

enumerated, several centered on affect (e.g., "reduction of anger," "affective stability," and "unaffected episodic anxiety"). Affect, then, was embedded within and
helped to constitute the "broad image" of behavior—and was addressed via patient
demeanor, gestures, actions, and expressions. It appeared in a variety of forms and
was underwritten by different kinds of evidence that was gathered via different kinds
of observations by different kinds of people. Those patients placed by Klein and Fink
in the reaction pattern group "Suppressive Denial," for example, were distinguished
by "a fearful suspiciousness accompanied by derogatory ideas of reference." Claims
that they are "fearful, agitated, and panicky" were grounded in references to their
speech being "evasive" or "guarded," and their social interactions being "hostile,
fearfully demanding, and leading to mutual withdrawal." Crucially, one source of observational evidence was the affective reactions in those staff interacting with them:
these patients were described as "engender[ing] uncomfortable feelings in staff personnel, with fears of assaultive behavior."[63] In comparison, the "somatizing" group
was characterized not only by patients' "chronic use of bodily complaints" but by
"manipulation as a basis for interpersonal relatedness." Manipulation was evidenced
by a fascinating range of affectively tinged behaviors that were interpreted through a
contrast between patients' outward expression and their "inner states." Before treatment with phenothiazines, for example, somatizing patients were deemed to be

> friendly during those interactions where they felt that they were about to get their way,
> and depressed, fearful, reproachful, sulky, and covertly angry when their demands were
> denied. Their symptoms and affective upheavals were most prominent in relation to the
> medical staff, appearing to be role-playing devices rather than expressions of inner
> states.[64]

After treatment with phenothiazines, these patients responded with "heightened manipulation": the evidence that Klein and Fink marshaled here included "histrionic
demonstrations of physical distress such as slumping slowly to the floor, wearing a
wet towel around the head, walking around the corridors leaning against or touching
the wall or using both hands on the stair bannister," and the abandoning of hospital
activities "as another gesture of helpless distress." Klein and Fink concluded that
these patients' somatic and affective complaints seemed "best understood as manipulative communications rather than the direct expression of anxiety or depression";
they claimed, furthermore, that "secondary gain is marked, and their illness is utilized
in an attempt to maintain a protected dependent status."[65]

What is noticeable in these descriptions is the range of different frameworks used
to characterize and interpret both patients' actions and their displays of affect. Descriptions of affect frequently embedded affect within an account of social interaction
(either between patients or between a patient and a member of clinical staff) or a verbal exchange (between a patient and his/her psychotherapist or with a member of
ward staff). Not infrequently, evidence was given that was not necessarily about affect witnessed in the patient, but that comprised feelings invoked by the patient in the
attending clinical staff. (Clinical staff might have been turning, here, to psychoana-

---

[63] Ibid., 451.
[64] Ibid., 454.
[65] Ibid.

lytic principles concerning countertransference, or to other models in which personal feeling was relied upon to assist with diagnosis—such as those indebted to Rümke's "Praecox Gefühl," which was used to identify schizophrenia.)[66] Sometimes there was the implication that interpretations were being made of patients' bodily and/or facial demeanor, or of the affective timbre of their speech (e.g., "patients now approached the interviewer in an ingratiating manner," or patients "expressed boredom with hospital routine").[67] Sometimes, an affectively tinged descriptor—a patient appearing "fearful" or "helpless"—was associated with (inauthentic) "role-play" that ran counter to the inner state, and at other times it was invoked as an apparent endorsement of the patient's authentic affective state. Affective displays were often linked to particular kinds of encounters in particular socio-spatial contexts (e.g., differences were noted between how the patient might behave in the context of a psychotherapeutic encounter, in comparison with social interactions on the wards). This array of frameworks and modes of gathering evidence makes us aware of how heterogeneous the practices of observing, assessing, and interpreting patient behavior and affect appear to have been within the psychiatric hospital at that moment.[68] Eric J. Engstrom, in his analysis of Kraepelin's late nineteenth- and early twentieth-century interest in the role of emotions in psychiatric illness, emphasizes Kraepelin's desire—as manifested in his diagnostic cards [*Zählkarten*]—to develop "reliable diagnostic techniques that, in turn, would lead the way toward greater prognostic certainty in day-to-day clinical practice."[69] How to identify and document details regarding a patient's behavior, emotions, and cognitive abilities—and how to relate these to a diagnosis—remained a challenge for psychiatry through the course of the twentieth century. Nosological schemas and modes of identifying and classifying symptoms, behaviors, and affects remained labile and heterogeneous. Indeed, the fact that there was no uniformly accepted method through which to evaluate the patient—and the effect of the drugs on him or her—was one strong impetus behind the 1956 conference on "The Evaluation of Pharmacotherapy in Mental Illness" discussed earlier.

We see how Klein's and Fink's published texts intermingle psychoanalytic principles and techniques (e.g., the concept of "secondary gain"), both so-called folk and scientific descriptors of affect, and various clinical frameworks used to describe phenomenology, symptomatology, and psychopathology (e.g., the concept of "ideas of reference"). They always inserted affect into a broader hermeneutic matrix through which to assess changes in the patient as a whole. Klein and Fink critiqued the use of "target symptoms" in relation to psychopharmacological research—arguing that such a model, by erroneously assuming that each manifestation of affect (in the context of a psychopathological symptom) was "identical in nature from patient to pa-

---

[66] H. C. Rümke, "Das Kernsymptom der Schizophrenie und das 'Praecox Gefühl,'" *Zentralbl. Gesam. Neurol. Psychiat.* 102 (1942): 168–9.

[67] Klein and Fink, "Behavioral Reaction Patterns" (cit. n. 31), 451.

[68] For additional insights into how Klein is likely to have been assessing patient behavior (including affective behavior), see Loring L. Burnett and Donald F. Klein, "A Guide for the Psychiatric Case Study," *J. Hillside Hospital* 14 (1965): 54–68. In this guide, "Affect" (which was considered under the heading "Direct Observations") included such diverse subheadings as "Tension: level, fluctuation, startle reactions"; "Anxiety: fidgeting, blushing, wet palms"; "Mood: apathy, expansiveness, depression, exhilaration, fluctuation"; "Expressiveness: inhibited, spontaneous, impulsive"; and "Genuineness vs. simulation: exaggerated, feigned."

[69] Eric J. Engstrom, "Tempering Madness: Emil Kraepelin's Research on Affective Disorders," in this volume.

tient,"[70] ended up "implicitly promot[ing] a mosaic view of psychopathology." They went to some lengths to convey how any particular affect might be exacerbated or attenuated in very different ways according to the distinctive behavioral typology in which it appeared. (For example, phenothiazine treatment had, they argued, very different effects on behaviors understood as "anxiety," depending on the overall behavioral reaction group of which anxiety was one part: in the "somatizing" group, anxiety after phenothiazine treatment became "markedly accentuated with much dramatic expressiveness, when dealing with psychiatric staff but it was not apparent during the patient's social intercourse," whereas "anxiety" in the "episodic anxiety" was entirely unchanged.)[71] They concluded that "each symptom represents a prominent facet of various complex adaptations . . . which can be most fruitfully described in a patterned multivariate context."[72]

What allowed the identification of those temporally and spatially patterned accounts of changes in patients' affective demeanors, gestures, and expressions was, Klein and Fink made clear, long-term, "expert observation" by psychiatric researchers who knew the patients. They disparaged the approach taken in many large hospital programs, where many patients were examined and tested by several raters who were not able to have prolonged clinical contact with patients, arguing that "the experienced clinician is our most sensitive cluster analytic device, given the opportunity to use his skills." ("Cluster analysis" emerged in the 1950s, and clustering algorithms began to be used in psychiatry in an attempt to cluster different groups of patients according to symptomatology. That Klein and Fink believed the individual, highly trained observer to trump the technological potency of clustering algorithms emphasized how sophisticated they believed that observer's techniques of parsing, amalgamating, and discerning to be.)[73] In contrast to the contemporaneous use of rating scales, such as Max Hamilton's "Assessment of Anxiety States," which was published in 1959, Klein and Fink's approach constituted the patient's body as a distinct and complex entity that existed in relation to other bodies in particular social settings: it was not something that could be dismantled and disaggregated into a tessellation of target symptoms.[74]

But not all "expert observation" was regarded with equal esteem by Klein and Fink. The observers and practices documented in their early publications made up a collectivity—including Klein and Fink themselves, the ward staff, the psychoanalytic psychotherapists and their supervisors, the nursing aides, and the patients—in which certain kinds of observation were privileged over others. The impact of this privileging became clear when Klein and Fink's favored approach and another mode of observation yielded different judgments about a patient. For example, Klein and Fink, in critiquing the reliance in many psychopharmacological evaluations on sim-

---

[70] Klein and Fink, "Behavioral Reaction Patterns" (cit. n. 31), 457.

[71] Ibid., 458.

[72] Ibid.

[73] Fionn Murtagh, "The History of Cluster Analysis," in *Visualization and Verbalization of Data*, ed. Jörg Blasius and Michael Greenacre (Boca Raton, Fla., 2014). See also Harvey A. Skinner and Roger K. Blashfield, "Increasing the Impact of Cluster Analysis Research: The Case of Psychiatric Classification," *J. Consult. Clin. Psychol.* 50 (1982): 727–35.

[74] Hamilton, "Assessment"; Worboys, "Hamilton Rating Scale" (both cit. n. 21). For a summary of the use of scales and checklists in the 1950s, see Richard L. Jenkins and Maurice Lorr, "Symptom Scales and Check Lists for Determining Symptomatic Improvement in Psychotic Patients," in Cole and Gerard, *Psychopharmacology* (cit. n. 18), 469–77.

plistic psychotherapeutic notions of "cured" or "improved," argued for a means by which the "rich complexity of behavioral change" might be registered. Such complexity would not, they implied, be recognized by psychotherapists, who might rate behavior change negatively because it interfered with the progress of psychotherapy, or by ward staff (who might rate the change as "positive" because they perceived the nursing burden to be alleviated).[75] Their research also downplayed the robustness of much of the evidence gleaned via linguistic utterances (remember Klein's account of the originary patient—in which the therapist discerned no difference in therapy after the ingestion of imipramine, and the patient himself was interpreted as providing erroneous explanations for his own actions). Theodore Porter's historical research on different forms of scientific objectivity is helpful here in allowing us to discern the professional and disciplinary jostling over when and how individual expertise ought to be trusted over forms of "mechanical objectivity" (such as scales or checklists). Klein and Fink are implicitly defending one kind of clinical observation as far more epistemologically robust than another kind of clinical observation (poor expertise in the form of psychoanalytic techniques of observation), and also more robust than the use of symptom checklists (a poor example of mechanical objectivity).[76]

Klein and Fink attempted to position observations of behavioral changes as a way of sidestepping some of the difficulties attendant upon observing and adjudicating changes in patients. But behavior was not as pellucid a means of capturing the potential effects of drug action as Klein and Fink might have wished it to have been. As we have already seen, the means by which they brought attention to particular kinds of behaviors rather than others was inflected by their interpretations of how behavior emerged in the context of particular kinds of communicative actions. Klein and Fink appeared to interpret some behaviors as not possessing ambivalent psychic overlays (e.g., the "helplessness" of the patients who ran to the nurses when beset by panic attacks), whereas other behaviors were associated with complex psychic motivations (the "dependent façade" of the somaticizing group, who engaged in "role-playing" with the nurses, and whose "helpless distress" was seen as a manipulative "gesture"). In short, Klein and Fink worked with a complex hermeneutics that ended up pulling particular affects and behaviors into analytical visibility and left others, no doubt, in the shadows. This is perhaps particularly striking in relation to the prowess of that "keen clinical observer"[77]—who was judged to have spotted the core of what was happening in relation to the "helpless" originary patient. For while there were surely multiple behavioral transformations that might have been noted after treatment with imipramine, what actually was foregrounded and endowed with the greatest significance was the fact that the patient was deemed to have stopped running to the nurses several times a day. It was the cessation of a particular kind of locomotor behavior— over and above the timbre and specifics of the affect of anxiety—that was privileged in Klein and Fink's account. Paroxysmal anxiety became newly visible as a distinct kind of psychopathological affect by dint of the removal (after imipramine treatment) of a particular kind of socially communicative locomotor behavior.

[75] Klein and Fink, "Behavioral Reaction Patterns" (cit. n. 31), 457.

[76] Theodore M. Porter, *Trust in Numbers: The Pursuit of Objectivity in Science and Public Life* (Princeton, N.J., 1995).

[77] Klein, "Anxiety Reconceptualized" (cit. n. 6), 238.

What were Klein and Fink actually seeing? What did they privilege in this scenario of "pure empiricism"? While they stressed the importance of attending to behavior, their enumeration of particular bodily actions was buttressed by a theoretical framework that underpinned their descriptive typology. After all, the patient's running was not documented simply as locomotor action but specifically as a manifestation of helplessness and "fearful clinging." And it was here that the concept of "separation anxiety," as formulated by the psychoanalyst and ethologist John Bowlby, haunted the empirical scene unfolding in Hillside Hospital.[78] Bowlby, dissatisfied with the accounts that Freud and later psychoanalysts had provided to explain the relationship between the child and mother, turned to ethology in order to frame attachment as serving a biological, prosurvival function of protection. For Bowlby, anxiety was "a primary response not reducible to other terms and due simply to the rupture of the attachment to [the] mother."[79] Klein, borrowing from Bowlby,[80] decided that early separation anxiety might be a particular kind of evolutionary process. Furthermore, in his seminal paper on separation anxiety, Bowlby had described conditions of isolation for the baby as activating both "crying" and "clinging" in relation to the mother figure: "until he is in close proximity to his familiar mother-figure these instinctual response systems do not cease motivating him," such that until this outcome is reached "his subjective experience is that of primary anxiety."[81] We can see here how Bowlby, in shifting the weight of interpretation away from agoraphobic anxiety concerning streets and squares, assisted in establishing a model of phobic anxiety in which attachment figures (particularly the mother) were equated with the environment of the home.[82] Klein and Fink superimposed Bowlby's small-scale dyadic scene featuring a crying, clinging child and a reassuring mother onto the figures of an adult male patient and a nurse within the space of the Hillside Hospital psychiatric ward.

---

[78] It is of course impossible to know from the published documentation whether Klein and Fink were already on the lookout for manifestations of "separation anxiety" as the early experiments began, or whether their theorizations took place subsequent to the "keen clinical observer" noticing the cessation of the running to the nurses' station. Klein and Fink's first paper on imipramine certainly referred to Bowlby's "separation anxiety," noting that for Bowlby, "separation anxiety has the biological function of evoking the retrieving and mothering response in a parent. . . . One may speculate that imipramine, in these patients, has some specific reparative effects upon this disordered emotion" (Klein and Fink, "Psychiatric Reaction Patterns" [cit. n. 29], 436).

[79] John Bowlby, "Separation Anxiety," *Int. J. Psychoanal.* 41 (1960): 89–113, on 93.

[80] Klein retrospectively described his turn to the work of Bowlby as follows: he had noticed that many of his agoraphobic patients manifested dependent behavior, and that many had been clinging children, fearful of going to school. Subsequently, he and his colleague Rachel Gittelman-Klein conducted a double-blind, placebo-controlled study of "school phobic" children whose central problem was deemed to be separation anxiety; the results indicated that imipramine was very successful in helping with school phobia. The same drug that apparently blocked panic attacks in adults seemed also to diminish separation anxiety in children. Klein, therefore, again working backward from drug responses, began to consider whether "in some sense, an outbreak of separation anxiety was at the root of agoraphobia" (Klein, "Anxiety Reconceptualized" [cit. n. 6], 245). See Rachel Gittelman-Klein and Donald Klein, "Controlled Imipramine Treatment of School Phobia," *Arch. Gen. Psychiat.* 25 (1971): 204–7; Gittelman-Klein and Klein, "School Phobia: Diagnostic Considerations in the Light of Imipramine Effects," *J. Nerv. Mental Disease* 156 (1973): 199–215.

[81] Bowlby, "Separation Anxiety" (cit. n. 79), 93.

[82] Bowlby's model of separation anxiety served several purposes for Klein. It manifested the same kind of distaste for "baroque structures" of symbolic interpretation as Klein's theory of panic. For Bowlby, separation anxiety functioned as a kind of unmediated protest mechanism whose form was very similar to Klein's understanding of panic as an autonomic discharge of paroxysmal anxiety.

CONCLUSION

Klein and Fink's early psychopharmacological experiments involved a small number of patients and a small number of "expert" clinical observers who inhabited a small psychiatric hospital away from the heft of mainstream large university research centers. In time, those experiments would come to have an impact that was both geographically and conceptually extensive, for Klein's work on pharmacological dissection in Hillside Hospital acted as the germinator for a diagnosis, panic disorder, that is now firmly embedded in multiple countries and across many psychiatric cultures.[83] Klein and Fink privileged particular observational practices as they traced a route through a hermeneutically dense terrain composed of heterogeneous patients, heterogeneous drugs, and all manner of "noise" vis-à-vis the behavioral features, linguistic utterances, and affective transformations that might, potentially, be of use in assessing whether and how imipramine acted. Those practices brought a particular manifestation of psychopathological anxiety to center stage. At a historical moment in which the clinical-evaluative drive was toward working with larger numbers of patients, larger research sites, and the use of target symptoms and the development of complex rating scales, Klein and Fink's experiments were characterized by their "intimate geographies."

Those geographies were centered on one research site and entailed the direct observation of patients' bodies—as entire, communicative, and spatially and temporally patterned entities—within the ward by clinical researchers and ward staff who knew those patients well. The intimacy of those geographies was perhaps dramatized most poignantly by the coming to life of Bowlby's separation-anxiety-disordered infant (in need of her mother) in the body of the "episodic anxiety"-disordered male adult patient (in need of the reassurance of the nurse). Whereas Westphal's case histories had referenced Berlin squares that piqued agoraphobics' fear, Klein's narrative of his originary patient was one that displaced the city and replaced it with the drama of a child-and-parent dyad. The agoraphobia of Westphal's patients was exemplified by their stuttering, stalled passage through the public sphere; the panic disorder of Klein's originary patient was exemplified by a frenzied running to the nurse/mother.

Klein and Fink mobilized a complex and creative experimental apparatus comprising heterogeneous bodies and heterogeneous drugs moving within a particular sociospatial setting. They rendered visible and validated particular interpretations of affective behavior in their consolidation of distinct behavioral reaction patterns—which, in turn, led to Klein's powerful elaboration of the logics of psychopharmacological dissection. David Healy has argued that for Klein and Fink,

> the new drugs were an experiment that would lead to new observations. The trick was to remain open-minded enough to see phenomena that available theories did not predict. New theories to explain these observations could be elaborated later. This was almost

---

[83] Admittedly, this was a long process. Many psychiatrists profoundly disagreed with Klein's interpretation of the imipramine findings. Some argued that he had mistaken his patients' symptoms and that the patients were actually suffering from depression; others (e.g., the behaviorist Isaac Marks who was based at the Maudsley Hospital in the United Kingdom) argued that Klein's panic disorder patients should instead be diagnosed as agoraphobic and saw no basis for Klein's new nosological category.

a new form of science, one that acknowledged that techniques drive progress as much as, if not more than, anything else.[84]

But Klein himself has not claimed, in fact, to have made any actual new observations. (He has emphasized that Freud had, in fact, described panic attacks "beautifully" in 1895 but argued that Freud's "theory prevented his observations.")[85] "So, it's not like it's a new observation," stated Klein: "What's new is that I put it together a different way."[86] What was new was a complex socio-spatial assemblage that Fink and Klein put together and set into motion. We need, I argue, to attend to the spatial as well as temporal specifics of this assemblage in order to discern how this new form of psychopathological affect—which would come to be termed panic disorder—emerged and then gained epistemological and ontological consistency. Through Klein and Fink's experiments, one drug (imipramine) operated in combination with one "good clinical observer"—to which was added the compelling overlay of Bowlby's figure of mother and child. Such were the elements that brought to center stage one small, affectively dramatic scene. Klein's analysis of the patient running to the nurse allowed him to "singl[e] out panic attack as being the key variable that was changing with imipramine"[87] and led, in time, to the consolidation of a new nosology of anxiety.

[84] Healy, *Creation* (cit. n. 16), 282.
[85] Klein, "Anxiety Reconceptualized" (cit. n. 6), 245; Klein, interview by Callard (cit. n. 46).
[86] Klein, "Anxiety Reconceptualized" (cit. n. 6), 245.
[87] Klein and Healy, "Donald Klein" (cit. n. 31), 331.

Cold War "Super-Pleasure":
Insatiability, Self-Stimulation,
and the Postwar Brain

by Otniel E. Dror*

ABSTRACT

In this contribution, I study the post–World War II discovery of a new "supra-maximal" "super-pleasure" in the brain. I argue that the excessiveness of the newly discovered supramaximal super-pleasure challenged existing models of organisms, of the self, and of nature and society, and that it prescribed a rethinking and a repositioning of pleasure. I reconstruct the laboratory enactments and models that constituted this new pleasure as "supramaximal," instant, and insatiable, suggest several postwar contexts that situate the new pleasure, and examine expert and vernacular reactions to the new super-pleasure. I also introduce and reflect on an approach that "sides with" emotion, and I present the notion of a "missed" emotion. I conclude with a brief consideration of "repetitions"—for science and for pleasure.

> A system (the fornix system) in the so-called "olfactory" part of the brain is in some way involved in intensely pleasurable activity.
>
> —John C. Lilly, 1955[1]

> Among the most recently discovered is a region within [the hypothalamus] . . . which on stimulation gives rise to a strongly pleasurable sensation. . . . Evidently all the desirable things in life are desirable only insofar as they stimulate the pleasure center. To stimulate it directly makes all else unnecessary.
>
> —Isaac Asimov, 1965[2]

> For the first time in the history of neuroscience, this arrangement enabled an animal to push a lever and deliver a stimu-

* History of Medicine, Hebrew University Medical Faculty, P.O. Box 12272, Jerusalem 91120, Israel; otnield@ekmd.huji.ac.il.

I wish to heartily thank Timothy J. Crow, Jaak Panksepp, Larry Stein, and Roy Wise for numerous conversations and e-mail exchanges regarding their scientific research and contributions during and following the period under study in this essay. I am also very grateful for the comments and suggestions of my fellow coeditors, the anonymous reviewers, and the editor of *Osiris*.

[1] "Analysis of NIH Program Activities: Project Description Sheet," p. 1. Principal Investigator: John C. Lilly, December 1955, folder 18, box 29, John C. Lilly Papers, Record ID M0786. Courtesy of Department of Special Collections and University Archives, Stanford University Libraries (hereafter JCL).

[2] Isaac Asimov, *The Human Brain: Its Capacities and Functions* (1963; repr., New York, 1965), 188.

lus—in this case, obviously, reinforcing and, presumably, pleasurable—to the interior of its own brain! The later popular term
for this behavior—"self-stimulation"—was introduced by Joseph V. Brady in 1958.

—H. W. Magoun, 1981[3]

In 1954, James Olds and Peter Milner discovered pleasure in the brain of a laboratory
rat. Pleasure, which had been ostracized as a nameable experience by the behaviorist
sciences, which had been de-ontologized during the early twentieth century as merely
aponia—as only an absence of pain—and which had been excluded from the repertoire of possible states of laboratory animals since the early inauguration of a laboratory science of emotions during the nineteenth century, returned with a vengeance.[4]
The return of pleasure introduced a major transformation whose repercussions and
offshoots are still very much with us today, including the development of a neurophysiology of decision making, risk taking, addiction, affective neuroscience, and more.

The history of the discovery of pleasure during the mid-twentieth century intertwined with numerous historical developments and contemporaneous contexts. Some
harked back to classical times, like the long history of Western hedonisms. Others
emerged in their modern guise during the eighteenth century: in Kantian aesthetics,
and in the aesthetics of Burke and Hume; in the hedonisms of Thomas Hobbes and
Robert Whytt; in Bentham's "pig's philosophy"; and in what Colin Mercer has defined as the eighteenth-century "*individualisation* of pleasure . . . the first stage in
the introduction of individual pleasures into the field of documentation and social
regulation."[5] Other developments occurred during the decades that immediately preceded the return of pleasure, like the rise of behaviorism and the praxis of (deep)
brain stimulation. Still others emerged contemporaneously, including the significant
and interrelated development of behavioral or psychopharmacology. The emergence
of pleasure also coincided with the anxieties of the Cold War and the development of
a unique "Cold War rationality" (which was distinct from "reason"), the negation and
"pathologization of emotions" in the post-Holocaust social sciences of liberal democratic societies, and the approaching horizon of the long sixties (1958–74). Freudianism was still thriving in the United States, and post-Freudian adaptations, reinterpre-

---

[3] For an explanation of the "arrangement" that Magoun was referring to, see below. H. W. Magoun,
"The Law of Effect: From Thorndike, to Skinner, to Olds" (1981), p. 10, folder 19, box 30, Magoun,
Horace Winchell Papers, Manuscript Collection no. 140, Neuroscience History Archives, History and
Special Collections for the Sciences, University of California, Los Angeles, Library Special Collec-
tions (hereafter HWM).

[4] Unlike "emotions," which were often studied in animals, the study of "pleasure" since the nineteenth century was confined almost exclusively to human subjects.

[5] On Kantian, Burkian, and Humean aesthetics, see Laura Frost, *The Problem with Pleasure: Mod-
ernism and Its Discontents* (New York, 2013), Kindle edition; Catherine Cusset, *No Tomorrow: The
Ethics of Pleasure in the French Enlightenment* (Charlottesville, Va., 1999); Regenia Gagnier, *The
Insatiability of Human Wants: Economics and Aesthetics in Market Society* (Chicago, 2000). On
Hobbes, Whytt, and the historical roots of modern hedonism, see Roy Porter, "Enlightenment and
Pleasure," in *Pleasure in the Eighteenth Century*, ed. Roy Porter and Marie Mulvey Roberts (London,
1996), 1–18; Ian Small, *Conditions for Criticism: Authority, Knowledge, and Literature in the Late
Nineteenth Century* (Oxford, 1991). On the individualization of pleasure, see Colin Mercer, "A Pov-
erty of Desire: Pleasure and Popular Politics," in *Formations of Pleasure*, ed. Formations Collective
(London, 1983), 84–100, on 91; emphasis in the original. See also Lynn Hunt and Margaret Jacob,
"The Affective Revolution in 1790s Britain," *Eighteenth-Cent. Stud.* 34 (2001): 491–521.

tations, and critiques appeared in the works of Herbert Marcuse, Norman O. Brown, Betty Friedan, Jacques Lacan, and Ernest (né Ernst) Dichter.[6] Into this melee of cultural forces, contradictions, and ideologies, scientists introduced "supramaximal" "super-pleasure."

In this contribution, I study the discovery of super-pleasure during the mid-twentieth century.[7] I partly adopt the suggestion made by Edward Boring in the mid-1920s, when the National Research Council convened a confidential Committee and Conference on the Experimental Study of Human Emotions: "the first thing to do," Boring suggested, "is to bring together all the people who are working on emotion and let them tell what they are doing. . . . Then you will find out what in the world emotion is that can be investigated."[8] In adopting Boring's suggestion, I follow experimenters who encountered pleasures inside their laboratories. Rather than providing a close reading of a particular development or laboratory, I present a broad perspective in characterizing the new brain super-pleasure of the immediate postwar period.[9] I focus on the first decade or decade and a half following the discovery of the new pleasure. This early period was characterized by multiple models and hypotheses, incongruent and conflicting terminologies and ideologies, and an intensive empiricism.

In describing the discovery of pleasure, I do not present a "balanced" history. Rather, I "side with" pleasure in a broader behaviorist context, in which "pleasure"—the very word—was often derided, obfuscated, and maligned. Pleasure often had to make itself heard, after an absence of several decades. My first intention in siding with pleasure is thus to empower this underdog emotion. This historiographical choice is significant,

---

[6] On Cold War rationality, see Paul Erickson, Judy L. Klein, Lorraine Daston, Rebecca Lemov, Thomas Sturm, and Michael D. Gordin, *How Reason Almost Lost Its Mind: The Strange Career of Cold War Rationality* (Chicago, 2013). On the "pathologization of emotions," see Frank Biess and Daniel Gross, eds., *Science and Emotions after 1945: A Transatlantic Perspective* (Chicago, 2014), 3. On Freudianism in the United States, see Alfred I. Tauber, "Freud's Social Theory: Modernist and Postmodernist Revisions," *Hist. Hum. Sci.* 25 (2012): 43–72; Dorothy Ross, "*AHR Roundtable*: American Modernities, Past and Present," *Amer. Hist. Rev.* 116 (2011): 702–14. On Dichter, see Stefan Schwarzkopf and Rainer Grie, eds., *Ernest Dichter and Motivation Research: New Perspectives on the Making of Post-War Consumer Culture* (Basingstoke, 2010). For the notion of the "long" 1960s, see John Carlevale, "The Dionysian Revival in American Fiction of the Sixties," *IJCT* 12 (2006): 364–91. Carlevale is following Arthur Marwick in adopting this periodization. On the contradictions and transformations of the 1960s, see Howard Brick, *The Age of Contradiction: American Thought and Culture in the 1960s* (Ithaca, N.Y., 1998), 1–22; Marianne DeKoven, *Utopia Limited: The Sixties and the Emergence of the Postmodern* (Durham, N.C., 2004).

[7] I bracket the ethical critiques that emerged in retrospect in regard to some of the experiments that were carried out in the context of the study of pleasure. For this latter ethical critique, see Christina Kathryn Fradelos, "The Last Desperate Cure: Electrical Brain Stimulation and Its Controversial Beginnings" (PhD diss., Univ. of Chicago, 2008).

[8] E. Boring to Stratton, 8 March 1926, NAS-NRC Archives Div A&P Rec Grp: DNRC: A&P: "Conf on Experimental study of Human Emotions: Second," 1926 March, Washington, D.C.

[9] Other historians have provided succinct summaries and reviews of some of the major scientific developments that led to the new discoveries. These included the research on deep brain stimulation by the Nobel laureate W. R. Hess, the anatomical and physiological contributions of S. W. Ranson, the discovery of the Reticular Activating System by W. H. Magoun and G. Moruzzi, and the elucidation of hypothalamic control of drives. For a well-informed history of reward, see Lawrence E. Marks, "A Brief History of Sensation and Reward," in *Neurobiology of Sensation and Reward*, ed. J. A. Gottfried (Boca Raton, Fla., 2011), http://www.ncbi.nlm.nih.gov/books/NBK92791/ (accessed 13 December 2013). For the historical background, see Alan A. Baumeister, "Serendipity and the Cerebral Localization of Pleasure," *J. Hist. Neurosci.* 15 (2006): 92–8; Henry J. de Haan, "Origins and Import of Reinforcing Self-Stimulation of the Brain," *J. Hist. Neurosci.* 19 (2010): 24–32; John Gardner, "A History of Deep Brain Stimulation: Technological Innovation and the Role of Clinical Assessment Tools," *Soc. Stud. Sci.* 43 (2012): 707–28.

since the allusion or nonallusion to "pleasure" was—and still is—highly contentious for many investigators. It reflected methodological and ideological distinctions that permeated the discovery of super-pleasure and also permeates my analysis below.

In siding with pleasure and presenting a narrative that takes off from "emotion," I also indicate that "pleasure" in my narrative was not the doppelgänger—or evil twin—of scientific "reward," as many investigators conceived—and still conceive—"pleasure." Pleasure was not an afterthought, a misspoken term and presence, a puerile throwback to be regretted in retrospect (by behaviorist scientists), or a layperson's misapprehension of science. "Pleasure" was significant for its scientific history, for the history of scientific "pleasure," from its very beginning. It was also highly significant for the scientific history of "reward."

Siding with pleasure also highlights the alternative narratives that become available from a history-of-emotions perspective. As I will argue below, the discovery of the new pleasure evoked and invoked a broad angst at the very moment of its emergence. "Reward," on the other hand, was exempt from these connotations. There was no angst associated with reward. There was no ostensive affective charge to the discovery of reward in the brain. Furthermore, and significantly, there was "super-pleasure" and "supramaximal" pleasure, but there was no "super-reward" or "supramaximal reward"—there was just "reward." In addition, "pleasure" and "reward" harked back to different histories and presented divergent meanings, associations, and connotations. A history in terms of "pleasure" and a history in terms of "reward" present two alternative narratives of the exact same developments—of the exact same discoveries, experiments, laboratories, technologies, investigators, and rats. Here, I let pleasure be heard.[10]

THE DISCOVERY OF "PLEASURE"

Dr. Marianne E. Olds remembered the day of the discovery of pleasure in the brain:

> At the time no one ever expected that the brain had reward sites, and not that reward was simply the absence of pain. My husband who came from a deeply religious family was not very familiar or had ever seen individuals taking rats, drilling holes in their brains and putting in twisted wires whose ends had been scraped very, very slightly. The result was very poor surgery, the electrode was placed at a site where the reward was very minimal which was lucky because the rat did not go crazy with pleasure but was sufficiently activated to come back, time and time again, to the site where my husband was testing him. . . . The upshot was Jim Olds coming home that Sunday and saying that he had discovered the pleasure center.[11]

The discovery of pleasure in the brain was the immediate effect of the rapprochement of behaviorism with neurophysiology and serendipity. There are several versions of

---

[10] However, I emphasize throughout my analysis that a history in terms of "reward" is essential for writing the complete history of "pleasure" (and reward). This essay is the first installment in a broader project on the scientific and cultural histories of postwar pleasure and reward in the sciences—*The Sciences and Cultures of Pleasures and Rewards*. The study of pleasure and reward generated an enormous number of publications. In the footnotes that follow, I provide only a few exemplary references out of many.

[11] Dr. M. E. Olds, e-mail message to Otniel E. Dror, 10 July 2014. I am grateful to Dr. M. E. Olds, James Olds's spouse and part-time collaborator, for her personal reminiscence of the fateful day of discovery. Dr. Olds passed away several weeks after our e-mail exchange.

the renowned story of the discovery of pleasure-reward in the brain. The main thread of the discovery story is as follows: James Olds, a recent Harvard PhD in social psychology and an advisee of Talcott Parsons, was a new postdoc in Donald O. Hebb's laboratory at McGill University. Hebb's laboratory was one of the first major centers that integrated psychology-behaviorism with neurophysiology.[12] One basic methodology in this emergent, bifurcated, and schizoid field was the electrical stimulation of deep—subcortical—regions in the brain, coupled with behaviorist methodologies and conditioning. The stimulation of deep structures in the brain reflected a major and significant shift from a previous focus on the neocortex to a new focus on subcortical structures in explaining complex behaviors.[13] Olds, who had no previous experience in brain physiology, was assisted by Peter Milner, a graduate student in Hebb's laboratory. One fateful day, after inserting electrodes into a rat's brain, Olds observed that his brain-stimulated rat seemed to seek the location in the cage where its brain had been stimulated: the "animal could be 'pulled' to any spot in the maze by giving a small electrical stimulus after each response in the right direction. This was akin to playing the 'hot and cold' game with a child."[14] Convinced that the rat was "rewarded," rather than punished, by the electrical shock to its brain, Olds concluded that he had discovered a reward-pleasure site in the brain. Olds immediately realized that his discovery directly challenged the reigning and dominant "drive-reduction" theory and paradigm of motivation (see below). "Jim had a movie made in case it never happened again" and refused to "sacrifice his rat to find out where the electrode had gone," fearing that his one-of-a-kind rat might be irreproducible. An X-ray of the rat's brain revealed the true (mis)placement of the electrodes. Together with Milner and Seth Sharpless, another graduate student in Hebb's laboratory, Olds eliminated alternative explanations for the rat's behavior in establishing the reality of a true reward-pleasure site in the brain.[15]

Reports of Olds's discovery spread quickly in the professional and vernacular literatures. Several major researchers were skeptical at first. Among them was Neal E. Miller of Yale University, perhaps the leading protagonist of the drive-reduction theory. Miller, for example, proposed that Olds had not observed a true reward on stimulating the rat's brain, but an "itch"-like addictive reaction. As Miller explained, each stimulus had "an aversive after effect—something like an itching," and the rats pressed the stimulating lever a "second time, and a third, and so forth, much in the

[12] Originally published in 1949, Hebb's *The Organization of Behavior* was a major contribution to the neurophysiology of behavior and to a more general attempt to neurophysiologize psychology. See D. O. Hebb's *The Organization of Behavior: A Neuropsychological Theory* (1949; repr., New York, 1959).

[13] This shift was emphasized by many investigators. See, e.g., Donald B. Lindsley, "Physiological Psychology," *Annu. Rev. Psychol.* 7 (1956): 323–48; Eliot Stellar, "Physiological Psychology," *Annu. Rev. Psychol.* 8 (1957): 415–36.

[14] James Olds, "Physiological Mechanisms of Reward," in *Nebraska Symposium on Motivation*, ed. Marshall R. Jones (Lincoln, Neb., 1955), 73–147, on 84.

[15] Some suggested that the stimulus was turning brain activity off rather than on. Others suggested that the stimulus was activating the viscera, and feedback from the viscera was "pleasurable." Still others proposed that the stimuli had an anesthetic effect and alleviated pain, rather than producing reward. Other investigators suggested that the stimulus induced a seizure in the brain, which in and of itself, irrespective of reward, increased the number of lever presses. For the citation, see Peter M. Milner, "The Discovery of Self-Stimulation and Other Stories," *Neurosci. & Biobehav. Rev.* 13 (1989): 61–7, on 63.

way one scratches a mosquito bite."[16] Within a very short period, these objections and alternative explanations were excluded until only pleasure-reward was left.[17]

The discovery of "pleasure" and its discovery in the brain reframed the neurological self.[18] The neurological self was a pleasure seeker, rather than a drive-pain-tension-excitation-need-desire reducer. The new science shifted to a framework of pleasure seeking, "excitement" seeking, and the "good life," in rejecting the prewar framework in which "pleasure ceased to be a drive in and of itself and was conceived as the experience which occurs when a drive, or motive, terminates, i.e., is satisfied, relieved, or fulfilled."[19] According to Abraham Maslow, "practically all historical theories and contemporary theories of motivation" regarded

> needs, drives, and motivating states in general as annoying, irritating, unpleasant, undesirable, as something to get rid of . . . need reduction, tension reduction, drive reduction, and anxiety reduction. . . . The drive or need presses toward its own elimination. Its only striving is toward cessation, toward getting rid of itself, toward a state of not wanting. Pushed to its logical extreme, we wind up with Freud's Death-instinct.[20]

This dominant ("drive-reduction") framework, in which pleasure was only a "negation of a negation" and whose roots hark back to Plato's *Philebus* and to Epicurian *aponia*, was championed by Freud and Hull during the first half of the twentieth century.[21]

Pleasure was (re)-endowed with an independent identity, history, and narrative. It was henceforth an ontological and independent principle in the materiality of the brain substance. It had its own special neurons, its distinct neural matrix, and its own "centers"—soon to be reconceived in terms of "pathways," and soon thereafter in terms of "circuits."[22] The brain, moreover, was wired for pleasure. Experimenters determined that the volume of brain stuff that embodied—or rather cerebrated—pleasure was much larger than the volume of brain stuff that cerebrated pain (35 percent for pleasure vs. 5 percent for pain, according to early measurements).[23] These cumulative discoveries also conferred an independent evolutionary history on pleasure, and they reconfigured the interrelationships between pleasure and pain. Pleasure had come into its own, although Gilles Deleuze (among others) would later lament

---

[16] James Olds, "Pleasure Centers in the Brain," *Eng. & Sci.* 33 (1970): 22–31, on 26. This was a criticism that had been suggested by Miller.

[17] Some, like Miller, continued to suggest models that retained drive reduction, without rejecting Olds's findings. See "Dr. Neal E. Miller, Comments on the Implications of the Olds Reward Effect for Theories of Reinforcement, 1956," folder 62, box 8, HWM.

[18] I emphasize here both the discovery of "pleasure" and its discovery in the brain.

[19] O. H. Mowrer, "Motivation," *Annu. Rev. Psychol.* 3 (1952): 419–38, on 419.

[20] Abraham Maslow, "Deficiency Motivation and Growth Motivation," in Jones, *Nebraska Symposium* (cit. n. 14), 1–30, on 10–3.

[21] On the history of pleasure, and for "negation of a negation," see Frost, *The Problem with Pleasure*, 8. See also Cusset, *No Tomorrow* (both cit. n. 5).

[22] For the shift from "pathway" to "circuit," see Anthony G. Phillips and Gordon J. Mogenson, "Brain-Stimulation Reward: Current Issues and Future Prospects," *Can. J. Psychol.* 32 (1978): 124–8, on 125. These shifts in the geometry of pleasure, from brain "center(s)" to "circuit(s)," were highly significant.

[23] The remaining 60 percent was neutral. For these estimates, see James Olds, "Differentiation of Reward Systems in the Brain by Self-Stimulation Technics," in *Electrical Studies on the Unanesthetized Brain*, ed. Estelle R. Ramey and Desmond S. O'Doherty (New York, 1960), 17–51, on 18.

its vacuousness, writing, "I cannot give any positive value to pleasure, because pleasure seems to interrupt the immanent process of desire."[24]

PLEASURING "REWARD"

The discovery of pleasure between behaviorism and neurophysiology bequeathed a split personality to pleasure and to reward. From the very moment of its inception, creation, and discovery, pleasure-reward was haunted by this tenuous duality. Pleasure-reward appertained to two distinct and (often) incongruent epistemologies and ideologies. Pleasure was anathema to behaviorists, like Miller, Larry Stein, and Roy Wise, yet it appeared in the writings of many prominent investigators, like Olds, John C. Lilly, and C. W. Sem-Jacobsen. This is one of the conundrums and ambiguities of referring to pleasure and/or to the behaviorist reward (and "positive reinforcement") in a semantic field whose members sometimes explicitly referred to pleasure, while at other times obstinately refusing to use the word "pleasure." Their work was referred to by other scientists and in numerous vernacular literatures in terms of "pleasure," while a significant, though far from absolute, number of its practitioners rejected "pleasure" and clearly distinguished between pleasure and the behaviorist's reward and positive reinforcement.[25] For this behaviorist group, reward and pleasure were not interchangeable: a reward was not necessarily pleasurable at all. Reward, moreover, indicated for many the energizing of a drive, in addition to its reinforcing dimensions. "Pleasure" talk, from their perspective, was a return to a preobjective and puerile phase in the development of modern objective psychology.[26]

Nevertheless, pleasure often trumped and usurped reward. John C. Lilly, one of the prominent early leaders of the new neurophysiology, referred to the discovery of brain reward in terms of the "intensely pleasurable."[27] Olds, who inaugurated the new field, referred time and again to pleasure, including in his 1958 *Science* article.[28] Other contemporary and prominent investigators, like Maslow, David C. McClelland, and Eliot Stellar, also referred to Olds's discovery in terms of pleasure.

This "pleasure" challenge was not the product of deliberate attempts to overthrow behaviorism, nor was it an outcome of the infiltration of behaviorism by the ideology and language of animal protection leagues. Rather, it was the result of multiple and local factors: the conjoining of a Skinnerian paradigm with the study of the internal physiology of the brain (Skinner had explicitly excluded a neurophysiology of behaviorism in his 1938 *The Behavior of the Organisms*);[29] the influx of investigators

---

[24] On Deleuze, see Frost, *The Problem with Pleasure*, 11. See also Cusset, *No Tomorrow* (both cit. n. 5).

[25] Skinner took the term "reinforcement" from Pavlov. See B. F. Skinner to Roy A. Wise, 2 December 1986. I thank Roy Wise for sharing this letter with me.

[26] Within the "reward" camp, there were also disagreements. Some investigators, e.g., argued that "reward" and "punishment" were already too value laden. See Howard S. Liddell, James Olds, and Roger W. Sperry, "Post-Pavlovian Development in Conditional Reflexes," in *The Central Nervous System and Behavior*, ed. M. A. B. Brazier (New York, 1959), 211–31.

[27] "Analysis of NIH Program Activities" (cit. n. 1).

[28] For "pleasure" in Olds's publications, see, e.g., Olds, "Physiological Mechanisms" (cit. n. 14); Olds, "Pleasure Centers in the Brain," *Sci. Amer.* 195 (1956): 105–16; Olds, "Self-Stimulation of the Brain," *Science* 127 (1958): 315–24; Olds, "Pleasure Centers," 1970 (cit. n. 16).

[29] "Neurones, synapses, or any other aspect of the internal economy of the organism . . . lie outside the field of behavior as here defined." See B. F. Skinner, *The Behavior of the Organisms: An Experimental Analysis* (New York, 1938), 418.

who had no prior commitments to a strict behaviorist paradigm into this emergent field, including Olds, its founder, and Lilly, one of its major pioneers (these pioneers sometimes used the terms "pleasure" and "reward" interchangeably); the close interactions and collaborations between behaviorists and investigator-clinicians who studied human subjects. These investigator-clinicians, including Robert G. Heath and C. W. Sem-Jacobsen, continuously referred to the subjective experiences of their human subjects, in comparing animal rewards or pleasure to human testimony. They infused and confused introspective human reports of "pleasure," "joy," "well-being," and so forth with the behaviorist (and animal-focused) language of "reward" and "positive reinforcement." The vernacular literature also converted "reward" talk into "pleasure" talk. In newspapers and magazines, popular-science publications, and (science) fiction, "rewarded" rats and the neurophysiology of "reward" became "pleasure" or "ecstasy."[30] Philosophical and legal interrogations of these discoveries also referred to the new science in terms of pleasure. In the postwar United States, the science of reward was usurped by pleasure—by the pleasure of (mis)-speaking of "pleasure" and sex and "orgasms" (see below), rather than in terms of the behaviorist's "positive reinforcement" and "reward."

Behaviorism played a crucial—even if unwitting—role in this turn to pleasure. Though behaviorism is often credited with (or discredited for) eliminating emotion talk, it provided a crucial ingredient for the science of pleasure and, in particular, for an animal science of "pleasure" (and emotions). Behaviorism authenticated by distinguishing, in terms of behaviorist precepts, between a "psychologically valid reward" and "merely a sham, having the appearance but not the substance of a positive emotional effect" in animals.[31] Behaviorism solved on its own terms the long-standing problem of studying emotions in animals: by getting rid of emotions, on the one hand, but also by formulating authenticating practices for its notion of emotion, on the other hand. Behaviorism provided the necessary tools for identifying and authenticating its version of real emotions (and distinguishing between real emotions and mere motor-reflex movements in animals). This distinction was crucial for the study of drives and rewards, including in the brain.[32] Many "pleasure" investigators depended on the behaviorist's proof of authenticity (of "reward" in animals) but substituted the behaviorist's reward for pleasure. Pleasure thus partly conformed to the logic of the behaviorist's reward—despite being pleasure—and behaviorism partly prescribed the

[30] See, e.g., D. B. Macfarlane, "McGill Opens Vast New Research Field with Brain 'Pleasure Area' Discovery," *Montreal Star*, 12 March 1954, 1–2; Robert Coughlan, "Part I: Behavior by Electronics," *Life* 64 (8 March 1963): 90–106; Olds, "Pleasure Centers," 1956 (cit. n. 28); Asimov, *The Human Brain* (cit. n. 2); J. Anthony Deutsch, "Brain Reward: ESB and Ecstasy," *Psychology Today* 6 (1972): 45–9.

[31] James Olds, "Commentary," in *Brain Stimulation and Motivation: Research and Commentary*, ed. Elliot S. Valenstein (Glenview, Ill., 1973), 81–99, on 85.

[32] This authentication already appeared in the first ever publication to demonstrate that direct stimulation of the brain produced an authentic "aversive" reaction, rather than mere motor-reflex movements. This first study, by Delgado, Miller, and Roberts of Yale University, set the stage for Olds's discovery of authentic brain-induced pleasure-reward. This first study was also significant since it challenged Karl Lashley's and Jules Masserman's determination that brain-induced aversive reactions and emotions were mere reflex motor movements, since, as they argued, they could not be conditioned and could not be used to motivate learning. Delgado, Roberts, and Miller demonstrated that direct stimulation of the brain could produce emotional conditioning and motivate learning of instrumental responses. See José M. R. Delgado, Warren W. Roberts, and Neal E. Miller, "Learning Motivated by Electrical Stimulation of the Brain," *Amer. J. Physiol.* 179 (1954): 587–93.

possibilities for and of pleasure in laboratories, despite being behaviorism. Behaviorism was thus also partly responsible for animal pleasure in the postwar history of animal emotions.[33]

The alter ego of pleasure-reward, neurophysiology, also partly prescribed the nature of reward-pleasure. Neurophysiological pleasure was studied in and confined to the nerves. It conformed to the physiological laws of nerve cells—like the all-or-none principle or the refractory period of nerve cells. These laws prescribed the interpretation of experiments on pleasure.[34] The shift to the brain also disembodied pleasure, but it did not spiritualize it; sometimes it aestheticized pleasure. The cerebration of pleasure, its materialization in terms of brain substance, stripped pleasure of its fleshiness, of its extensions in the body, and of its sensuousness. Pleasure was devoid of drools, secretions, contractions, and spasms, and of their temporal logic, progression, and reverberations in the experienced-felt matter of the body.

Pleasure was studied henceforth in its point of origin in the brain: before it was inhibited, controlled, civilized, and socialized by the internal mechanisms of the brain and of society; before it was distorted and corrupted; before it corrupted and distorted—pleasure in its "pure," "supramaximal," and unadulterated form.[35] This ostensibly unnatural laboratory pleasure, which was immediately recognized as such, was the basis for the study of all natural pleasures.

PERPETUUM-PLEASURE-*MOBILE*: "PURE" "SUPRAMAXIMAL" "SUPER-PLEASURE"[36]

Pure supramaximal super-pleasure was the immediate effect produced by experimental enactments. These enactments made pleasure, and they made it supramaximal. They drew on the behaviorist's toolbox. They included crossing electric-pain grids, solving mazes, making preference choices, and pressing levers in order to receive an "electrical brain stimulation" (EBS). Their multiplicity thwarted attempts to establish a common denominator or even an agreed-upon value of and for pleasure-reward.[37] Despite their multiplicity, the new pleasure-reward materialized most ostensibly and predominantly in the image and laboratory enactment of the insatiable self-stimulating

---

[33] The pleasure of laboratory rats immediately preceded the imminent emergence of an ethics of pain and pleasure in animal ethics during the late 1960s and early 1970s. The new animal ethics was distinguished from previous animal ethics in emphasizing and valuing the pleasure/well-being of animals, beyond the elimination/reduction of animal pain and suffering, which had dominated previous approaches to nonhuman animals. See Peter Singer's *Animal Liberation* (New York, 1975) as the most obvious example. See also the important publications on pleasure in animals by both Michel Cabanac and Jaak Panksepp.

[34] The all-or-none principle of nerve action undergirded the interpretation of electrical brain stimulation particularly and crucially during the early cartographic-mapping phase of pleasure in the brain. The study and identification of different neurotransmitters would be crucial from the 1960s onward. The firing rates of neurons would be important for pinpointing the neurons that were significant for pleasure/reward and excluding others.

[35] For "purely rewarding," see, e.g., John C. Lilly, "The Psychophysiological Basis for Two Kinds of Instincts: Implications for Psychoanalytic theory," *J. Amer. Psychoanalyt. Assoc.* 8 (1960): 659–70, on 667.

[36] For "supramaximal," see John C. Lilly, "LSD: Reward and Punishment," undated, folder 3, box 28, JCL. For "super-pleasure," see William H. Davis, "The Pleasure Helmet and the Super Pleasure Helmet," *J. Thought* 10 (1975): 290–3; Sandor Rado, "Fighting Narcotic Bondage and Other Forms of Narcotic Disorder," *Comprehens. Psychiat.* 4 (1963): 160–7.

[37] On this problem, see, e.g., Elliot S. Valenstein, "Problems of Measurement and Interpretation with Reinforcing Brain Stimulation," *Psychol. Rev.* 71 (1964): 415–437.

rat. The insatiable self-stimulating (male) rat pressed a lever that stimulated its own brain's "pleasure centers" at a rate of up to 8,000 (sometimes even 10,000) presses per hour, forsaking food, water, sleep, and sex (a female rat in heat)—until it collapsed from exhaustion, convulsed, or even died.[38] "He" was soon joined by the insatiable self-stimulating monkey, dolphin, rabbit, goldfish, dog, chicken, and human: "If the tip of a fine wire electrode is placed in a specific area of the brain and a brief, weak electrical current introduced . . . [a monkey] will respond as though he liked the sensation. In fact, if the circuitry is arranged so that he can press a bar to induce the current for a moment, he will press the bar steadily 3 times a second for 16 hours a day, seven days a week for years. Special arrangements must be made to keep the animal from starving or thirsting to death."[39] There was a report of a dolphin that pushed the lever "too rapidly, [which] caused a seizure, [and then the dolphin] became unconscious, respiration failed, and he died."[40]

The new super-pleasure was self-perpetuating, self-rewarding, and self-reinforcing. It was produced instantly, on demand, and it dissipated instantaneously. It did not have a refractory period, and it did not lead to satiation. In terms of the "newer information theories," it was a "positive feedback mechanism" and therefore demanded "external control" (in contrast to the "negative feedback" of drive reduction).[41] Pleasure had to be controlled from outside and beyond itself since pleasure begot pleasure begot pleasure. Science had designed/discovered a *perpetuum*-pleasure-*mobile*.

The *perpetuum*-pleasure-*mobile* drew directly on previous conditioning experiments. In these experiments, an animal in a Skinner box would receive, for example, a food pellet after pressing a lever. As Olds explained, "By putting the animal in the 'do-it-yourself' situation (i.e., pressing a lever to stimulate its own brain) we could translate the animal's strength of 'desire' into response frequency, which can be seen and measured."[42]

The scientific and social power and potency of the *perpetuum* derived foremost from its numerical, experiential, and cultural excessiveness. On the shop-floor level, the excessiveness of the *perpetuum* challenged investigators, who initially failed to keep up with the rats. Counting individual lever presses of the indefatigable rats "would have taken months, possibly years," Milner recalled of the first discovery, so the measurement of "the self-stimulation performance [was] by using the length of time a rat was pressing at more than a criterion rate."[43] This partly explains why "pleasure"-"rewardiness" was not represented in absolute and discrete counts of lever presses, but in terms of rates. The uncountable nature of a super-pleasure on the shop-floor level collapsed a discrete number system for counting the *perpetuum*. Elliot S. Valenstein, one of the leading investigators of reward and EBS, also presented the excess of the

---

[38] For "pleasure centers," see Olds, "Pleasure Centers," 1956 (cit. n. 28); Olds, "Pleasure Centers," 1970 (cit. n. 16).

[39] Lilly, "LSD" (cit. n. 36), 2.

[40] John C. Lilly, "Some Considerations Regarding Basic Mechanisms of Positive and Negative Types of Motivations," *Amer. J. Psychiat.* 115 (1958): 498–504, on 501. As Lilly explained, dolphins are a species that must remain conscious in order to continue breathing. For extensive background on Lilly and his research on dolphins, see D. Graham Burnett, *The Sounding of the Whale: Science and Cetaceans in the Twentieth Century* (Chicago, 2012).

[41] James Olds, "Pleasure and Value," XVth International Congress of Psychology, *Acta Psychol.* 15 (1959): 76–7, on 76. This discovery, Olds noted, would also establish "an experimental psychology of positive motivation" (77).

[42] Olds, "Pleasure Centers," 1956 (cit. n. 28), 110.

[43] Milner, "The Discovery" (cit. n. 15), 64.

perpetuum in recalling an experiment in which a "rat self-stimulat[ed] for 21 days during which time it operated the switch about 850,000 times. . . . There was no indication that the animal would ever break the pattern, and it was only the fatigue of my associate, Dr. Bernard Beer, and myself—we were taking turns observing the animal around the clock—that finally forced us to end these observations."[44]

These failures to count in the face of the *perpetuum* were partly products of the design of the lever press itself.[45] Because the experiments depended on escalating numbers, rats were compelled to release the lever within 0.5 second in order to receive the next train of brain stimulations. If the rat persisted and continued to press the lever, the current was automatically turned off. This guaranteed that the rat would have to press again in order to receive brain stimulation—providing the experimenters with a response rate as a measure of its pleasure-reward. The 0.5-second limit also guaranteed that the rats (or dolphins) would not stimulate themselves unto death: "The story is told of a dolphin who, inadvertently left in a pool with the switches connected, delighted himself to death after an all-night orgy of pleasure."[46] These precautions against "Death by Ecstasy," as Larry Niven titled a fictional account of a murder, in which the "weapon" was a "pleasure plug," accentuated the phantasmagoric image of the *perpetuum*.[47]

In attempting to place a value on self-stimulation and to incorporate EBS into existing scales, investigators matched it against hunger, thirst, sex, pain, fatigue, and more. Self-stimulation often prevailed over its competitors. It was "more than," on the edge, off the grid, rather than in between the supposedly stable and coherent outliers against which it was matched. Animals preferred to starve to death rather than rationally calculate their pleasures and pains. These observations and modes of valuing EBS accentuated the extremity of the *perpetuum* by emphasizing its perverse pathological deviations from any notion of a rational felicity calculus.

The *perpetuum* was also a product of the unique relationships that were forged between the hypothalamus and human-induced artificial electrical stimulation of the brain. The *perpetuum* depended on the indefatigability of the hypothalamus, which received thousands of electrical stimulations and continued to react, unlike the motor cortex, which would enter a refractory period of unresponsiveness.

The investigators were partly cognizant of these—their own—constructions of the enacted phenomena. Some expressed the gap between an imagined pleasure and the pleasure that was necessarily constituted in laboratories. Any setup and measurement of pleasure construed pleasure and its measurement in particular ways. Trivial details, like the size of the lever and its relative size with respect to the overall size of the Skinner box partly determined the frequency of lever presses, as Olds observed.[48]

---

[44] Elliot S. Valenstein, *Brain Control: A Critical Examination of Brain Stimulation and Psychosurgery* (New York, 1973), 67–8.

[45] For a different example of how "failure" can contribute to the evolving understanding of affect, see Eric J. Engstrom, "Tempering Madness: Emil Kraepelin's Research on Affective Disorders," in this volume.

[46] Roderic Gorney, "The New Biology and the Future of Man," *UCLA Law Rev.* 15 (1967–8): 273–356, on 338. There is no explanation for why the dolphin died.

[47] Larry Niven, "Death by Ecstasy," *Galaxy Magazine* 27 (1969), ftp://ftp.seti.org/phillips/scifi /Death%20by%20Ecstasy%20-%20Larry%20Niven.txt (accessed 1 June 2016).

[48] J. Olds, R. P. Travis, and R. C. Schwing, "Topographic Organization of Hypothalamic Self-Stimulation Functions," *J. Comp. Physiol. Psychol.* 53 (1960): 23–32, on 27.

Lilly, who studied monkeys and dolphins, reported that in monkeys the rate depended on whether the monkey was required to press the lever with its hand (3 presses/second), tongue (2 presses/second), or foot (1 press/second).[49] Other investigators observed that when the stimulating electrodes were positioned in the substantia nigra the rat turned itself around after every lever press. As Roy Wise reminisced, "there was a pronounced motor effect, forcing the animal to turn 360° from and back to the lever between lever-presses. . . . St. Laurent, a post-doc of Olds made a movie of the different forms that self-stimulation could take."[50] Still others reported that some rats stayed on the lever when given the opportunity to receive continuous brain stimulations, while others preferred short trains of electrical stimulations and turned off the current after a short burst. These and numerous other variations depended on the location of the stimulation and on the rat. These variations also prompted a variety of hypotheses. While some argued that rats self-limited their own EBS because prolonged stimulation became aversive, others explained that the rats were maximizing their pleasure. The latter group reasoned that, because the continuous stimulus might be habituating, the marginal returns—to rephrase the investigators—decreased over time. By getting on and off the lever, the rats increased their rewards.

There were also reports that could temper the dominant vision of the *perpetuum*. For example, in his observations of monkeys, Lilly noted that if "you make him [a monkey] push the lever with his tongue, he then reaches over and picks up a piece of apple with his mouth and stores it in the pouches on either side of his face, goes on pushing with his tongue and chews the food between pushes."[51] Reports from human subjects and from dolphins described this resistance and refusal to press the lever if the experimenters made it too difficult or uncomfortable.[52]

In addition, self-stimulation was criticized as an absolute measure of pleasure-reward by prominent insiders, including Olds. The *perpetuum* provided more of an illusion or semblance of absolute quantification. Though many early publications implicitly vaunted the excess of numbers, Olds himself presented self-stimulation only as a comparative tool, objecting to its use as an absolute quantitative measure of reward-pleasure. This comparative perspective did away with any notion of absolute pleasure quantities. However, even this numerical compromise was criticized by some investigators, who established that self-stimulation did not correlate with other measures of reward. Most distinctly, when animals were given the choice to lever press and stimulate differently positioned electrodes in their brains, they often chose to press a lever that produced lower rates of lever pressing. This observation, however, was not indicative of the true measure of reward-pleasure, since, as some argued, the lower rates could simply indicate that the effects of the stimulation persisted for several seconds, rather than declining rapidly. Despite the criticisms, alternatives, and complications that haunted the *perpetuum*, the *perpetuum* largely overwhelmed its detractors both inside and outside the scientific community.

---

[49] PHS-NIH, Individual Project Report, "Mapping the Behavior Elicitable by Electrical Stimulation of the Brain," 1957, folder 18, box 29, JCL.

[50] Roy Wise, e-mail message to Otniel E. Dror, 19 September 2014.

[51] "Discussion between Dr. Sidney Cohen and Dr. Lilly," p. 2, 10 February 1967, folder 6, box 44, JCL.

[52] See, e.g., Valenstein, *Brain Control* (cit. n. 44); Coughlan, "Behavior by Electronics" (cit. n. 30). Some, but not all, of these reports should be read in the context of attempts to assuage public fears—by showing that humans, e.g., can resist this type of control.

THE SUPER-PLEASURE PRINCIPLE VERSUS THE REALITY PRINCIPLE

The image of an instantaneously produced, insatiable, self-perpetuating super-pleasure captured the imagination of contemporaries and of generations to come. Philosophers, legal scholars, psychologists, journalists, fiction writers, movie producers, and investigator-clinicians drew on the pleasured rat in challenging entrenched philosophical views (e.g., Gilbert Ryle's language-based analysis of pleasure as a nonsensation);[53] in "query[ing] whether a constitutional issue, not to say a basic ethical issue, would arise if the offender's submission to such 'brave new-world' wiring of his brain were made a legal condition for his being returned to society";[54] in voicing concerns over governmental mind control and brainwashing; in seeking inspiration for artistic and cultural productions of a utopian or dystopian nature; and in suggesting new disease models and therapeutic interventions.

One major preoccupation was where to position the newly discovered super-pleasure in the social and natural orders. As Carol Tavris put it, "we hear of mysterious pleasure centers of the brain that turn the surliest male ecstatic when they are stimulated. Enthusiasts hail ESB [electrical stimulation of the brain] as a salvation for the sick and an antidote for the aggressive. Skeptics see visions of 1984: electrodes implanted in everyone, controlled by a computer that assures happiness and political docility."[55] This quandary was relevant to Nature and humans alike. In terms of the natural order, investigators presented a nature that was both dependent on and wary of the new super-pleasure. Previous naturalists and investigators had already incorporated (normal) pleasure into the natural history of animals or, after Darwin, into evolutionary theory. These latter incorporations presented a pre-Darwinian Nature, in which "the Creator has given to man the two faithful guards of pleasure and pain for his preservation; the one to avert evil, the other to invite him to useful actions," as Albertus Haller put it in the eighteenth century; or, as the post-Darwinian perspective was summarized by Charles Scott Sherrington in the early twentieth century: "The concomitance between certain nervous reactions and psychosis seems an alliance that strengthens the . . . continuance of existence."[56]

The post-Olds interpretations emphasized and took off from the super-pleasure. Pleasure was no longer just a motive or spring for action, but a positive feedback *perpetuum* loop that could disrupt even a minimal version of the felicity calculus of organisms. This image framed the animal kingdom in terms that were akin to human society. Both the animal kingdom and human society had to come to terms with their super-pleasure principle within a reality principle of the struggle for existence (animals) or the reality principle of society (humans). According to investigators, in the animal kingdom, the management of super-pleasure was built-in and wired into the hypothalamus of organisms and had to be so for the very possibility of their ex-

---

[53] For a direct critique of Ryle's argument, which drew directly on Olds's experiments, in arguing against Ryle's assertion that "pleasure is not a sensation because, inter alia, it is not separable from its source, a cause or effect, clockable or locatable or describable the way pains are," see, e.g., Roland Puccetti, "The Sensation of Pleasure," *Brit. J. Phil. Sci.* 20 (1969): 239–45, on 239.

[54] Paul E. Meehl, "Psychology and the Criminal Law," *Univ. Richmond Law Rev.* 5 (1970): 1–30, on 13.

[55] This quotation from Carol Tavris appeared in Deutsch, "Brain Reward" (cit. n. 30), 45. Deutsch was one of the leading investigators of EBS and reward.

[56] The quotations are, respectively, from Albertus Haller, *First Lines of Physiology*, ed. Lester S. King (1786; repr., New York, 1966), 80; Charles S. Sherrington, *The Integrative Action of the Nervous System* (New Haven, Conn., 1906), 333.

istence. They presented the following model: the architecture of the hypothalamus was made up of discrete centers of "pure" aversive-pain and "pure" reward-pleasure. These two principles interacted in the hypothalamus and were ultimately integrated into one output. These positive and negative centers—two for each drive—were in hyper-mode and were reciprocally interrelated and inhibitory.

At the atomic level of the hypothalamus, the brain was not a moderate consumer: its hunger or its pleasure or its sex was not construed on a model of moderation, but on extremes that were held in check and "released" when the conditions were appropriate. This basic model of "release" harked back to the nineteenth century, but the new model introduced significant distinctions. In the new hypothalamic model, the inhibiting centers were not exclusively "higher" centers that released "lower" centers. Investigators did not present a model of the "super-ego" that controlled the "id"—to transpose this new science into a different domain; an apt transposition, since various twentieth-century investigators explicitly positioned the id in the hypothalamus, both prior to and following the new discoveries.[57] The id was made up of drive-inhibit dyads, whose reciprocal and integrated interactions were the id. These inhibitors did not inhibit the id. They were part of the id itself. The higher centers indeed exerted control over the lower hypothalamus, as they had done before, but they were not the "stop" button in a model in which the lower id was exclusively "go." They exerted their control over the go-stop dyad of the now dual/Janus-faced id, and they were demoted to (just) one more type of input among several types of inputs into the hypothalamus, which now also included direct inputs from the internal environment of the body.

For human society, the implications of super-pleasure were more complex and knotty. Isaac Asimov, Abraham Maslow, Sandor Rado, John C. Lilly, Sidney Cohen, and several philosophers, each in his own way, presented the potential dangers of a society of pleasure or, rather, of a society of insatiable self-pleasuring individuals, of individualized pleasure gone mad. In this dystopia, the "invisible hand" collapsed from a pleasure that was way off the grid. The humanistic psychologist Maslow expressed this more general angst about super-pleasure in his comments on a 1955 presentation of the pleasured rat by James Olds, shortly after its initial 1954 discovery:

> Supposing the subjective pleasure state brought on by septal stimulation turns out to be the end or goal of *many* kinds of motivated behavior . . . which has its goal, and therefore its justification, in a subjective, pleasurable, conscious state. Then what would happen if we could short-circuit or bypass the whole complex of troublesome . . . behaviors which have in the past been the only ways of achieving this pleasurable end-experience? . . . Would we eat? So for drinking, copulating, and *temperature regulation*? So also perhaps for love. . . . If complex civilizations are based upon forced, sometimes unpleasant work in order to get . . . a pleasurable subjective state, what would happen to work (and to civilization) if this same end could be achieved by plugging oneself into a nearby Olds-intermittent-stimulator-socket?[58]

---

[57] For the (hypo)thalamus as the locus of the "id," see Walter B. Cannon to Carl A. L. Binger, 24 October 1934, folder 1529, box 110, Walter Bradford Cannon Papers (H MS c40), Harvard Medical Library in the Francis A. Countway Library of Medicine, Boston, Mass.

[58] "Temperature regulation" is not italicized in the text. I emphasize it here in order to stress Maslow's perceptive nondistinction between the super-pleasure of copulating, eating, and temperature regulation. I will return to this below in discussing "missed" polymorphous pleasures. For the citation, see Olds, "Physiological Mechanisms" (cit. n. 14), 145–6.

Maslow had presented these apocalyptic visions of the super-pleasure in arguing for and insisting on an ethics of higher versus lower pleasures. The affirmation of distinct—higher versus lower—pleasures would safeguard against the breakdown of a civilization that had discovered the super-pleasure and made it technologically available. Similar concerns and analogous resolutions would appear in philosophical interrogations of the EBS. These philosophical commentaries, like "Pleasure Helmet and Super Pleasure Helmet," ethicized and/or rationalized the new super-pleasure away by imagining and/or construing and/or explicitly arguing for a hierarchy of lower versus higher motivations qua ethical and/or rational objectives in and meanings of life.[59]

Experimenters, however, had not discovered, nor had they proffered, an imagined "higher" pleasure in the brain. This higher pleasure would provide an easy and illusory solution for the predicaments of a society that came face-to-face with "supra-maximal" pleasures, which were the source of all pleasures. Super-pleasure was, indeed, celebrated, but it was distinctly non-"cerebral." It was not a pleasure that assumed distinctions between a "low"-sensuous-embodied pleasure and a "high"-aesthetic-cerebral pleasure. These distinctions were established most significantly and pertinently in Kant's work, when he differentiated between "rational, universal taste and appreciation" and "embodied, instinctual, voluptuous stimulation."[60] Cerebrated (noncerebral) pleasure presented a model of the brain in which the "higher"—neocortical—"cerebral" pleasures had their source in the cerebrally "low" subcortical brain centers or circuits of pleasure. This topology of the brain also ensued from earlier observations during the 1940s and early 1950s, in which experimenters discovered that the neocortex was devoid of pleasure neurons and that its direct stimulation did not yield pleasure (in humans). All pleasures were low pleasures. These discoveries implicitly democratized pleasure by leveling all pleasures. They also undermined ideologies and cultures that were invested in these distinctions.[61] This implicit challenge from the laboratory resonated, but it did not seem to reverberate, with significant analogous and contemporaneous shifts in pleasure beyond the laboratory.

These particular concerns with higher and lower pleasures, ethics, meanings, and objectives were largely absent from much of the vernacular literature. It was not the frenzied hedonism of these rats, their unbridled self-stimulation, and their disruption of a rational and calculating self that evoked widespread concerns. It was the specter of mind control. Midcentury commentators were captivated by the Cold War era literature on "mind control" and "brainwashing," including *Hidden Persuaders* (1957).[62] Several investigators further accentuated these vernacular concerns, notably José M. R. Delgado. Delgado, who was one of the pioneers and a well-respected researcher from

[59] Davis, "The Pleasure Helmet" (cit. n. 36). I note that Robert Nozick's contemporaneous and renowned "Experience Machine" did not allude to Olds's rats. See Nozick, "The Experience Machine," in *Anarchy, State, and Utopia* (1974; repr., Oxford, 1999), 42–5.

[60] Kant distinguished between high and low pleasures, high and low aesthetics, and the sensual vs. the reflective. For these distinctions, see Noel B. Jackson, "Rethinking the Cultural Divide: Walter Pater, Wilkie Collins, and the Legacies of Wordsworthian Aesthetics," *Mod. Philol.* 102 (2004): 207–34; for the citation, see Frost, *The Problem with Pleasure* (cit. n. 5), 9.

[61] For a concerted attempt to come to terms with the neurophysiological discovery/argument that all pleasures are "low," see H. J. Campbell, *The Pleasure Areas* (London, 1973).

[62] On the emergence, contexts, and history of "brainwashing," see David Seed, *Brainwashing: The Fictions and Mind Control: A Study of Novels and Films* (Kent, Ohio, 2004). "Brainwashing" was coined by Edward Hunter in 1950. Huxley's *Brave New World* (*Revisited*) figured prominently in the Cold War literature on brainwashing. See Vance Packard, *The Hidden Persuaders* (New York, 1977).

Yale at the time, performed various controversial demonstrations of brain control in bulls and monkeys, which led to the publication of his widely read monograph, *Physical Control of the Mind*, in 1969.[63]

These widespread public concerns with direct mind control by despotic overseers, however, failed to internalize the true meaning of the super-pleasure. Invoking Aldous Huxley, Lilly perspicaciously observed of the new technologies, "The Brave New World will not be imposed upon us by Big Brother. We will beg for the raptures of Soma *and* call *it* a sacrament."[64] Skinner would make a similar astute observation in the course of his testimony during the 1973 Kennedy hearings on human experimentation: "Punishment has at least the merit that it generates countercontrol. The punitive controller eventually runs into trouble, but the positive controller may achieve a new and frightening kind of despotism."[65]

Yet even these two astute observers seem to have underappreciated the full meaning of the *perpetuum*. Several post-"Olds-intermittent-stimulator-socket" authors— as Maslow referred to Olds's new technology—seem to have grasped and internalized the meaningful distinctions between the new EBS "super-pleasure" and Huxley's obsolete model of "Soma," which was often invoked as prescient in the brain-control literature.

Larry Niven's fictional account of "Death by Ecstasy" (1969) exemplifies this new work. Niven drew on the new super-pleasure in constructing a crime scene in which the deceased, "Owen," died from an "ecstasy plug" that was wired into his "pleasure center." Death ensued from starvation and "lack of will power." As one of the protagonists explained, Owen died after a "month of ecstasy, a month of the highest physical pleasure man can attain . . . [though] food [was] only a few footsteps away," Owen would "have [had] to pull out the droud [electronic brain implant] to reach it."[66]

"Soma," Huxley's drug of blissful happiness, projected an image of a pleasure of contentment and conformity, of incorporating the dominant ideology of a society and the order of things—of a productive pleasure, in terms of the totalitarian regime of the *Brave New World*. Soma was overtly functional for the *Brave New World*, rather than disruptive of the social order. The new super-pleasure was not a pleasure of comfort and acquiescence, nor of docility, nor of acceptance. It was a super-pleasure that overwhelmed; it disrupted production and the reproduction of obedient subjects and society. It was a pleasure that killed. These distinctions between Huxley's food pellet, ingested Soma, and the new direct brain-stimulated super-pleasure are perhaps best captured by and/or are analogues to the suggestive distinctions between *plaisir*-pleasure and *jouissance* that emerged in Lacanian theorizing during the same historical moment.[67] Moreover, these distinctions, as I will argue below, also capture and reflect the shift from an older social order, which was embodied in a lever-pressing rat that

---

[63] José M. R. Delgado, *Physical Control of the Mind: Toward a Psychocivilized Society* (Evanston, Ill., 1969).

[64] Lilly, "LSD" (cit. n. 36), 7; emphasis in the original.

[65] *Quality of Health Care—Human Experimentation, 1973 Hearings before the Subcommittee on Health of the Committee on Labor and Public Welfare*, United States Senate, Ninety-Third Congress, First Session, 370.

[66] Niven, "Death by Ecstasy" (cit. n. 47).

[67] On *plaisir* and *jouissance* and their respective conformity with and disruption of the dominant ideology, see, in particular, Brian L. Ott, "(Re)Locating Pleasure in Media Studies: Toward an Erotics of Reading," *Comm. Crit. Cult. Stud.* 1 (2004): 194–212.

was rewarded by a food pellet, to the new social order, which was embodied in a lever-pressing rat that was rewarded by direct brain stimulation.

MISSED EMOTIONS: POLYMORPHOUS PLEASURES[68]

In the Lacanian concept of *jouissance*, and in current scholarship on pleasure, much attention is focused on sex and the erotic body—"How difficult it is to uncouple the terms pleasure and sexuality," Cora Kaplan rightfully observed. For mid-twentieth-century investigators, however, there was much more to pleasure than sex (or drugs).[69] During the initial phase of the science of pleasure, sex and the erotic writ large were conspicuously absent.[70] The brain science of pleasure was initially and predominantly a study of hunger pleasure-reward, and less so of thirst pleasure-reward. Sex pleasure-reward appeared soon after. Thermoregulation pleasure-reward, aggression pleasure-reward, and drug pleasure-reward would also join the fray.

These distinctions between drive-specific rewards-pleasures emerged from the very beginning of the new discoveries. Investigators discovered a gamut of drive-specific rewards-pleasures, one for each drive: a feeding-drive reward-pleasure, a sex-drive reward-pleasure, a thirst-drive reward-pleasure, and so forth. Each drive-specific reward-pleasure was independent of the others, was semidistinct anatomically, and was sensitive to different experimental manipulations.

The iconic, self-stimulating, insatiable super-pleasured rat that was invoked in numerous publications and images was more often than not stimulating its feeding-center pleasure site (or its drinking-center pleasure site), and not its sex-center pleasure site or its orgasm pleasure site. In terms of laboratory enactments, there was nothing unique about sex pleasure/reward. Its mode of production, measurement, and reward/pleasure value were indistinguishable from other pleasures-rewards. There was no distinction between the stimulation of the pleasure-reward zones of the thirst drive, the feeding drive, the thermoregulation drive, or the sex drive. The self-stimulating rat was also indifferent to these distinctions, and during the protocol, the experimenters were often ignorant of and indifferent to the drive specificity of the "reward" electrode prior to the dissection of the rat's brain.[71]

The innumerable allusions to and depictions of the insatiable ecstatic rats as inspirations for varieties of orgasm-inducing—"Orgasmatron"-like—brain machines ought to have been framed and named in terms of "drinkotrons" (thirst-drive super-pleasure) or "feedotrons" (feeding-drive super-pleasure) or even "thermotrons" (thermoregulation-drive super-pleasure). These names would have more realistically reflected the true nature of these early devices. The pleasure-frenzied rat was not experiencing a barrage of orgasms, but a barrage of feeding super-pleasures, drinking super-pleasures, and so

---

[68] This subtitle invokes Herbert Marcuse's contemporaneous "polymorphous perversity" or "polymorphous sexuality" in *Eros and Civilization*. The term originated in Freud's studies of infantile sexuality. See Marcuse, *Eros and Civilization: A Philosophical Inquiry into Freud* (1955; repr., Boston, 1966).

[69] Cora Kaplan, "Pleasure/Sexuality/Feminism," in Formations Collective, *Formations of Pleasure* (cit. n. 5), 15–35, on 15.

[70] This was in stark contrast to the Freudian-inspired motivation research of Ernest Dichter.

[71] This nondistinction is obvious in numerous publications from this earlier period. Larry Stein, one of the leading investigators during this early period, affirms that in numerous experiments the investigators were indifferent to the drive specificity of the "reward." Stein, telephone conversation with Otniel E. Dror, 10 October 2014.

forth. In fact, and as experimenters clearly showed, when given the choice between the stimulation of their thirst-center reward-pleasure site by EBS and real sex with a real female in heat, male rats preferred the thirst-drive super-pleasure over copulation.

These varieties of nonsexual super-pleasures presented a gamut of polymorphous pleasures. But these polymorphous possibilities were lost on contemporaries and on the cultural imagination of future generations. They represent "missed" pleasures and missed paradigms and templates for and of pleasure.[72] Super-pleasures were orgasms, which were now streamlined. The orgasm usurped super-pleasure and framed for many the 1960s pleasure revolution—and for some, women's liberation—in terms of a sexual-pleasure revolution.[73] The orgasm hijacked (and high jacked) the *perpetuum*.[74]

Scientists were also partly hijacked by the orgasm conception, despite their enactments of polymorphous pleasures. Some, like Lilly and Cohen, explicitly referred to the frenzied rats (monkeys and dolphins) in terms of orgasms, despite knowing better. For others, the basic stipulations of the laboratory and of scientific protocols partly prescribed the orgasm as an implicit model for studying pleasure. Instantly reproducible, rapidly climaxing, and immediately dissipating—the orgasm template offered a discrete moment-phenomenon, which was clearly distinguishable from background fluctuations, could serve as a focal point for research, and could be conditioned, in contrast to the hunger-satiety template, which was spread over time, was not a moment in time, and progressively and insidiously metamorphosed from hunger into satiation, with no clear demarcations between its presence and absence.

The requirement for instantaneity in reward studies was ostensibly presented by Neal E. Miller, for whom EBS was a godsend, since it allowed him to produce instants: instant hunger, instant satiety, and instant thirst. EBS transformed hunger, satiety, and thirst—which were intractable—into phenomena that were instantaneous and brisk. Miller redesigned hunger, thirst, and satiety in these terms, since his conditioning experiments demanded a definite and brisk change in the animal economy.[75] EBS bypassed the temporal order and logic of the body and its materialized-embodied reverberations. It redesigned super-pleasure in terms of the momentary, climaxing, and fleeting orgasm moment. The science of pleasure-reward thus offered polymorphous pleasures, but it also implicitly designed them on an orgasm template.

The orgasm template was also partly prescribed by the English language. There seems to be no specific word in English for a "supreme pleasure moment" of eating-satiation or drinking-quenching. There seems to be no analogue to orgasm in speaking of a gastronomic experience (other than the derived and contemporary colloquialism—"foodgasm"). The seeming absence of a way to speak (and think) about extreme pleasure moments of biological drives other than in terms of sex positioned orgasm as the only available template for imagining and speaking these moments.

---

[72] For an exemplary account of an alternative, pre- and non-sex-dominated conception of pleasure during the Enlightenment, see Emma Spary, *Eating the Enlightenment: Food and the Sciences in Paris, 1670–1760* (Chicago, 2012), esp. chap. 5.

[73] On the association between women's sexual-pleasure revolution and women's liberation, see, famously, Anne Koedt, "The Myth of the Vaginal Orgasm," in *Notes from the Second Year* (New York, 1970), 37–41. See also Jane Gerhard, "Revisiting 'The Myth of the Vaginal Orgasm': The Female Orgasm in American Sexual Thought and Second Wave Feminism," *Feminist Stud.* 26 (2000): 449–76.

[74] I thank Robert Gregory Boddice for introducing me to the meaning of "high jack."

[75] E.g., according to the drive-reduction theory, the instantaneous production of satiety via EBS should work like a reward.

There were, nevertheless, other contemporaneous models for imagining and framing extreme pleasures: drugs (LSD and narcotics) and nirvana. Rado, Lilly and Cohen, and the vernacular literatures proffered these other possibilities. In respect to drugs, the associations with the *perpetuum* were already evident by the early 1960s. These drug associations were for the most part absent from the publications of the broader community of expert ESB reward investigators. The associations between EBS reward and drugs were of two kinds. For Rado and for Lilly and Cohen, the major thrust of the links was in terms of shared hyper-pleasure. Drugs and the *perpetuum* were instances of "supramaximal" pleasure (Lilly, in speaking of LSD) or "super-pleasure" (Rado, in speaking of narcotic drugs and addiction—"bondage"). The link with EBS was in terms of the unique and shared experience of excessive pleasure (which for Rado also explained the addiction). This link via excessive pleasure, and the modeling of addiction in terms of an addiction to pleasure, explains why the broader community of EBS experimenters ignored both of these possible connections. Many EBS investigators did not speak (or think) in terms of pleasure. The EBS rats, moreover, were not "addicted" to the EBS, at least not in terms of the physiological meaning of addiction, which framed addiction primarily in terms of withdrawal, rather than in terms of being addicted to the "pleasure" of drugs.[76] In the vernacular literature, the preoccupation with mind control forged immediate associations between EBS and drugs. Both represented different methods for brain control: electrical and chemical, respectively.[77]

The nirvana model potentially offered a real alternative to the EBS-orgasm model. As Lilly explained in retrospect, the science of the orgasm and the template of the fleeting climaxing moment were incompatible with nirvana, which was an extended and sustained pleasure state. EBS-orgasm failed as a model for this different super-pleasure possibility and cosmology. For Lilly, moreover, nirvana could be achieved by extremes of pain, rather than only in terms of pleasures.[78]

For Maslow, in contrast, the relationships between the *perpetuum* and nirvana, or rather, between the *perpetuum* and "mystic or oceanic experiences," were not oppositional, as they were for Lilly. Rather than focusing on the pleasure of super-pleasure, addiction, or alternative cosmologies of nirvanic—nonorgasmic—pleasures, Maslow emphasized that the rats were oblivious to the outside world—to the external environment—during lever pressing. This withdrawal into the self, which could model various pathologies, like schizophrenia, was also typical of "beatific states of concentration on 'higher' conscious experiences reported from the East, as well as with the concentrated fascination of aesthetic and cognitive insight experiences."[79] The *perpetuum* was turned on its head. Rather than modeling a frenzied pleasure and its dystopic possibilities, as Maslow had done several paragraphs above in this same commentary, he presented the frenzied rats as existing on or modeling higher states of consciousness—as, in fact, modeling nirvana.

---

[76] For the community of experimenters, the more significant direct links with drugs would be forged during the 1970s.

[77] See the references in n. 30.

[78] Judith Hooper, "John Lilly: Altered States," *Omni Magazine* (January 1973), https://www.erowid.org/culture/characters/lilly_john/lilly_john_interview1.shtml (accessed 5 July 2014).

[79] Maslow continued: "I suppose Olds and Hebb will be horrified by this kind of speculation." See Olds, "Physiological Mechanisms" (cit. n. 14), 147.

THE *PERPETUUM* SOCIETY

There was one other possibility for why the orgasm was more appealing than nirvana in modeling super-pleasure. This possibility was not articulated explicitly, but it emerged from an interchange between Lilly and Cohen, in which they discussed, among other things, what a society of nirvana would have to look like. If the nirvana model of EBS were to function as a general model for Western society, if Western societies were to adopt EBS-nirvana as a common practice (rather than EBS-orgasm), then it would require the imagining of a different social organization. This social organization was imagined to exist in the East by Lilly and Cohen, and to be incompatible with the Western social order. The Western model of democratized access to super-pleasure and ecstasy could not tolerate, afford, or support a nirvana model, a model of instantly produced extended nirvanas, in which a large segment of the population would enter and sojourn in prolonged states of nirvana-induced EBS. The EBS-as-nirvana society was thus at odds with the cultural imagination and possibilities of Western EBS investigators and their vernacular audiences. The EBS-orgasm model, on the other hand, was instantly appealing. It was instantly produced, immediately climaxed, and it dissipated instantaneously. It could easily be incorporated and integrated into, and managed within, Western society, and it was easily imagined by Western minds in terms of their society.

It was also good for business. This latter "good" of the EBS-orgasm brings us back to one unresolved issue in respect to the *perpetuum*, which links several threads of this history together. Considering all the negative press of super-pleasure, why did Nature select for a super-pleasure that continuously positioned organisms on the edge of their own *perpetuum*? A full discussion of this significant question is beyond the scope of this essay. Below, I briefly outline in broad strokes one interpretation that took shape during this initial period.[80]

Throughout the period under study, EBS-rewarded rats and the *perpetuum* were often under suspicion. The preternatural presentation and materialization of EBS continuously undermined its status as a model of nature.[81] The naturalization of EBS entailed crucial and fascinating developments and experiments in a broader postwar cultural context that was itself rapidly shifting. In the context of this broader transformation, one discovery encapsulates the significance of the shift to super-pleasure, the evolutionary logic of the *perpetuum*, and the emerging postwar consumer market. This discovery was the perplexing finding that stimulation of the hunger center was rewarding. Experimenters observed that the same electrical stimulation that was rewarding for the rat also activated the rat's hunger drive. Why would a rat frantically press a lever that made it hungry, and why was it rewarding for the rat?

This discovery was inexplicable in terms of the drive-reduction theory, in which hunger was an aversive state of deprivation that activated the animal into a state of high drive. The driven animal found food, ate, and was satiated. Satiety, not hunger, was thus rewarding (pleasure), since it reduced the hunger (drive). In evolutionary

---

[80] For a fuller discussion, see Otniel E. Dror, *The Sciences and Cultures of Pleasures and Rewards* (manuscript, History of Medicine, Hebrew University).

[81] EBS reward had an extremely fast extinction rate, it often required "priming" the animal, it was debatable whether it could be used for secondary conditioning, and more.

terms and as Miller succinctly put it: "Animals who are not pleasantly rewarded by substances that reduce their drives come from a long line of extinct ancestors."[82]

This basic scheme of the drive-reduction theory also implied an economy of scarcity: it was enacted inside the laboratory by depriving animals prior to the experiment and provisioning food, sex, water, and more as the "reward" during the experiment. This implied scarcity economy, its enactment inside the laboratory, and its affective logic of an aversive hunger and a rewarding satiety were turned upside down with the discovery of super-pleasure in the brain. The new super-pleasure-in-the-brain cerebrated and embodied an economy of affluence and of pleasure seeking: of consumption in satiety—of consumption that was driven not by deprivation, but in satiation and within an economy of abundance, since EBS-rewarded rats were satiated; and of an insatiable "hunger" that was, moreover, rewarding, rather than aversive(!). This was the enacted discovery of the new super-pleasure: a satiated, nondeprived rat that continued to press the lever insatiably and was continuously "hungry" for more reward—for, in fact, more "hunger." "Eating begets eating," as Olds succinctly put it.[83] This was the meaning of the *perpetuum* in its real embodied and natural form, that is, in terms of the drives to which it appertained (since all rewards and pleasures appertained to drives and were never abstract).

In trying to make sense of these discoveries, Olds proposed a new hypothetical and tentative model of the relationships between hunger, satiety, and starvation, which could be broadened to all other drives, and in which he accounted for the EBS-rewarded behavior, evolutionary theory, and the perplexing (for the older model) finding that hunger was rewarding.[84] First, he analogized between the albino rat—"having been bred in a laboratory"—and the average American college student—"having been bred in America"—neither of which "has ever experienced a need in its life." What, then, he wondered, "might drive behavior in the absence of needs"? The drive-reduction theory, he continued, was a "Procrustean bed" for "an organism that seeks novelty, ideas, excitement, and good-tasting foods." He then surveyed some of the major scientific findings since his initial discovery, and then he turned to consumption. He emphasized the positive feedback loop of consumption—"Once the eating mechanism has been triggered, it moves forward under its own power and would go on indefinitely if other extraneous control devices were not brought to bear." But "why does the animal keep on eating? . . . Why does the animal start eating even if he is not starving, but is only stimulated by the sight or smell or taste of food?"

Evolution made sense of these seemingly inexplicable behaviors:

> [the] mechanisms for hoarding promoted the survival of the species. The abstract animal (who never lived so far as I can make out in phylogenetic history) who waited until demise was imminent and then began looking to satisfy his need was on the edge of demise at all times. You might say he was "just" living. His lucky cousin, who is no abstraction, stocked up his larder during the fat years, in preparation for the lean ones, and you might say he enjoyed it. Instead of "just" living, the positive reinforcement creature was really living.[85]

[82] Miller, "Comments" (cit. n. 17).
[83] Olds, "Pleasure Centers," 1970 (cit. n. 16), 30.
[84] Ibid.
[85] For the various citations, see ibid., 22, 31.

The relationships between consumption, affect, and drive had shifted from an economy of (wartime) scarcity and food rationing and from an experimental program that
took off from deprivation and "drive reduction" to a (postwar) political and experimental economy of abundance and consumption, in which satiated humans were
driven insatiably to consume, and in which satiated animals inside laboratories were
driven insatiably to self-stimulate their brain pleasure circuits.[86]

REPETITIONS: REVOLUTION THROUGH REVOLUTIONS[87]

Roy Wise still remembers the fascination with and the gravitational pull of the *perpetuum* during his graduate studies in the 1960s, walking down the hall and hearing
the noise of the clicking of relays of electrical stimulation, and the abundant "pleasure"-
talk, in which he, in his youth, was also caught up during one instance.[88]

There was (and still is) something irresistible about the lever-pressing, self-
stimulating, insatiable rat. The pleasure revolution of the mid-twentieth century was
a revolution by revolutions—a simple rattine lever press repeated hundreds of thousands of times. Perhaps it was only as a *perpetuum* that pleasure could be heard in
the context of the behaviorism from and in which it initially emerged. Would there have
been "pleasure" if it were not for the *perpetuum*, if it had emerged in terms of maze-
running or electric-grid-crossing rats? The repetition of these repetitions in a growing
number of laboratories was at the heart of the scientific revolution in pleasure. The behaviorist's toolbox of repetitions and frequencies of repetitions materialized "pleasure"
and made it supramaximal.[89]

As supramaximal, as a *perpetuum*, pleasure was perhaps too "super" for a postwar
society that wanted its super-pleasures but still retained a powerful notion of self-
mastery and of "healthy" hygienic pleasure. As Maslow said of "peak experiences"
(which he had on previous occasions explicitly compared to the *perpetuum* rats),
"They would kill us if they lasted too long or came too often. (Supposing a great orgasm lasted for 15 minutes instead of 10 or 15 seconds! The organism couldn't stand
it. Surely the heart would collapse.)"[90] In the mid-twentieth-century United States,
rats, monkeys, dolphins, and (institutionalized) patients were the sole beneficiaries—
or subjugated victims—it seems, of supramaximal, super, pure pleasure.[91]

---

[86] This postwar shift to abundance-affluence was highly differentiated and discriminatory in post–
World War II United States society. There is an abundant literature on the shift from scarcity to abun-
dance and its implications for numerous post–World War II developments. See, e.g., Brick, *The Age of
Contradiction* (cit. n. 6).

[87] I am drawing here on Raymond Williams's pointed analysis of the historical conjunction between
"revolution" in terms of revolving in space and "revolution" in terms of a political rebellion. See Williams, *Keywords: A Vocabulary of Culture and Society* (1976; repr., New York, 1983), 270–4.

[88] Roy Wise, telephone conversation with Otniel E. Dror, 5 September 2014.

[89] For an analysis that also emphasizes practices in consolidating a new affective ontology, see Felicity Callard, "The Intimate Geographies of Panic Disorder: Parsing Anxiety through Psychopharmacological Dissection," in this volume.

[90] For the citation, see Jessica Lynn Grogan, "A Cultural History of the Humanistic Psychology
Movement in America" (PhD diss., Univ. of Texas at Austin, 2008), 105–6.

[91] Pleasure, however, was never really pure. Even at its source, it was corrupted and contaminated,
since it was always embodied in distinct drives that diminished its true virtuous purity. The *perpetuum*
itself could often be manipulated at its source through these specific drives—e.g., by changing levels
of androgen (sex-reward *perpetuum*) or levels of satiety (feeding-reward *perpetuum*).

The endlessly repetitive, mechanical, and factory-like lever presses of the hard-working rat were neither alienating, nor tedious, nor boring for the rat.[92] Tedium, boredom, habituation—these ostensible effects of repetition were absent from the literature on repetitions in lab rats and in lab "hermits," as G. Stanley Hall christened the new breed of late nineteenth-century laboratory-based researchers. These hermits were the human version and equivalent of lab rats.[93] They were looped in a *perpetuum* of their own: the *perpetuum* of knowledge. They engaged in endless repetitions and replications (though they were beaten by the rats), had an insatiable appetite (for knowledge), were indefatigable, and like the rats, they renounced food, sleep, sex, and sensual gratifications (for the sake of knowledge)—for the "reward," perhaps even (super) pleasure, of science.[94]

This parallelism between rats and scientists can be taken one step further. Like a lab rat that accidentally and serendipitously pushed the lever of pleasure-reward and went into a frenzy of repetitions, the science of pleasure-reward writ large was a science that literally began with—and was often described in terms of—accidental serendipitous discoveries, which were followed by a frenzy of repetitions and replications.[95] This core narrative and formula, in which the effect of an accidental and contingent behavior of an organism leads to an exponential increase in the frequency of repetitions of the (accidental) act that had preceded the effect, constituted the primary and basic definition of the behaviorist-Skinnerian "positive reinforcement" and "reward."

[92] I note that in transposing this design to humans in terms of, e.g., "pleasure helmet" or otherwise, the workaholic rat was replaced by an automated super-pleasure generator that automatically stimulated the human brain and provided for "easy" passive pleasures.

[93] For G. Stanley Hall and "laboratory hermits," see Laurence R. Veysey, *The Emergence of the American University* (Chicago, 1965), 151.

[94] For further elaborations, see Otniel E. Dror, *Repetitions: For Science and for Pleasure* (manuscript, History of Medicine, Hebrew University).

[95] During its formative years, this was a science of inserting electrodes into brains and then stimulating and recording behaviors. This partly explains the relatively frequent reports of serendipity: one was never absolutely sure what the behavior of the animal would be during stimulation and where the electrode ended up in the brain until the post hoc dissection of the brain. This empiricism was accentuated by the fact that in the rat's small brain, major distinct function centers were located in very close proximity to one another.

Notes on Contributors

Damien Boquet is Senior Lecturer in History of the Middle Ages at the University of Aix-Marseille, and author of *L'ordre de l'affect au Moyen Âge: Autour de l'anthropologie affective d'Aelred de Rievaulx* [The order of affect in the Middle Ages: Around the affective anthropology of Aelred of Rievaulx] (Caen, 2005).

Felicity Callard is Reader in Social Science for Medical Humanities at the Department of Geography and Centre for Medical Humanities, Durham University. She has wide-ranging research interests in the historical geography of twentieth- and twenty-first-century psychiatry, psychology, psychoanalysis, and cognitive neuroscience. She is Editor-in-Chief of *History of the Human Sciences*, and Group Leader of "Hubbub at Wellcome Collection"—an interdisciplinary research program on rest and its opposites that brings together humanists, social scientists, life scientists, and artists.

Naama Cohen-Hanegbi is a lecturer in the History Department of Tel Aviv University. She has published articles on the medical and religious treatment of emotions and on medicine within interreligious encounters. Currently she is preparing a manuscript entitled "Practices of Care: Medical and Pastoral Writings on the Accidents of the Soul, 1200–1500."

Otniel E. Dror is the Joel Wilbush Chair in Medical Anthropology and Head of the Section for the History of Medicine in the Medical Faculty at the Hebrew University of Jerusalem. He recently published *Knowledge and Pain* (coedited with Esther Cohen, Leona Toker, and Manuela Consonni; New York, 2012), and his *Blush, Flush, Adrenaline: Science, Modernity and Paradigms of Emotions, 1850–1930* is under revision for the University of Chicago Press. He is currently working on the history of the study of pleasure and reward during the post–World War II period.

Eric J. Engstrom is a Research Associate in the Department of History at Humboldt University in Berlin and a member of the working group on the history of psychiatry at the Max Planck Institute of Psychiatry in Munich. He has published widely on the history of psychiatry, including a monograph, *Clinical Psychiatry in Imperial Germany* (Ithaca, N.Y., 2004). He is also coeditor of the multivolume edition of the papers and correspondence of the German psychiatrist Emil Kraepelin. He is currently researching and writing a

book on the history of forensic cultures in Wilhelmine Berlin.

Anne Harrington is the Franklin L. Ford Professor of the History of Science at Harvard University, specializing in the history of psychiatry, neuroscience, and the other mind and behavioral sciences. Her publications include *Medicine, Mind and the Double Brain* (Princeton, N.J., 1988), *Reenchanted Science* (Princeton, N.J., 1997), and *The Cure Within* (New York, 2007), and a forthcoming book, *The Biological Turn in Psychiatry*.

Bettina Hitzer is Minerva Research Group Leader at the Max Planck Institute for Human Development, Berlin. Her research interests range from the history of migration and the history of religion to the history of medicine and the history of emotions. She is a coauthor of *Learning How to Feel: Children's Literature and Emotional Socialization, 1870–1970* (Oxford, 2014); coeditor, with Pascal Eitler and Monique Scheer, of "Feeling and Faith—Religious Emotions in German History," special issue, *German History*, volume 32, number 3 (2014); and author of "Oncomotions: Experiences and Debates in West Germany and the United States after 1945," in *Science and Emotions after 1945: A Transatlantic Perspective*, edited by F. Biess and D. M. Gross (Chicago, 2014).

Anja Laukötter is a researcher in the Department of the History of Emotions at the Max Planck Institute for Human Development in Berlin, where she is working on a book concerning the history of knowledge and emotions in health education films in the twentieth century. After earning her PhD in the history of anthropology and anthropological museums at Humboldt University in Berlin, she researched the history of human experiments at the Institute for the History of Medicine, Berlin. She is the author of various publications in the field of the history of science and knowledge, postcolonial studies, visual media (including the history of educational films), and history of emotions throughout the nineteenth and twentieth centuries. She is codirecting with Christian Bonah the ERC Advanced Grant *BodyCapital, The Healthy Self as Body Capital: Individuals, Market-Based Societies and Body Politics in Visual Twentieth Century Europe*.

Pilar León-Sanz is Associate Professor of the History of Medicine and Medical Ethics at the University of Navarra and a member of the project "Emotional Culture and Identity" at the Institute of Culture and Society (UN). Her research interests include medicine in eighteenth-century Spain, especially music therapy, and the practices of health care professionals during the nineteenth and twentieth centuries; she is also studying the evolution of the concept of emotion and its place in medical knowledge. Her publications include *La Tarantola Spagnola: Empirismo e tradizione nel XVIII secolo* (Lecce, 2008); "From Claims to Rights: Patient Complaints and the Evolution of a Mutual Aid Society (*La Conciliación*, 1902–1936), in *Complaints, Controversies and Grievances in Medicine: Historical and Social Science Perspectives*, edited by Jonathan Reinarz and Rebecca Winter (New York, 2015); "Music Therapy in Eighteenth Century Spain: Perspectives and Critiques," in *Music and the Nerves 1700–1900*, edited by James Kennaway (London, 2014); "Resentment in Psychosomatic Pathology (1939–1960)," in *On Resentment: Past and Present*, edited by Bernardino Fantini, Dolores Martín Moruno, and Javier Moscoso (Newcastle upon Tyne, 2013); and "Evolution of the Concept of Emotion in Medicine: A Music-Therapy Approach," in *The Emotions and Cultural Analysis*, edited by Ana Marta González (New York, 2012).

Rafael Mandressi is a historian and researcher at the Centre Alexandre-Koyré d'Histoire des Sciences et des Techniques at the Centre National de la Recherche Scientifique in Paris. His main field of research is the history of medicine in early modern Europe. He has published *Le Regard de l'anatomiste: dissections et invention du corps en Occident* (Paris, 2003), and about sixty articles, mainly on the history of anatomy, on printed medical images, and on the history of cerebral functions.

Dolores Martín Moruno works at the Institute for Ethics, History, and the Humanities (iEH2) at the University of Geneva. After earning a PhD in philosophy from the Universidad Autónoma (Madrid) and in the history of science from the Centre Alexandre-Koyré (Paris), she was a Visiting Research Fellow at the Centre de Recherche en Histoire des Sciences et des Techniques (Paris), the Bakken Museum (Minneapolis), and the Queen Mary University of London. She is the author of various publications on the history of emotions and specifically about the changing cultural perceptions of resentment, melancholy, and love between the eighteenth and the nineteenth centuries.

Piroska Nagy is Professor of History of the Middle Ages at Université de Québec à Montréal and author of *Le Don des larmes: Un instrument spirituel en quête d'institution, V–XIIIe s.* [The gift of tears: A spiritual instrument seeking institution, 5th–13th c.] (Paris, 2000). In 2006, she and Damien Boquet founded the research program EMMA (Emotions in the Middle Ages; http://emma.hypotheses.org). They have codirected several publications: *Le Sujet des émotions au Moyen Âge* (Paris, 2009), *Politiques des émotions au Moyen Âge* (Florence, 2010), and "La Chair des émotions au Moyen Âge," special issue, *Médiévales*, volume 61 (2011). Her most recent book, coauthored by Boquet, is *Sensible Moyen Âge: Une histoire culturelle des émotions et de la vie affective dans l'Occident médiéval* [Emotional Middle Ages: A cultural history of the emotions and affective life in the medieval West] (Paris, 2015).

Index

SUGGESTIONS FOR CONTRIBUTORS TO OSIRIS

OSIRIS is devoted to thematic issues, conceived and compiled by guest editors who submit volume proposals for review by the OSIRIS Editorial Board in advance of the annual meeting of the History of Science Society in November. For information on proposal submission, please write to the Editor at osiris@etal.uri.edu.

1. Manuscripts should be submitted electronically in Rich Text Format using Times New Roman font, 12 point, and double-spaced throughout, including quotations and notes. Notes should be in the form of footnotes, also in 12 point and double-spaced. The manuscript style should follow *The Chicago Manual of Style*, 16th ed.

2. Bibliographic information should be given in the footnotes (not parenthetically in the text), numbered using Arabic numerals. The footnote number should appear as superscript. "Pp." and "p." are not used for page references.

   a. References to books should include the author's full name; complete title of book in *italics*; place of publication; date of publication, including the original date when a reprint is being cited; and, if required, number of the particular page cited (if a direct quote is used, the word "on" should precede the page number). *Example*:

   [1] Mary Lindemann, *Medicine and Society in Early Modern Europe* (Cambridge, 1999), 119.

   b. References to articles in periodicals or edited volumes should include the author's name; title of article in quotes; title of periodical or volume in *italics*; volume number in Arabic numerals; year in parentheses; page numbers of article; and, if required, number of the particular page cited. Journal titles are spelled out in full on the first citation and abbreviated subsequently according to the journal abbreviations listed in *Isis Current Bibliography*. *Example*:

   [2] Lynn K. Nyhart, "Civic and Economic Zoology in Nineteenth-Century Germany: The 'Living Communities' of Karl Möbius," *Isis* 89 (1999): 605–30, on 611.

   c. All citations are given in full in the first reference. For succeeding citations, use an abbreviated version of the title with the author's last name. *Example*:

   [3] Nyhart, "Civic and Economic Zoology" (cit. n. 2), 612.

3. Special characters and mathematical and scientific symbols should be entered electronically.

4. A small number of illustrations, including graphs and tables, may be used in each volume. Hard copies should accompany electronic images. Images must meet the specifications of The University of Chicago Press "Artwork General Guidelines" available from the Editor.

5. Manuscripts are submitted to OSIRIS with the understanding that upon publication copyright will be transferred to the History of Science Society. That understanding precludes consideration of material that has been previously published or submitted or accepted for publication elsewhere, in whole or in part. OSIRIS is a journal of first publication.

OSIRIS (ISSN 0369-7827) is published once a year.

Single copies are $33.00.

Address subscriptions, single issue orders, claims for missing issues, and advertising inquiries to *Osiris*, The University of Chicago Press, Journals Division, PO Box 37005, Chicago, IL 60637.

Postmaster: Send address changes to *Osiris*, The University of Chicago Press, Journals Division, PO Box 37005, Chicago, IL 60637.

OSIRIS is indexed in major scientific and historical indexing services, including *Biological Abstracts, Current Contexts, Historical Abstracts*, and *America: History and Life*.

Paperback edition, ISBN 978-0-226-39204-2

 **A RESEARCH JOURNAL DEVOTED TO THE HISTORY OF SCIENCE AND ITS CULTURAL INFLUENCES**

A PUBLICATION OF THE HISTORY OF SCIENCE SOCIETY

W. Patrick McCray
Co-Editor, Osiris
Department of History
University of California, Santa Barbara
Santa Barbara, CA 93106-9410 USA
pmccray@history.ucsb.edu

Suman Seth
Co-Editor, Osiris
Department of Science & Technology Studies
306 Rockefeller Hall
Cornell University
Ithaca, NY 14853 USA
ss536@cornell.edu